Topics in Industrial Mathematics

Applied Optimization

Volume 42

Series Editors:

Panos M. Pardalos
University of Florida, U.S.A.

Donald Hearn
University of Florida, U.S.A.

The titles published in this series are listed at the end of this volume.

Topics in Industrial Mathematics

Case Studies and Related Mathematical Methods

by

Helmut Neunzert

Kaiserslautern, Germany

and

Abul Hasan Siddiqi

Aligahr, India
and Dhahran, Saudi Arabia

KLUWER ACADEMIC PUBLISHERS

DORDRECHT / BOSTON / LONDON

A C.I.P. Catalogue record for this book is available from the Library of Congress.

ISBN 0-7923-6417-1

Published by Kluwer Academic Publishers,
P.O. Box 17, 3300 AA Dordrecht, The Netherlands.

Sold and distributed in North, Central and South America
by Kluwer Academic Publishers,
101 Philip Drive, Norwell, MA 02061, U.S.A.

In all other countries, sold and distributed
by Kluwer Academic Publishers,
P.O. Box 322, 3300 AH Dordrecht, The Netherlands.

Printed on acid-free paper

Printed in the Netherlands.

To Renate Neunzert and Azra Siddiqi

without whose patience and cooperation this work would not have been possible.

Contents

Preface

Industrial Mathematics is a relatively recent discipline. It is concerned primarily with transforming technical, organizational and economic problems posed by industry into mathematical problems; "solving" these problems by approximative methods of analytical and/or numerical nature; and finally reinterpreting the results in terms of the original problems. In short, industrial mathematics is modelling and scientific computing of industrial problems.

Industrial mathematicians are bridge-builders: they build bridges from the field of mathematics to the practical world; to do that they need to know about both sides, the problems from the companies and ideas and methods from mathematics. As mathematicians, they have to be generalists. If you enter the world of industry, you never know which kind of problems you will encounter, and which kind of mathematical concepts and methods you will need to solve them. Hence, to be a good "industrial mathematician" you need to know a good deal of mathematics as well as ideas already common in engineering and modern mathematics with tremendous potential for application. Mathematical concepts like wavelets, pseudorandom numbers, inverse problems, multigrid etc., introduced during the last 20 years have recently started entering the world of real applications.

Industrial mathematics consists of modelling, discretization, analysis and visualization. To make a good model, to transform the industrial problem into a mathematical one such that you can trust the prediction of the model is no easy task. One needs plenty of experience because modelling is mainly learnt by doing. A nice approach would be to pose real-world problems to the students who should work on them under the guidance of an experienced modeler. In international programmes in Kaiserslautern, "modelling seminars" are organized each semester along these lines. They are proving an important tool for the education of industrial mathematicians. The problems are mainly supplied by an "Institute for Industrial Mathematics" which cooperates with industry on a large scale, doing about 40 different projects every year. This institute is a very important source of interesting problems. But not every university has such a source. How do others get appropriate problems? Again, by searching them, where they are – in industry. Staff members have to visit companies, discuss their problems – they will find a variety of good projects.

This book is designed to help the beginners, to show what we have experienced during our interaction with industry and teaching industrial mathematics. It tries to teach modelling by reading. It may not be the best solution – learning by doing is clearly preferable. We have, however, tried to maintain the flavour: first, by presenting five case studies and then adding some background material related to the theories used for the case studies.

The case studies which make up the first chapter of the book have been taken from a modelling seminar in Kaiserslautern and handle problems of molecular alignment in drug design, acoustic identification of vehicles, the security of air bag sensors, quality control of fabrics and fatigue life analysis. The subsequent chapters provide the reader with mathematical concepts and methods which are essential for a proper analysis of these models and for exploration of related new areas.

For example, a problem of fatigue life analysis dealing with the estimation of the lifetime of critical car components is presented in Chapter 1, while a mathematical formulation based on the concept of hysteresis is given in chapter 6. Airbag sensors need Maxwell's equations whose basis and relevant literature is given in Chapter 3. Optimization is needed in drug design and the acoustic identification of vehicles – Chapter 2 discusses some important algorithms in continuous optimization. Random numbers and so-called Monte Carlo methods help to evaluate very complex integrals (as in drug design) and to solve high-dimensional kinetic equations needed for nuclear reactors, space flight, semiconductors. Chapter 4 is devoted to these methods. Image processing is an emerging field, where a whole bunch of new mathematical ideas are used; our problem deals with quality control of fabrics and uses the fundamental concepts of multiscale analysis. But other methods like wavelets, fractals, energy model, etc. may also be equally important and we describe some of them in Chapter 5.

In the appendices, we have provided discussion on certain topics which are essential for understanding of the main text as well as some results which could not find an appropriate place in a particular chapter. At the end of each chapter, we have given some problems, some of which may lead to research problems, especially in Chapters 3 and 5. Hints of some of these problems are mentioned there. At the end, we have provided an extensive bibliography.

The book addresses several types of readers. We hope it to be useful for all those who have genuine curiosity to know about Industrial Mathematics. It is intended as a handy manual of Mathematical Methods for current industrial and technological problems which may be very useful for engineers and physicists. It is also intended to serve as a lucid commentary on most applied methods which are likely to attract more attention in years to come. Finally, it can be used for a course on Mathematical Methods of current real life problems/industrial problems at graduate and advanced undergraduate levels. A deeper insight, if needed, can be obtained through updated and appropriate references mentioned in the text. Proof of theorems like Theorems 3.2, 3.3 and 6.3 may be omitted by the readers who are not interested in a rigorous analysis.

We are grateful to a number of persons who have provided valuable help in the completion of the book. In particular, we wish to express our thanks to Martin Brokate, Axel Klar, Michael Hack, Franz-Josef Pfreundt, Sergej Rjasanow, Joachim Weickert, Pammy Manchanda, Kalimuddin Ahmad, Firozzaman, Günter Gramlich, Sudarshan Tiwari and Ingeborg Woltman. Abul Hasan Siddiqi would also like to express his thanks to the German Academic Exchange Service, the University of Kaiserslautern and the International Centre of Theoretical Physics, Trieste, Italy, for providing excellent opportunities to acquire knowledge for this emerging area. He would also like to express his gratitude to the founder director of the ICTP late Prof. Abdus Salam, Nobel Laureate, who motivated him to work in applied areas of mathematics.

Abul Hasan Siddiqi would like to thank King Fahd University of Petroleum and Minerals, Dhahran 31261, Saudi Arabia for providing excellent facilities during the preparation of the final manuscript.

Helmut Neunzert
Kaiserslautern University
Kaiserslautern, Germany

Abul Hasan Siddiqi
King Fahd University of Petroleum
and Minerals
Dhahran, Saudi Arabia
and
Aligarh Muslim University, Aligarh, India

Chapter 1

Case Studies at Kaiserslautern

As mentioned in the introduction, modelling can only be learnt by doing. A textbook can only show examples how others have done it – it gives the spirit of what is industrial mathematics, modelling or scientific computing, but it cannot teach it.

However, in order to remain as much realistic as possible, we present some paradigmatic case studies in the way they came to us. Each problem is first posed in a nonmathematical way; so everybody may feel free to start modelling the problem by himself. In a subsequent section, we present our modelling; of course, there is no unique model; others may have other and even better ideas. Finally, we discuss how we have evaluated the model; i.e., we give algorithms to solve the mathematical problem posed by the model. Although different approaches are possible, we can claim that our solutions are good ones in the sense that they give answers to the original problems. We should mention that all the five problems were posed by companies in Germany; some small, others rather big. All these problems were treated by mathematicians from the "Laboratory for Technomathematics" in Kaiserslautern, Germany. The "we modelled", will clearly identify the name in each case. It may be mentioned that the companies paid for the work done by the mathematicians involved – a good indicator that the work was useful. As mentioned in the introduction, the five problems chosen from different industrial areas asking for different mathematical methods have been used as a guideline in the book. Mathematics needed in these problems will be presented in detail in the following chapters.

1.1 Molecular alignment

1.1.1 The problem

Pharmaceutical industry uses extensive computer simulations in drug design. One aspect is molecular alignment. Some drugs are molecules which may bond themselves to a macromolecular receiver, e.g., a protein. These molecules operate like keys which must fit into a key-hole of the protein. However, the exact structure of this key-hole is often unknown; what is known, however, is another matching key which we want to replace by the newly synthesized drug. This new key should be similar to the old key. The problem is: find a model for the similarity of two molecules which allows us to find new keys and develop algorithms to decide how similar in this newly defined sense the two given molecules are.

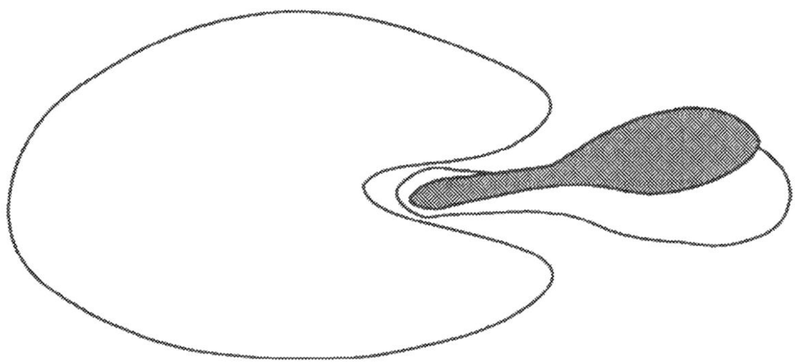

Figure 1.1.1

Such modelling problems have to be handled in close cooperation with chemists. The mathematics part was done by Dr. F. J. Pfreundt of the Technomathematics Lab, Kaiserslautern; the partners were Dr. G. Kleib and Th. Rietznet, BASF.

1.1.2 The model

Molecules are given by their geometric as well as their electromagnetic structure. Geometrically, we consider **molecules** as a rigid structure of balls (atoms) of different radii and different charges. **Rigidity** means that the distance between the N atoms forming a molecule is fixed. The radius of each atom is normally called the **van der Waals radius** and is the minimal distance between this molecule and any other.

In a fixed position, we may therefore represent a molecule by the N positions of the centers of the balls $\underline{x}_1, ..., \underline{x}_N$, by the N corresponding radii $r_1, ..., r_N$ and by N charges $q_1, ..., q_N$; i.e.

$$M = \{\underline{x}_1, ..., \underline{x}_N; r_1, ..., r_N; q_1, ..., q_N\}.$$

To compare the shapes of molecules and to define the **similarity**, the molecules must be translated and rotated. Since the structure is rigid, translating a molecule by a vector $\underline{a}$ and rotating it by a rotation A means to translate and rotate any atom by $\underline{a}$ and A, respectively; i.e.

$$AM + \underline{a} = \{\underline{a} + A\underline{x}_1, ..., \underline{a} + A\underline{x}_N; r_1, ..., r_N; q_1, ..., q_N\}.$$

A rotation is given by Euler's angles Θ, φ and Ψ; i.e.

$$A = A(\Theta, \varphi, \Psi).$$

Any motion of $M \to AM + \underline{a}$ is therefore characterized by 6 parameters $\underline{a} \in \mathbb{R}^3$, $\Theta \in [0, \pi]$, $\varphi, \Psi \in [0, 2\pi]$. Assume that we have already defined a distance d of 2 molecules; i.e., we know what

$$d(M_1, M_2),$$

for given M_1, M_2 should be. ***Similarity*** would then mean "small distance". But then the distance would depend on the relative positions of the molecules. To define similarity, we would have to move one of the molecules until it best fits the other. Similarity should therefore be measured by

$$\mathrm{Sim}\,(M_1, M_2) = \min_{\underline{a}, A} d\,(\underline{a} + AM_1, M_2).$$

To find the best possible position for M_1, i.e., to find the optimal $\underline{a}$ and A is called **alignment**. It is clear that this alignment depends on d, the distance between the molecules we choose. We want to remind that the main purpose of the entire investigation is to substitute one key M_1 by another key M_2. The similarity should be a similarity of keys with respect to one lock! We speak about distances, but we do not expect to get distances in the sense of a (mathematically correct) metric. d must be a functional on pairs of molecules with $d\,(M_1, M_2) = 0$ if and only if $M_1 = M_2$. We do not insist on having a triangle inequality – it is easily obtained, and not even required. As mentioned before, one has to take geometrical aspects into consideration as well as the electrostatic situation.

Typical attempts in the past (see, for example, the so-called SEAL-metric, developed by Kearsley and Smith [1990]) defined $d\,(M_1, M_2)$ for

$$M_1 = \{\underline{x}_1, ..., \underline{x}_N; r_1, ..., r_N; q_1, ..., q_N\},$$

and

$$M_2 = \left\{\underline{y}_1, ..., \underline{y}_M; \widetilde{r}_1, ..., \widetilde{r}_M; \widetilde{q}_1, ..., \widetilde{q}_M\right\},$$

by

$$d\,(M_1, M_2) := C - \sum_{i=1}^{N} \sum_{j=1}^{M} \omega_{ij} e^{-\alpha \|x_i - y_j\|^2},$$

where $\omega_{ij} = \omega_E q_i \widetilde{q}_j + w_s r_i^\beta \widetilde{r}_j^\beta$. C is chosen such that $d(M_1, M_1) = 0$. Clearly C depends on M_1. The sum measures the overlap of the two molecules and is maximal and equal to C if both molecules are the same.

There are several free parameters in the game: α, w_E, w_s and β; they were found by fitting to experiments. One may also do some geometric considerations: if we have two balls of the same radius R, the volume of the intersection of the 2 balls having a distance r, $V(r)$, is 0 for $r > 2R$; for $0 \leq r \leq 2R$, we compute by elementary geometry

$$V(r) = \frac{2\pi}{3}\left(\frac{r^3}{8} - 3r\frac{R^2}{2} + 2R^3\right),$$

which is maximal for $r = 0$: $V(0) = \frac{4\pi}{3}R^3$. We shall now choose w_s and α, β such that

$$\widetilde{V}(r) = w_s R^{2\beta} e^{-\alpha r}$$

fits optimally to $V(r)$. Again, "optimal" is not defined and one may think of $\widetilde{V}(0) = w_s R^{2\beta} = \frac{4\pi}{3}R^3$, i.e. $w_s = \frac{4\pi}{3}$ and $\beta = \frac{3}{2}$. Then α may be chosen such that $V'(0) = \widetilde{V}'(0)$, i.e. $+\alpha = +\frac{3}{4R}$. Other ideas are possible.

Let us now try to do the modelling differently. The domain "fitted" by the molecule $M = \{\underline{x}_1, ..., \underline{x}_N; r_1, ..., r_N; q_1, ..., q_N\}$ is, of course,

$$\Omega = \{\underline{x} : \|\underline{x} - \underline{x}_1\| \leq r_1 \& \underline{x} : \|\underline{x} - \underline{x}_2\| \leq r_2 \& ... \& \|\underline{x} - \underline{x}_N\| \leq r_N\}.$$

Hence, the "surface" of M is $\partial\Omega$. It is quite a complex structure. The atoms carry the charges q which define an electrostatic Coulomb potential

$$\phi(\underline{x}) = \sum_{j=1}^{N} \frac{q_j}{\|\underline{x} - \underline{x}_j\|}.$$

This is repulsive. We may think of taking the distance between the potentials of two molecules, ϕ_1 and ϕ_2, on the surfaces of both molecules, i.e., on $\partial\Omega_1$ and $\partial\Omega_2$ and may try to define

$$d(M_1, M_2) = \left(\int_{\partial\Omega_1} |\phi_1 - \phi_2|^p \, d\omega + \int_{\partial\Omega_2} |\phi_1 - \phi_2| \, d\omega\right)^{\frac{1}{p}};$$

($d\omega$ is the surface measure on $\partial\Omega$). This would lead to a symmetric d, but practically it would be enough to consider only one integral.

Experimental validation, however, shows that what counts is the electric field and not the potential; i.e.,

$$\underline{E} = -\nabla \phi.$$

Moreover, it is neither the flux $\langle \underline{E}, \underline{n} \rangle$ nor the electric energy at the surface; i.e., $\int_{\partial\Omega} \|\underline{E}\|^2 \, d\omega$, which plays the most important role, but rather the direction

$$\underline{e} = \frac{E}{\|\underline{E}\|},$$

which determines similarity.

The electrostatic component of our d may, therefore, be modelled by

$$\int_{\partial\Omega_1} \|\underline{e}_1 - \underline{e}_2\|^2 \, d\omega.$$

What remains is the geometrical part. The most natural choice from the point of view of a mathematician would be the **Hausdorff distance**

$$\delta(\Omega_1, \Omega_2) = \max_{x \in \Omega_1} \min_{y \in \Omega_2} \|x - y\| \, .$$

We first take the shortest distance of a point x to Ω_2; i.e.,

$$d_{\Omega_2}(x) = \min_{y \in \Omega_2} \|x - y\| \, ,$$

and then use the maximum of all these distances with respect to Ω_1

$$\delta(\Omega_1, \Omega_2) = \|d_2\|_{\mathbb{L}^\infty(\Omega_1)} = \|d_2\|_{\mathbb{L}^\infty(\partial\Omega_1)} \, .$$

Since $\mathbb{L}^\infty$-norms are unpleasant to compute, we choose

$$\widetilde{\delta}(\Omega_1, \Omega_2) = \|d_2\|^2_{\mathbb{L}^2(\partial\Omega_1)} \, .$$

Modelling is always a balance between the correctness of the model and the complexity of its evaluation. Hence, there is no need to choose a more complex model. **Hausdorff** is mathematically a very pleasant metric, but what about computationally! O.K., says the chemist, but then begins again to play (having the old SEAL-model in mind): Let us be more flexible by using

$$\int_{\partial\Omega_1} \left(1 - e^{-\beta d_2^2(x)}\right) d\omega(x),$$

(it is almost $\beta \|d_2\|^2_{\mathbb{L}^2(\partial\Omega_1)}$, at least, if d_2 is small – and it is more flexible).

Here we are, at the moment

$$d(\Omega_1, \Omega_2) = \alpha \int_{\partial\Omega_1} \|\underline{e}_1 - \underline{e}_2\|^2 \, d\omega + \int_{\partial\Omega_2} \left(1 - e^{-\beta d_2(x)^2}\right) d\omega,$$

and we may play with α, β controlling the weight of the electrostatic and the geometric aspects.

Let us stop here with modelling; one is never at the end of a modelling task. There is no "the model" – it is either better or worse. To know how good a model is, it must be evaluated, d must be computed, alignment performed, and comparisons with experiments done.

1.1.3 The evaluation

There are 2 numerical problems to be solved, if one wants to use the model given above.:

(a) One has to evaluate integrals over quite complicated surfaces $\partial\Omega$, surfaces of domain Ω formed by the union of balls centered at different positions and of different radii.

(b) One has to minimize

$$d(M_1, A(\theta, \varphi, \Psi)M_2 + \underline{a})$$

with respect to $\theta, \varphi, \Psi, \underline{a}$.

One should realize that the evaluation of the function d is quite an elaborate task. This means that we should look for algorithms which need minimal evaluation.

Considering these 2 problems, one may realize one "principle" of industrial mathematics, which we claim to hold almost always. It is a brother of "Murphys principle": what can go wrong, goes wrong. Of course, we know how to evaluate surface integrals: we approximate the surfaces by piecewise polynomial surfaces etc. How can we do that, if the surface is as complicated as that of a molecule? Of course, we know many optimization strategies: those which need derivatives and others which do not. Which one should we choose here?

We come back to these questions in Chapters 4 and 2, respectively.

1.1.4 The interpretation of the results

What we have done are only the first steps; the problem is not yet solved. But what means "solved"? We expect that molecules, which are known to be similar in the sense defined above, have a small distance – small compared to a typical distance between two arbitrary molecules. By our normalization, it means to be near 1 or even less than 1. We took 4 pairs of rather similar molecules; to explain their structures is beyond the scope of this book. We compare the SEAL-distance with our distance d:

$\mathbf{M_1}$	$\mathbf{M_2}$	$\mathbf{d}_{SEAL}$	$\mathbf{d}$
trypsin 1	trypsin 2a	0.58	0.56
ltmn	ltmp	1.16	0.59
abs	tapap	3.16	1.18
napap	lapap	1.47	1.04

So, it seems that our d discovers similarities better than the former distance concepts.

However, the time to compute d is far too long to be used in practice. The most important task, therefore, is now to accelerate the algorithms. To develop faster algorithms is the genuine work of a numerical analyst, i.e., a genuine mathematical task.

1.2 Acoustic identification of vehicles

1.2.1 The problem

An observer should watch movements of ships around him. He has only ears; he does not see the ships (there may be fog); he has neither radar nor other sensors. He knows from which direction the sound reaches him and the intensity with which the sound reaches him. But he has no information about the distance of the ship and the intensity of the sound emitted. The first question is: how can he use the information available to him? What does it tell him about the movements of the ship and the kind of ship under consideration?

Moreover, whatever he may get out of it, he wants it as quickly as possible. Therefore, he needs on-line algorithms to determine what can be determined. The problem was investigated by Dr. S. Rjasanow of the Technomathematics group at Kaiserslautern (the customers want to remain anonymous).

1.2.2 The model

We put the observer at the origin of the (x, y)-plane, in which the ships are supposed to move in straight lines with uniform speed. These assumptions seem to be justified in "our" situation (says the problem poser!). Therefore,

$$\underline{x}(t) = \underline{a} + t\,\underline{v}$$

is the trajectory of the ship, where we do not know $\underline{a}$ and $\underline{v}$. What we observe is the angle between the position of the ship and (say) the x-axis. We call it $\alpha(t)$ and we measure $\alpha_j = \alpha(t_j)$ at times $t_1, ..., t_k$.

Moreover, we measure the intensity of the incoming signal, $d(t)$. Since it is inversely proportional to $||x(t)||^2$, we get

$$d(t) = \frac{\delta}{||x(t)||^2},$$

where δ denotes the unknown intensity of the sound emitted from the observed vehicle. It is not sufficient to measure α_j and d_j and $d_j = d(t_j)$ in order to determine $\underline{a}, \underline{v}$ and δ, even if we are very diligent and do it quite often. A more distant, but faster and noisier vehicle, may create the same signals at the origin and may not be distinguishable from the nearer but slower and more silent vehicle. We need more information in order to determine δ and motion $(\underline{a}, \underline{v})$, and we may pose different problems depending on the information available.

__Problem 1__. We know the speed $||v|| = \sigma$ (probably since there is a maximal speed which ships normally use) and we measure only α_j, $j = 1, ..., k$. We want to know the motion $\underline{a}$ and $\underline{v}$.

Problem 2 uses, in addition to the information provided in Problem 1, measurements of d_j, $j = 1, ..., k$ and asks for better results. Apart from the motion, we want to know the type, i.e., δ.

Problem 3 does not assume that we know σ, but measures δ and is again interested in the motion. All measurements carry an element of error.

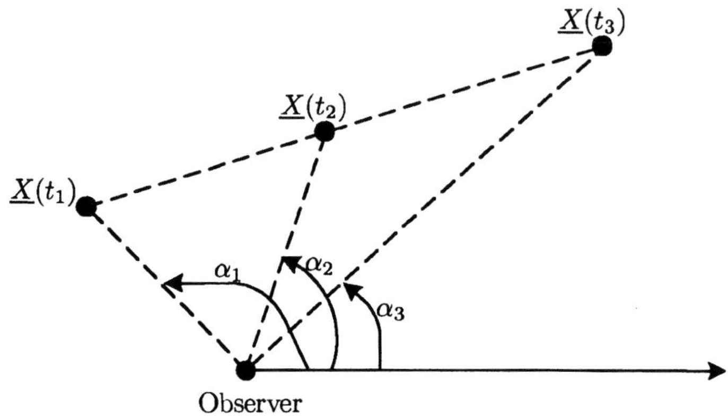

Figure 1.2.1

We know that

$$\underline{x}_j := \underline{x}(t_j) = \underline{a} + t_j v = r(t_j) \begin{pmatrix} \cos \alpha_j \\ \sin \alpha_j \end{pmatrix}, \; j = 1, ..., k.$$

Since we do not know $r(t_j)$, we express the fact that $\underline{x}_j$ points in the direction $\begin{pmatrix} \cos \alpha_j \\ \sin \alpha_j \end{pmatrix}$ by saying that

$$\underline{w}_j = \begin{pmatrix} \sin \alpha_j \\ -\cos \alpha_j \end{pmatrix}$$

is orthogonal to $\underline{x}_j$

$$\langle \underline{x}_j, \underline{w}_j \rangle = 0.$$

If we assume that the measurement error of $\langle x_j, w_j \rangle$ is normally distributed, we would make a regression to get $\underline{a}$ and $\underline{v}$ by minimizing

$$\sum_{j=1}^{k} \left(\langle \underline{x}_j, \underline{w}_j \rangle - 0 \right)^2 .$$

If we think that the cosine of α is normally distributed, we minimize

$$\sum_{j=1}^{k} \left(\cos \angle \left(\underline{x}_j, \underline{w}_j\right)\right)^2 = \sum_{j=1}^{k} \frac{\left\langle \underline{x}_j, \underline{w}_j \right\rangle^2}{\left\| x_j \right\|^2}.$$

Since we did not get the information about the instrument measuring the angles, we choose these two functionals for our further investigation; they are quite good to handle. It is an "economics" principle in modelling (formulated by the Austrian philosopher Ernst Mach during the last century) to choose the simplest model coinciding with the given information.

For Problems 2 and 3, where d_j is used, we need another regression and, most likely, the functional to be minimized should be

$$\sum_{j=1}^{k} \left(\frac{\delta}{\left\| \underline{x}_j \right\|^2} - d_j \right)^2.$$

Although

$$\sum_{j=1}^{k} \left(\delta - d_j \left\| x_j \right\|^2 \right)^2$$

is simpler but harder to justify, we shall still use it.

1.2.3 The evaluation

Keeping in mind that $\underline{x}_j = \underline{a} + t_j \underline{v}$ depends on the parameters $\underline{a}$ and $\underline{v}$ to be estimated by regression, we write

$$\phi_1(\underline{a}, \underline{v}) := \sum_{j=1}^{k} \left\langle \underline{x}_j, \underline{w}_j \right\rangle^2$$

$$\phi_2(\underline{a}, \underline{v}) := \sum_{j=1}^{k} \frac{\left\langle \underline{x}_j, \underline{w}_j \right\rangle^2}{\left\| \underline{x}_j \right\|^2},$$

and

$$\Psi(\underline{a}, \underline{v}, \delta) := \sum_{j=1}^{k} \left(\delta - d_j \left\| \underline{x}_j \right\|^2 \right)^2.$$

Since $\left\langle \underline{x}_j, \underline{w}_j \right\rangle^2 = \left(\left\langle \underline{a}, \underline{w}_j \right\rangle + t_j \left\langle \underline{v}, \underline{w}_j \right\rangle \right)^2$, we get, for example,

$$\operatorname{grad}_a \phi_1 = 2 \sum_{j=1}^{k} \left(\left\langle \underline{a}, \underline{w}_j \right\rangle + t_j \left\langle \underline{v}, \underline{w}_j \right\rangle \right) w_j.$$

In order to simplify, we use the matrices

$$W^{(n)} = \sum_{j=1}^{k} t_j^n \cdot \underline{w}_j \cdot \underline{w}_j^T,$$

where we notice that

$$\begin{aligned}\underline{w}_j \cdot \underline{w}_j^T &= \begin{pmatrix} \sin\alpha_j \\ -\cos\alpha_j \end{pmatrix} \begin{pmatrix} \sin\alpha_j & -\cos\alpha_j \end{pmatrix} \\ &= \begin{pmatrix} \sin^2\alpha_j & -\sin\alpha_j\cos\alpha_j \\ -\cos\alpha_j\sin\alpha_j & \cos^2\alpha_j \end{pmatrix}.\end{aligned}$$

With this notation we easily get

$$\begin{aligned}\mathrm{grad}_a\phi_1 &= 2\left(W^{(0)}\underline{a} + W^{(1)}\underline{v}\right) \\ \mathrm{grad}_v\phi_1 &= 2\left(W^{(1)}\underline{a} + W^{(2)}\underline{v}\right).\end{aligned}$$

In the same way, we may express the gradients of the other functionals.
We now turn to.

<u>**Problem 1**</u>. We do not need Ψ; moreover $\|\underline{v}\| = \sigma$ is given. If we choose ϕ_1 for regression and use $\|\underline{v}\|^2 = \sigma^2$ as constraint, we get by using the Lagrange multipliers

$$F(\underline{a}, \underline{v}, \mu) = \phi_1(\underline{a}, \underline{\,}) + \mu\left(\|\underline{v}\|^2 - \sigma^2\right),$$

as functional to be minimized; necessary conditions for extrema are given by

$$\begin{aligned}\frac{1}{2}\mathrm{grad}_a F &= \frac{1}{2}\,\mathrm{grad}_a\phi_1 = W^{(0)}\underline{a} + W^{(1)}\underline{v} = 0; \\ \frac{1}{2}\mathrm{grad}_v F &= \frac{1}{2}\,\mathrm{grad}_v\phi_1 + \mu\underline{v} = W^{(1)}\underline{a} + W^{(2)}\underline{v} + \mu\underline{v} = 0,\end{aligned}$$

and

$$\|\underline{v}\|^2 = \sigma^2.$$

If $W^{(0)}$ is a regular matrix, we get $\underline{a} = -W^{(0)^{-1}}W^{(1)}\underline{v}$ and therefore an eigenvalue problem for μ and $\underline{v}$ is

$$\left(W^{(1)}W^{(0)^{-1}}W^{(1)} - W^{(2)}\right)\underline{v} = \mu\underline{v},$$

for which the condition $\|v\| = \sigma$ is only a normalization of the eigenvector. (We mention that a general treatment of "normal system of equations", which originate from regression, are treated in Ciarlet and Lions [1991]).

Since $W^{(i)}$ are symmetric matrices, the matrix of our eigenvalue problem is again symmetric and we get 2 real eigenvalues and 2 orthogonal real eigenvectors. The length of the eigenvectors is σ and therefore we get by using $a = -W^{(0)^{-1}}W^{(1)}\underline{v}$, the following 4 solutions

$$\underline{v}_1, \underline{a}_1; \; -\underline{v}_1, -\underline{a}_1; \; \underline{v}_2, \underline{a}_2; \text{ and } \; -\underline{v}_2, -\underline{a}_2$$

with eigenvalues μ_1, μ_2. It is not difficult to decide which sign is the correct one; but which eigenvalue, μ_1 or μ_2, should we choose? This may lead to a maximum instead of a minimum, or to no extremal value of F at all!

To proceed further, we consider the Hessian of F with respect to $(\underline{a}, \underline{v})$

$$H_\mu := 2\begin{pmatrix} W^{(0)} & W^{(1)} \\ W^{(1)} & W^{(2)} + \mu E_2 \end{pmatrix}.$$

It is easy to see that H_0 is positive definite. If we use the following four k-dimensional vectors:

$$\underline{S}^{(1)} = \begin{pmatrix} \sin\alpha_1 \\ \vdots \\ \sin\alpha_k \end{pmatrix}, \underline{S}^{(2)} = \begin{pmatrix} -\cos\alpha_1 \\ \vdots \\ -\cos\alpha_k \end{pmatrix}, \underline{S}^{(3)} = \begin{pmatrix} t_1 \sin\alpha_1 \\ \vdots \\ t_k \sin\alpha_k \end{pmatrix},$$

$$\underline{S}^{(4)} = \begin{pmatrix} -t_1 \cos\alpha_1 \\ \vdots \\ -t_k \cos\alpha_k \end{pmatrix},$$

where the matrix $\frac{1}{2}H_0$ is the Gramian of these vectors; i.e.,

$$\frac{1}{2}H_0 = \left(\langle \underline{S}_i, \underline{S}_j \rangle, 1 \leq i, j \leq 4\right).$$

Gramians are always positive definite, if the vectors S_j form a linear independent set, since

$$\sum_{i,j=1}^{4} \xi_i \xi_j \langle \underline{S}_i, \underline{S}_j \rangle = \left\langle \sum_{i=1}^{4} \xi_i \underline{S}_i, \sum_{j=1}^{4} \xi_j \underline{S}_j \right\rangle \geq 0,$$

with equality only if $\sum_{i=1}^{4} \xi_i \underline{S}_i = 0$.

But even in this case, H_0 is semidefinite. We search for solutions of

$$\frac{1}{2}\,\mathrm{grad}_{(a,v)} F = 0,$$

and see that this means solutions of

$$\frac{1}{2}H_0 \begin{pmatrix} \underline{a} \\ \underline{v} \end{pmatrix} = -\mu \begin{pmatrix} 0 \\ \underline{v} \end{pmatrix}.$$

Since H_0 is positive semidefinite, both eigenvalues, μ_1 and μ_2, are non-positive if

$$\mu_1 \leq \mu_2 \leq 0.$$

But F cannot have a local extremum at $\mu = \mu_1$. Consider H_{μ_1} and

$$\begin{aligned} \left\langle H_{\mu_1}\begin{pmatrix}\underline{a}_2\\ \underline{v}_2\end{pmatrix}, \begin{pmatrix}\underline{a}_2\\ \underline{v}_2\end{pmatrix}\right\rangle &= 2\left\langle \begin{pmatrix} W^{(0)}\underline{a}_2 + W^{(1)}\underline{v}_2 \\ W^{(1)}\underline{a}_2 + W^{(2)}\underline{v}_2 + \mu_1\underline{v}_2\end{pmatrix}, \begin{pmatrix}\underline{a}_2\\ \underline{v}_2\end{pmatrix}\right\rangle \\ &= 2\left\langle \begin{pmatrix} 0 \\ (\mu_1-\mu_2)\underline{v}_2\end{pmatrix}, \begin{pmatrix}\underline{a}_2\\ \underline{v}_2\end{pmatrix}\right\rangle \\ &= 2(\mu_1-\mu_2)\sigma^2 < 0\,. \end{aligned}$$

At the same time we get

$$\left\langle H_{\mu_1}\begin{pmatrix}\underline{a}_2\\ 0\end{pmatrix}, \begin{pmatrix}\underline{a}_2\\ 0\end{pmatrix}\right\rangle = 2\left\langle W^{(0)}\underline{a}_2, \underline{a}_2\right\rangle > 0,$$

since $W^{(0)}$ is the Gramian of $\underline{S}_1$ and $\underline{S}_2$ and, as we assumed, regular. We see that H_{μ_1} cannot be definite and therefore $F(\underline{a}, \underline{v}, \mu_1)$ cannot have a local extremum. Therefore, $\mu = \mu_1$ can be excluded. The only chance is $\mu = \mu_2$; and H_{μ_2} positive definite or at least semidefinite! Certainly not definite, since

$$H_{\mu_2}\begin{pmatrix}\underline{a}_2\\ \underline{v}_2\end{pmatrix} = 0.$$

The proof that D is semidefinite for "generic" measurements is still not available – it should not be too hard. The numerical results point to the fact that μ_2 is a good choice.

What remains is the question of orientation: Is $+\underline{v}_2$ or $-\underline{v}_2$ the correct motion, $\underline{a}_2$ or $-\underline{a}_2$ to starting point?

The angle between $\underline{x}(0)$ and the other directions $\begin{pmatrix}\cos\alpha_j\\ \sin\alpha_j\end{pmatrix}$ should be between $-\frac{\pi}{2}$ and $\frac{\pi}{2}$. If therefore $\left\langle a_2, \begin{pmatrix}\cos\alpha_j\\ \sin\alpha_j\end{pmatrix}\right\rangle > 0$ for all j, we choose the plus sign; if not, we choose $-\underline{a}_2$ and $-\underline{v}_2$.

A final remark to Problem 1 concerning the regularity of $W^{(0)}$: If the ship is on path hitting us directly, then all angles are equal if the measurements are exact. With noisy data the matrix $W^{(0)}$ might not be singular, but is certainly ill-conditioned and the algorithm does not work.

Now we turn to **Problem 2**, but we will shorten the presentation.

We combine ϕ_1 and Ψ using a parameter λ, which might be chosen according to the relative weight of direction and intensity measurements. Instead of F we get

$$G_\lambda(\underline{a}, \underline{v}, \delta; \mu) = \lambda\phi_1(\underline{a}, \underline{v}) + (1-\lambda)\Psi(\underline{a}, \underline{v}, \delta) + \mu(\|\underline{v}\|^2 - \sigma^2).$$

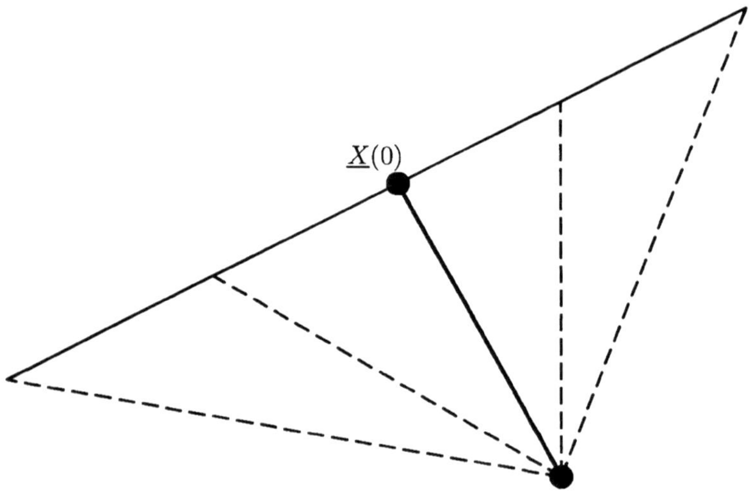

Figure 1.2.2

Regression with G_λ gives necessary conditions

$$\begin{aligned} \lambda\ \mathrm{grad}_a\phi_1 + (1-\lambda)\ \mathrm{grad}_a\Psi &= 0 \\ \lambda\ \mathrm{grad}_v\phi_1 + (1-\lambda)\ \mathrm{grad}_v\Psi + 2\mu\underline{v} &= 0 \\ (1-\lambda)\Psi_\delta &= 0 \\ \|\underline{v}\|^2 &= \sigma^2. \end{aligned}$$

Eliminating δ, we get

$$\begin{pmatrix} S_{11} & S_{12} \\ S_{21} & S_{22} + \mu E_2 \end{pmatrix} \begin{pmatrix} \underline{a} \\ \underline{v} \end{pmatrix} = \underline{0}\,,$$

with

$$\begin{aligned} S_{11} &= \lambda W^{(0)} + 2(1-\lambda) f_{11}(\underline{a}, \underline{v}) E_2 \\ S_{12} &= \lambda W^{(1)} + 2(1-\lambda) f_{12}(\underline{a}, \underline{v}) E_2 = S_{21} \\ S_{22} &= \lambda W^{(2)} + 2(1-\lambda) f_{22}(\underline{a}, \underline{v}) E_2, \end{aligned}$$

where f_{ij} denote functions quadratic with respect to $\underline{a}$ and $\underline{v}$ and depending on d_j.

We again get an eigenvalue problem, but now a nonlinear one. We rewrite it as a problem of determining the zeros $\underline{u}^T = \left(\underline{a}^T, \underline{v}^T, \mu\right)$ of

$$\underline{F}(\underline{u}) = 0,$$

where we include in $\underline{F}$ the component $\|\underline{v}\|^2 - \sigma^2$. This nonlinear system is solved by a several-dimensional **Newton method**, choosing a starting value $\mathring{\underline{u}}$, linearizing $\underline{F}$

at $\mathring{\underline{u}}$ to get

$$\underline{\widetilde{F}}(\underline{u}) = \underline{F}(\mathring{\underline{u}}) + D_F(\mathring{\underline{u}})(\underline{u} - \mathring{\underline{u}}),$$

where $D_F(\mathring{\underline{u}})$ is the Jacobian of $\underline{F}$ at $\mathring{u}$ and then iterating

$$\underline{u}^1 = \mathring{\underline{u}} - D_F(\mathring{\underline{u}})^{-1}\underline{F}(\mathring{\underline{u}}).$$

More generally,

$$\underline{u}^{j+1} = \underline{u}^j + \Delta^j,$$

with Δ^j solving

$$D_F(\underline{u}^j)\Delta^j = -\underline{F}(\underline{u}^j).$$

(It is not really necessary to compute $D_F(\underline{u}^j)^{-1}$; it is enough to solve this equation for a given right hand side).

In our case the **Jacobian** D_F has a special structure

$$D_F(\underline{u}) = \begin{pmatrix} J_{11} & J_{12} & 0 \\ J_{21} & J_{22} & \underline{v} \\ 0 & \underline{v}^T & 0 \end{pmatrix}.$$

The equation is solved by any iteration method such as Gauß–Seidel or SOR; we get an "inner" iteration in that way, being performed in any step of the "outer" j-iteration. We may remind that λ has to be chosen by the user.

Problem 3 is a bit simpler since there is no constraint; δ is given and we start with

$$D_\lambda(\underline{a}, \underline{v}) = \lambda\phi_1(\underline{a}, \underline{v}) + (1-\lambda)\Psi(\underline{a}, \underline{v}, \delta).$$

Again, we may write the regression as a nonlinear equation for $\begin{pmatrix} \underline{a} \\ \underline{v} \end{pmatrix}$, which we try to solve by a **Newton method**. We are not completely in command of the domain of convergence. Although, practically, everything looks fine yet, in reality, since we do not require that $\|\underline{v}\| = \sigma$, we get a new critical point $\underline{a} = \underline{v} = \underline{0}$ and we cannot exclude that our method just converges to this trivial solution. To explain the situation, we study a similar one-dimensional problem with $x \in \mathbb{R}$ instead of $\begin{pmatrix} \underline{a} \\ \underline{v} \end{pmatrix}$; G consists of a quadratic and a fourth order term

$$G(x) = \alpha x^2 + \beta(x^2 - \delta)^2,\ \alpha, \beta, \delta > 0.$$

If $\alpha \geq 2\beta\delta$, g has only one minimum: the trivial one $x = 0$;
if $\alpha < 2\beta\delta$, g has a maximum at $x = 0$ and two minima at
$x = \pm\sqrt{\delta - \frac{\alpha}{2\beta}}$

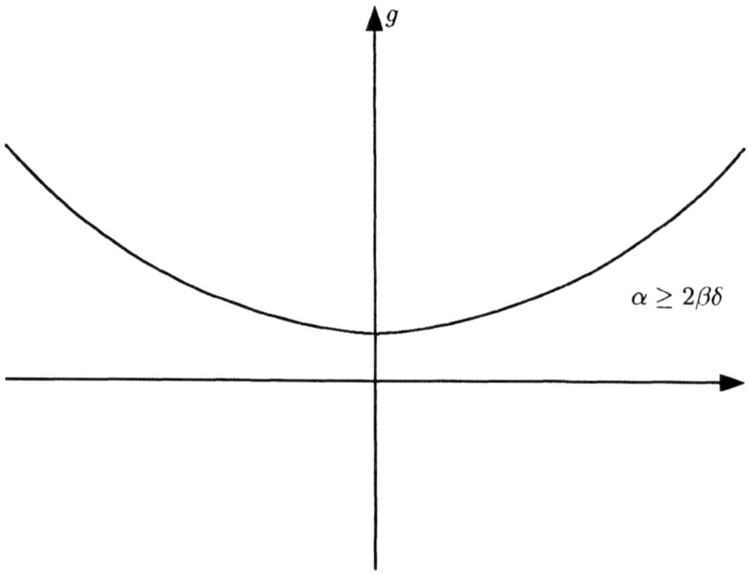

Figure 1.2.3

We are interested in the second situation only; the one-dimensional example suggests that we get an interesting solution if at $\underline{a} = \underline{v} = 0$ the function G has a maximum. To see when this is the case, let us consider the Hessian of G at $\underline{a} = \underline{v} = 0$. We get

$$DG(\underline{0},\underline{0}) = \lambda \begin{pmatrix} W^{(0)} & W^{(1)} \\ W^{(1)} & W^{(2)} \end{pmatrix} - 2(1-\lambda) \begin{pmatrix} f_{11}(\underline{0},\underline{0})E_2 & f_{12}(\underline{0},\underline{0})E_2 \\ f_{21}(\underline{0},\underline{0})E_2 & f_{22}(\underline{0},\underline{0})E_2 \end{pmatrix}.$$

The last matrix is again a Gramian of the two vectors $\left(\sqrt{d_1},...,\sqrt{d_k}\right)^T$ and $\left(\sqrt{d_1}t_1,..., \sqrt{d_k}t_k\right)^T$ and therefore positive definite. For sufficiently small λ, the matrix $DG(\underline{0},\underline{0})$ is therefore negative definite, so that we are in the (hopefully) interesting situation. Therefore we try.

Remark. Until now, we have only used ϕ_1. Let us at least look at the simplest problem 1 with ϕ_2 instead of ϕ_1. We want to see how sensitive the solution is with respect to the chosen functional. Moreover, we try another method of solution.

We want to minimize $\phi_2(\underline{a},\underline{v}) = \sum_{j=1}^{k} \frac{<\underline{x}_j,\underline{w}_j>}{||\underline{x}_j||^2}$ with the constraint $||\underline{v}||^2 = \sigma^2$. We avoid the Lagrange multipliers by putting

$$\underline{v} = \sigma \begin{pmatrix} \cos\varphi \\ \sin\varphi \end{pmatrix},$$

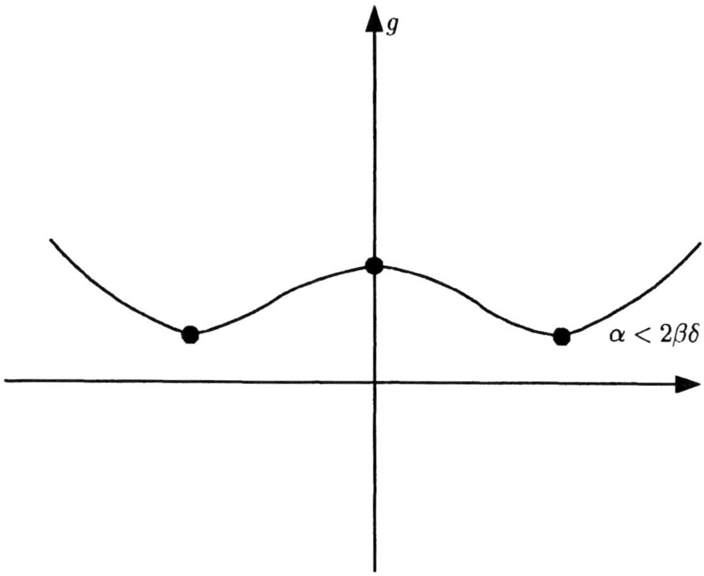

Figure 1.2.4

and writing ϕ_2 as a function of $\underline{a}$ and φ: $\phi_2(\underline{a}, \varphi)$. The price we pay is the shape of the normal equations. $\text{grad}_a \widetilde{\varphi}$ is a rational function of (a_1, a_2) and has many zeros. $\widetilde{\phi}$ oscillates near the origin and at infinity. Newton's method cannot be used for that reason–the domain of convergence is very small. We need other optimization methods to approach the problem, but we did not find methods which were absolutely reliable. In 95% of all practical problems they work - but clients in industry want software which never fails. To many exceptions, to interactivity with the user means to need perturbations. We therefore abandoned these ideas.

1.2.4 The interpretation of the results

We are not allowed to present real data. Therefore we show a constructed example which is not far from reality.

We assume that the "truth" is

$$\underline{x}(t) = \begin{pmatrix} -10 \\ 25 \end{pmatrix} + t \begin{pmatrix} 0 \\ -6 \end{pmatrix}, \; 0 \le t \le 50,$$

and $\delta = 0.1$.

We make the data very noisy by adding normally distributed errors to w_j with a variation w_r and also normally distributed errors to d_j with a variation d_r. For t_j, we take 100 equidistant points. For functional ϕ_1, Problems 1, 2 and 3 give the following estimates for $\underline{a}$ and $\underline{v}$, which are correctly $\begin{pmatrix} -10 \\ 250 \end{pmatrix}$ and $\begin{pmatrix} 0 \\ -6 \end{pmatrix}$:

$\mathbf{w}_r$	$\mathbf{d}_r$	Problem 1		Problem 2		Problem 3	
		$\underline{a}$	$\underline{v}$	$\underline{a}$	$\underline{v}$	$\underline{a}$	$\underline{v}$
0.1	0.0	-8.75	-0.03	-8.83	-0.03	-9.56	-0.01
		245.00	-5.99	245.00	-5.99	249.50	-5.98
0.3	0.0	-4.11	-0.14	-4.21	-0.14	-8.13	-0.05
		220.00	-5.99	221.00	-5.99	239.00	-5.73
0.5	0.5	0.36	0.23	0.43	0.25	-5.36	-0.12
		192.0	-5.99	197.0	-5.99	239.00	-5.73

Problem 1 takes less computing time, Problems 2 and 3 are solved more slowly, but with approximately equal effort. Problem 3 delivers clearly the best solution. Our results show that the functional ϕ_2 cannot compute – it is slower and less precise.

Of course the computation is done on-line – one uses the data at hand at a certain time. Each newly incoming measurements improve the result; the "old" result is used as starting value for the Newton iteration. In that way, this iteration needs fewer steps.

In the beginning, the results are less encouraging, especially if the course of the ship is directly pointed at the observer. However, they soon get better if the path does not go exactly through the origin.

The client accepted the results but soon asked for higher computational speed. A problem for industry is never solved: it is like in a German fairy tale of a little boy with the strange name Häwelmann. He was never content with what he got; he always shouted "more, more".

1.3 The Airbag-sensor

1.3.1 The objective

Usually, in an **airbag system** of a car, the blowing up of the airbag is controlled by an electronic and an electromechanical sensor. The airbag will be released if both sensors react on an outer influence closing an electronic circuit.

In this section, we deal with the electromechanical sensor, the so-called **Safing sensor.** Basically, it consists of a magnet along with a metallic ring which can move within certain boundaries. The electronic circuit will close if the ring moves over a certain distance. The sensor has to be built in such a way that the circuit is not closed if the force is small (say, in case of braking only), but is closed in case of a strong force (crash case).

We discuss here a model for the magnetic force and other influences on the ring.

1.3.2 The modelling project

Model a **Safing sensor**, that is, the movement of a ring under the action of an external force and the magnetic field.

1.3.3 The algorithmic project

Give a correct computational method for the model. This does not come immediately. No on-line answer is required. This project was carried out by Dr. Axel Klar, Martin Braun and by a group of students during the ECMI-Modelling week 1994.

1.3.4 The modelling of Safing sensor

Usually, in the airbag system of a car there are two sensors controlling the blowing up of the airbag. The first sensor is a purely electronical sensor. This sensor may also react on electromagnetic influences not caused by a car crash. To avoid the blowing up of the airbag in such a case, a second sensor is used. This is the so-called Safing sensor. It is built on an electromechanical basis. The airbag will blow up only if both sensors react.

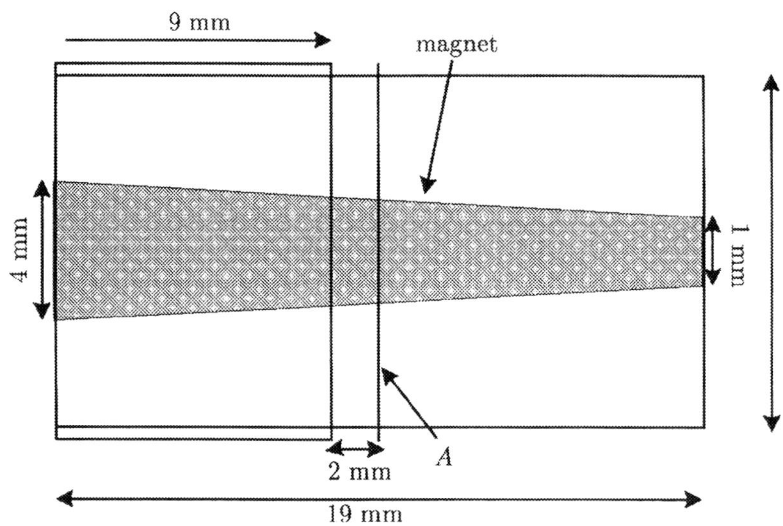

Figure 1.3.1

In this project, we deal with the Safing sensor only. It consists of a conical magnet in a cylinder. Around this cylinder, there is a metallic ring with mass m that can move along the cylinder. Initially, the ring must be at one side as shown in Figure 1.3.1. If it hits one side with a certain velocity v, then it bounces back with a velocity $c \cdot v$, where c is the elasticity coefficient (in our models $c = 0.3$).

Whenever there are forces on the car, there is also an acceleration of the ring in relation to the cylinder. So the ring starts moving. If it moves over a certain point, say A, the electronical circuit closes. However, this should only happen if the forces on the car are very large, i.e., if there is a crash. If these forces are only small, for example, if one brakes, the circuit should not close. In that case, the ring should not move over the point A. Further, after the braking, the ring must move back to its initial position. All this can be accomplished by the magnet and its magnetic field and force. This force causes the ring not to move too far if there is only a small force on the car.

In the project, we wanted to determine the movement of the ring. The main problem was the calculation of the magnetic field and the magnetic force. First, we assumed the magnetic force to be a constant or a linear function of x, where x is the direction of the movement. Then, we worked out the real magnetic field and the magnetic force theoretically and found a numerical solution for this case.

It is important to get an estimation for the accelerations of the ring in case of a crash and in case of braking only. Then, the required magnetic force and, thus, the magnetic field can be determined. When all forces are known, the movement of the ring can be modelled as a function of time.

In this section, realistic values for the acceleration of the car (and the ring in the sensor) will be determined for the crash case and the braking case. We will consider only a little crash and very strong braking. If the ring moves a suitable distance in these cases, then it will also move for a strong crash or little braking.

Suppose a car crashes a wall at a speed of 36 km/h; i.e., 10 m/s, and suppose the crash lasts 100 minutes. Then the acceleration is $(a = v/t)100\ \text{m/s}^2 \approx 10\,g$ where g is the earth acceleration. This is not very accurate, but from experiments it follows that this value is realistic, although it is very small.

Suppose a car that is driving at a speed of 36 km/h brakes suddenly and stands still after half a second (this is very fast). Then, the mean acceleration is $20\ \text{m/s}^2 \approx 2g$. This value is also realistic although it is very large.

From this, the magnitude of the magnetic force can be determined such that the circuit closes if the car crashes and does not close if the car only brakes. The acceleration of the ring due to the magnetic force should be within $20\ \text{m/s}^2$ and $100\ \text{m/s}^2$ and acting against the crash and braking force.

Governing equations

The physics of our problem can be described by Newton's second law:

$$m\ddot{\vec{x}} = \sum_{i=1}^{n} \vec{F}_i.$$

Choosing the positive x-direction towards the right of the initial position of the ring, we obtain

$$m_R\ddot{x}(t) = F_{\text{car}} + F_{fric}(\dot{x}(t)) - F_{\text{mag}}(t, x(t))F_{\text{ind}}(\dot{x}(t)),$$

where

- F_{car} is the force which is exerted on the car. It can either be the crash force F_{crash} or the brake force F_{brake}.

- F_{fric} is the friction force between the ring and the cylinder. It always acts against the movement; i.e.,

$$F_{\text{fric}} = \left\{ \begin{array}{ll} -\mu m_{\text{Rg}} & \text{for } \dot{x}(t) > 0, \text{ i.e, the ring moves} \\ & \text{towards the position } x\text{-direction} \\ \mu m_{\text{Rg}} & \text{for } \dot{x}(t) < 0, \text{ i.e., the ring moves} \\ & \text{towards the negative } x\text{-direction} \end{array} \right\},$$

  where $m_R = 0.002$ kg is the mass of the ring, $\mu = 0.5$ is the friction coefficient and $g = 9.81$ m/s^2 is the earth acceleration.

- F_{mag} is the magnetic force on the ring.

- F_{ind} is the induced induction force.

From now on, we consider only the different kinds of accelerations on the ring which can be obtained from the corresponding forces divided by the mass of the ring. Then,

$$\ddot{x}(t) = a_{\text{car}}(t) + a_{\text{fric}}(\dot{x}(t)) - a_{\text{mag}}(t, x(t)) - a_{\text{ind}}(\dot{x}(t))$$

holds. This equation is a non-linear ordinary differential equation which can be solved by numerical methods only. As an integration method, we used the trapezoidal rule which is applied twice to get the velocity $\dot{x}(t)$ and then the movement of the ring $x(t)$.

First model

In the first model, the magnetic force which is exerted on the ring is assumed to be constant in time and independent of the position of the ring. The acceleration of the ring due to this magnetic force is $a_{\text{mag}} = \text{const}$. Furthermore, the acceleration due to the force on the car is taken as

$$a_{\text{car}}(t) = \begin{cases} (1 - \cos\omega\ t) \cdot\ a_c & \text{for } t \leq T \\ 0 & \text{for } t > T, \end{cases}$$

where a_c is the amplitude of the acceleration, $\omega = \frac{2\pi}{T}$, and T is the time that the force lasts. This cosine function appears to be a rather good approximation of the real acceleration if there is a crash or a brake force. The acceleration due to the friction between the ring and the cylinder is as described above. To approximate the movement of the ring in case of a real crash with this model, the following values are used:

$$\begin{aligned} a_c &= 100 \text{ m/s}^2 \\ a_{\text{mcg}} &= 20 \text{ m/s}^2 \\ a_{\text{fric}} &= 5 \text{ m/s}^2 \\ T &= 25 \text{ ms}. \end{aligned}$$

To approximate the movement in case of braking, the same values are used except for a_c which is taken as 40 m/s^2.

Second model

In the second model, the acceleration due to the magnetic force and the friction force is the same as in the first model, but the acceleration due to the force which is exerted on the car is taken from a real crash data. A picture of these values is shown in Figure 1.3.2.

Again $a_{\text{mag}} = 20\text{m/s}^2$ and $a_{\text{fric}} = 5\text{m/s}^2$. The crash time T is approximately 90 ms whch can be easily calculated.

With these values, it appears that the ring stays a longer time at the right hand side of the cylinder than in the previous model. This is because the crash lasts a longer time. Therefore, the closing time of the electronical circuit is longer (about 90 ms). (The closing time is the time during which the ring is beyond point A, i.e., the time between two passings of points A).

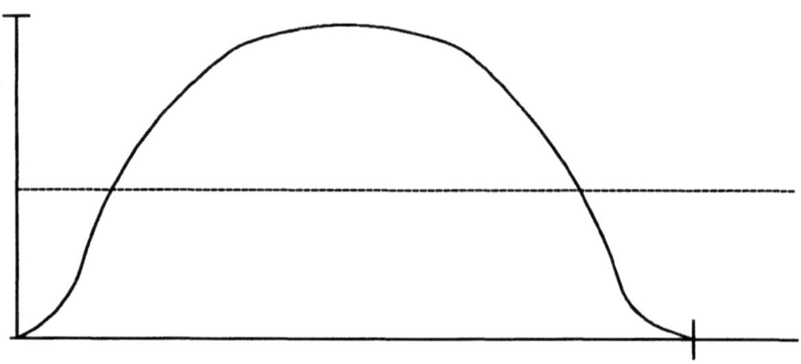

Figure 1.3.2

Third model

In the third model, again the acceleration due to the forces on the car is taken from a real crash, and the friction force is the same as in the two previous models. In this model, the magnet in the cylinder is replaced by a spring which gives the following linear acceleration of the ring as a function of its position:

$$a_{\text{spring}}(x) = C_{\text{spring}} \cdot x + c_0,$$

where C_{spring} is the spring constant and c_0 is the spring force on the ring, if it is in its initial position divided by the mass of the ring.

C_{spring} and c_0 are chosen $2000\,\text{s}^{-2}$ and $5\,\text{m/s}^2$, respectively. The acceleration c_0 is introduced to make sure that the ring moves back to its initial position after braking. The closing time with the approximation is 95 ms.

1.3.5 The advanced model

Here, we have to calculate the force the conical permanent magnet exerts on the ring. However, this includes magnetization in an indirect manner only: magnetization is generated by the microscopic current described by the Maxwell equation. For this we require a book on classical electro-dynamics, for example, by J.D. Jackson or by Sommerfeld or the comparatively current book by Dautry-Lions, see references in Chapter 3. The main tool is always the system of **Maxwell equations**. We are mainly interested in macroscopic effects, that is, in the **magnetizations** and the forces induced by it. Here one has to be careful: macroscopic currents also play a role when we note, for example, that the movement of the ring also implies a movement of charges and thus a current; the resulting so-called electromotive force is not taken it into account initially.

We observe that the movement of the ring is also a movement of the weight and that creates a current and we consider this electric force in the next step. Let us describe the magnetization of the ring and the force which results from it by formulating the Maxwell equation in terms of the magnetization instead of currents. **Magnetization** is a vector field $\underline{M}(\underline{x})$ which we may consider as a sum of moments of elements of magnetic field per unit volume, that is, as a density.

Between magnetic field $\underline{H}$, magnetic induction $\underline{B}$ and magnetization $\underline{M}$, the following relation holds:

$$\underline{H} = \frac{1}{\mu_0}\underline{B} - \underline{M} \text{ or } \underline{B} = \mu_0 \underline{H} + \mu_0 \underline{M},$$

where μ_0 is a well known constant.

$\underline{M}$ is the magnetization of the cone K, $\underline{M}_K$ and the induced magnetization of the ring $\underline{M}_R$ which is not known. However, $\underline{M}_K$ is given and known. It is constant in K and is in the direction of the axis of the magnet, that is, of the x-axis

$$\underline{M}_K(x, y, z) = (M_1, 0, 0).$$

The stationary Maxwell equation is given below

$$\text{div } \underline{B} = 0,$$

or

$$\text{div } \underline{H} = -\text{div } \underline{M}.$$

If we could determine H, we would get

$$\underline{B} = \mu_0 \underline{H} + \mu_0 \underline{M}.$$

$\underline{B}$ now generates the magnetization of the ring and

$$\underline{M}_R = \frac{\mu_R - 1}{\mu_0 \cdot \mu_R} \cdot \underline{B}/R = \mu \underline{B}/R.$$

In principle, we have to solve the following system of equations:

$$\begin{aligned} \operatorname{div} \underline{H} &= -\operatorname{div} \underline{M}_K - \operatorname{div} \underline{M}_R \\ \underline{B} &= \mu_0 \underline{H} + \mu_0(\underline{M}_K + \underline{M}_R) \\ \underline{M}_R &= \mu \underline{B}. \end{aligned}$$

For determining H and $\underline{B}$, we must know $\underline{M}_R$ which again depends on B. In reality, M_R has less influence than B at least compared to $\underline{M}_K$. Here, one can think of an iteration, say,

$$\begin{aligned} \operatorname{div} \underline{H}^{(i+1)} &= -\operatorname{div} \underline{M}_K - \operatorname{div} \underline{M}_R^i \\ \underline{B}^{(i+1)} &= \mu_0 \left(\underline{H}^{(i+1)} + \underline{M}_K + \underline{M}_R^{(i)} \right) \\ \underline{M}_R^{(i+1)} &= \mu \underline{B}^{(i+1)}, \quad i = 0, 1, 2, \cdots \end{aligned}$$

One can start with $M_R^{(0)} = 0$; so

$$\begin{cases} \operatorname{div} \underline{H}^{(1)} = -\operatorname{div} \underline{M}_K \\ \underline{B}^{(1)} = \mu_0(\underline{H}^{(1)} + \underline{M}_K) , \end{cases} \tag{$*$}$$

and finally

$$\underline{M}_R^{(1)} = \mu \underline{B}^{(1)}.$$

One can hope that $\underline{B}^{(1)}$ and $\underline{M}_R^{(1)}$ are sufficiently accurate and the following iterations do not yield a big change. In any case, we shall try to do this without having estimated the error so far.

How does one solve $(*)$? The first problem only appears because $\underline{M}_K$ is not differentiable. It jumps at the surface ∂K and is otherwise constant. This problem is, like several others in the field of differential equations, self-fabricated. Really, the equation is as follows:

The flux of $\underline{B}^{(1)}$ through an arbitrary surface of a bounded smooth body is always zero; that is,

$$\int_S \langle \underline{B}^{(1)}, \underline{n} \rangle dw = 0,$$

for all closed surfaces S; of course, the Gauss theorem changes this into div $\underline{B} = 0$ as:

$$\int_V \operatorname{div} \underline{B}^{(1)} d\underline{x} = \int_{\partial V} \langle \underline{B}^{(1)}, \underline{n} \rangle dw = 0,$$

for all sets V with sufficiently smooth boundary and from this it directly follows that

$$\operatorname{div} \underline{B}^{(1)} = 0 .$$

But, of course, the theorem is valid for differentiable $\underline{B}^{(1)}$ only– and this is not the case here, at least not on ∂K. Away from ∂K, we have div $\underline{M}_K = 0$; i.e., div $\underline{H}^{(1)} = 0$; on ∂K, and it follows from $\int_S \langle \underline{B}^{(1)} . \underline{n} \rangle dw = 0$, that

$$\langle \underline{B}^{(1)}, \underline{n} \rangle_{\partial K_+} = \langle \underline{B}^{(1)}, \underline{n} \rangle_{\partial K_-},$$

where $\underline{n}$ is the outer normal vector on ∂K and ∂K_+ denotes the boundary limit when we approach the ∂K from the interior, where ∂K_- is the exterior limit. This can be seen from Figure 1.3.3.

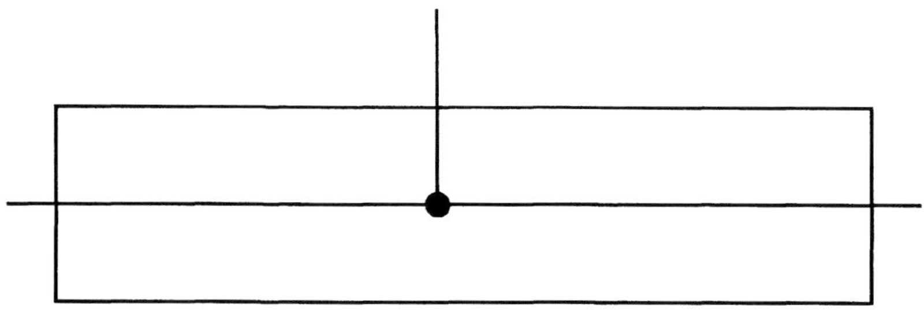

Figure 1.3.3.

From $\int_S \langle \underline{B}^{(1)}, \underline{n} \rangle dw = 0$, we have $\langle \underline{B}^1, \underline{n} \rangle_{\partial K_-} + \langle \underline{B}^{(1)}, \underline{n} \rangle_{\partial K_+} = 0$. (When the closed sets are very thin, the side face does not play any role). This is precisely the continuity of $\langle \underline{B}^{(1)}, \underline{n} \rangle$. Thus we have

$$\underline{B}^1 = \mu_0 (\underline{H}^{(1)} + \underline{M}_K),$$

which implies that

$$\langle \underline{H}^{(1)}, \underline{n} \rangle_{\partial K_-} = \langle \underline{H}^{(1)}, \underline{n} \rangle_{\partial K_+} + [\langle \underline{M}_K, \underline{n} \rangle],$$

where

$$[\langle \underline{M}_K, \underline{n} \rangle]$$

is the jump of the normal component of $\underline{M}_K$.

Our problem to determine $\underline{H}^{(1)}$ now reads div $\underline{H}^{(1)} = 0$ away from ∂K. On ∂K, the above jump condition must hold.

In order to calculate $\underline{H}^{(1)}$, we need the following Maxwell equation:

$$\text{rot} \underline{H}^{(1)} = \underline{j}\,.$$

Here, j denotes the microscopic and macroscopic currents. Our interest is $\underline{H}^{(1)}$ in the ring R, where we started with $\underline{M}_R^{(0)} = 0$. Out of K and specially in the ring R, the j is equal to zero, i.e., $\underline{H}^{(1)} = 0$. These are the integrability conditions for $\underline{H}^{(1)}$ and in view of the Poincaré Lemma, we can suppose a potential ϕ of $\underline{H}$

$$\underline{H}^{(1)} = -\nabla\phi.$$

This gives us

$$\Delta\phi = 0,$$

and

$$\frac{\partial\phi}{\partial\underline{n}}\Big/\partial K_{+} = \frac{\partial\phi}{\partial\underline{n}}\Big/\ \partial K_{-} - \langle\underline{M}_1, \underline{n}_1\rangle.$$

One observes how the special conical form of the magnet comes into play.

If the magnet were of cylindrical form, we would have $n_1 = 0$ on the surface of the cylinder and there would be no jump in the normal derivative; the field only appears at the side face. Now, we utilize our knowledge of the solution of Laplace equation and jumps of the solution. One needs the well-known theory of "single layer" which together with "double layers" is the main tool for the boundary integral method. We describe quickly all the relevant material.

Let $N(\underline{x}, \underline{y}) = \frac{1}{4\pi\|\underline{x}-\underline{y}\|}$ be the kernel of the Laplace equation, then we have, for given $m \in C^0(\partial K)$: The function

$$\phi(\underline{x}) = \int_{\partial K} N(\underline{x}, \underline{y})m(\underline{y})dw(\underline{y})$$

satisfying the Laplace equation outside the ∂K is continuous in the whole space and its normal derivative $\frac{\partial\phi}{\partial n}$ has a singularity of order m.

$$m = \frac{\partial\phi}{\partial\underline{n}}\Big/\partial \underline{K} - \frac{\partial\phi}{\partial\underline{n}}\Big/\ \partial K_{+}\ .$$

The desired potential is

$$\phi(\underline{x}) = \int_{\partial K} N(\underline{x}, \underline{y})\langle\underline{M}, \underline{n}\rangle dw.$$

We look at the ring R. Having ϕ, we get

$$\underline{B}^{(1)} = -\mu_0\nabla\phi\ \ (\underline{M}_K\ \text{ is in the zero ring}),$$

and therefore we have

$$\begin{aligned} \underline{M}_R^{(1)} &= \mu\underline{B}^{(1)} = -\mu_0\mu\nabla\phi \\ &= \frac{\mu_0\mu}{4\pi}\int_{\partial K}\langle\underline{n}_1, \underline{M}_1\rangle(\underline{y})\frac{\underline{y}-\underline{x}}{\|\ \underline{y}-\underline{x}\ \|^3}dw(\underline{y}), \qquad \underline{x} \in R. \end{aligned}$$

The conical surface consisting of side, top and bottom must be parametrized and integrated. It goes straight since M_1 is constant and $n_1(y)$ is fixed on any one of the three sub-surfaces.

Now it is nearly done: we have the magnetization of the ring from which we can calculate the force K on R. In principle, this is done again with the underlying micro-currents and associated Lorenz force $\underline{j} \times \underline{B}$. There are currents inside R and on the surface. We have

$$\text{Force} = \int_R (\underline{j}_i \times \underline{B}) d\underline{x} + \int_{\partial R} (\underline{j}_0 \times \underline{B}) dw.$$

We are in the "Macro-theory" with magnetization instead of current and we must know that

$$\underline{j}_i = \text{rot}\, \underline{M}_R \quad \text{and} \quad \underline{j}_0 = \underline{M}_R \times \underline{n} \ \text{ holds}\,.$$

With this we have

$$\text{Force} = \int_R (\text{rot}\, \underline{M}_R) \times \underline{B} d\underline{x} + \int_{\partial R} (\underline{M}_R \times, \underline{n}) \times \underline{B} dw.$$

By the identity $\underline{a} \times (\underline{b} \times \underline{c}) = \langle \underline{a}, \underline{c} \rangle \underline{b} - < \underline{a}, \underline{b} > \underline{c}$ and the relation div $\underline{B} = 0$ we get

$$\text{Force} = \int_{\partial R} \langle \underline{M}_R, \underline{n} \rangle \underline{B} dw.$$

We are again interested in the component of the force in the direction of the axes of the cylinder, that is, $\underline{B}_1$. This means the evaluation of another surface integral over the surface of the ring which has 4 parts: the inner and outer cylinders, and both frontal sides. This is also simple, although $\underline{M}_R$ in R is not constant in R.

The movement of the ring under this magnetic force can be seen in Figure 1.3.4.

This is what has been done up to now. However, there is still much to do, for example:

Clarify whether the iteration of $\underline{H}^{(1)}$ can really be terminated. Estimate how the electromotive forces influence the behaviour. Incorporate the switching into the investigation. It looks as if the third point will lead to a continuation of the project. But if we want to have simulation results which are really accurate, we have to look at many points in a more precise manner.

The fundamental equations (Maxwell's equations) are completely reliable, classical electrodynamics is quite a mature theory. Moreover, the material constants are well known. What needs to be examined here are the simplifications and neglections: problems which are clearly formulated mathematically but may sometimes be difficult to solve. This is a very rewarding area of research where, surprisingly, few mathematicians are working!

In Chapter 3, the fundamental results concerning Maxwell's equations are presented indicating the current techniques and methodology for their solutions.

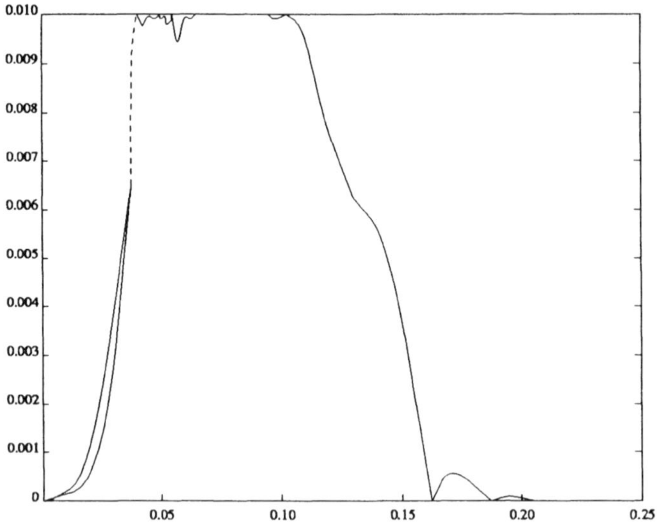

Figure 1.3.4. Movement of the ring with the real magnetic force.

1.4 How to judge the quality of a nonwoven fabric

1.4.1 The problem

A variety of industrial products, ranging from carpets to baby napkins, make use of artificial fabrics. Nonwoven fabrics and agglomeration of (plastic) fibres are mainly used. They are sometimes called "**fleeces**". The visual and mechanical properties of these fabrics depend on the homogeneity of these fibres – homogeneity with respect to the local density of the fibres and homogeneity with respect to their directions. If there are areas of different density, they look like dark or light clouds; the defect of nonuniform density is therefore called "cloudiness". If many parallel fibres stick together, the fabric shows anisotropic or "ships". **Clouds** and **ships** reduce the quality of a fabric; traditionally this quality was judged through visual inspection – experts looked at the fabric and ranked its quality. To get a more objective judgment, one needs a measure for **nonuniformity**. The company takes on-line images of the fabric; ca 1 m long pieces of the fabric ribbon are mapped and should be evaluated on-line in order to change or stop the production process if the quality decreases.

The problem consists in developing a mathematical model evaluating the fabric images which describes the nonuniformity of these images in a way corresponding to the judgment of quality by experts. The calculation of the quality given by the model should be fast enough to enable an on-line judgment.

The project has quite a long history, in which several people worked successfully but only for a short while, when further improvements were requested ("More,

more", shouts little Hävelmann!). With respect to cloudiness, it seems that we got a saturation and there is now even a software product judging the cloudiness in a proper way. Ships are still under consideration.

The long history has led to a long list of contributors: Dr. Brian Wetton, Dr. H. G. Stark (both are meanwhile professors, in Canada and in Germany respectively), Dr. P. Hackh, Dr. R. Rösch, Dr. J. Weickert (and one of us). We shall describe some attempts in detail; others just by mentioning them.

1.4.2 The models: A first approach

At the end of the production process, a piece of the fabric is screened by a laser and the intensity of the transparent light is measured "pointwise". Such a point has a diameter corresponding to that of the laser beam and there are ca $N = 5000$ points in each image. The points are called pixels, and to each pixel a grey value, i.e., the measured intensity, is assigned. Since an image is a two-dimensional object, we use double indices (i,j) to address an individual pixel and the grey value of pixel (i,j) is denoted by $\mu_{i,j}$ and assumed to be nonnegative. We are interested in intensity fluctuations and therefore only in relative values. Hence, we normalize μ by

$$\sum_{i=1}^{N_1}\left(\sum_{j=1}^{N_2}\mu_{i,j}\right) = 1,$$

where $N_1 \cdot N_2 = N$ is the total number of pixels. As a positive, normalized function on a finite domain, it may be interpreted as a probability density and this in fact helps us in getting ideas in spite of the fact that there is no randomness involved.

Absolute uniformity would mean $\mu_{i,j} = \frac{1}{N}$ for all (i,j); we denote this uniform distribution by $\mathring{\mu}_{i,j}$.

Our modelling task seems to be clear: if we want one number to characterize the quality, it should be something like a distance between a measure μ and $\mathring{\mu}$. The larger the distance, the poorer the quality.

1) Since μ is given by an N-tuple, i.e., $\mu \in \mathbb{R}^N$, we may use any distance known in $\mathbb{R}^N$, for example

$$d_p(\mu,\mathring{\mu}) := \left(\sum_{i,j}|\mu_{i,j} - \mathring{\mu}_{i,j}|^p\right)^{\frac{1}{p}}.$$

This is a simple but a wrong idea. Why is it wrong, we understand by looking at a "one-dimensional fabric" (it prevents us only from elaborate index handling – the idea is, of course, independent of the dimension): A cloud is a section $\alpha \le i \le \beta$, in which μ_i is larger (light cloud) or smaller (dark cloud) than the average $\mathring{\mu}_i = \frac{1}{N}$.

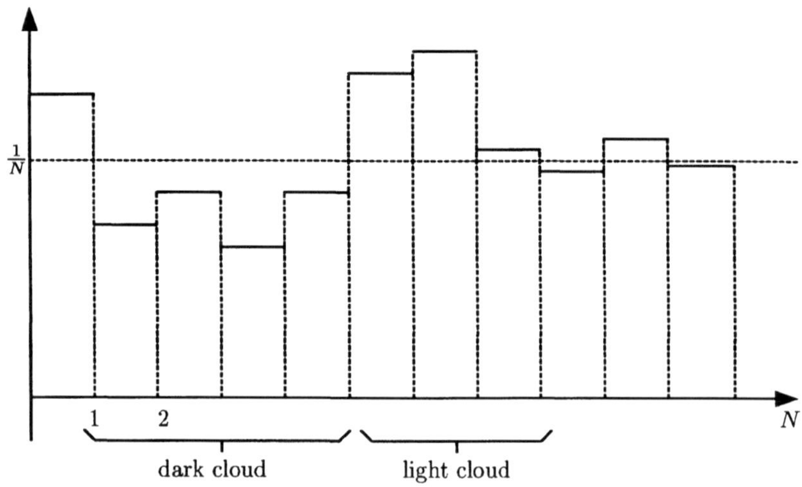

Figure 1.4.1

Now assume that we have a fabric with a large hole of length, say, $\frac{1}{10}N$ in the middle. The corresponding μ could be

$$\mu_{\text{hole}} = \left(\frac{10}{9}\frac{1}{N}, \ldots, \frac{10}{9}\frac{1}{N}, \underbrace{0, \ldots, 0}_{\text{length } \frac{N}{10}}, \frac{10}{9}\frac{1}{N}, \ldots, \frac{10}{9}\frac{1}{N} \right);$$

realize that μ_{hole} is normalized to 1: We have $\frac{9}{10}N$ pixels of intensity $\frac{10}{9}\frac{1}{N}$ and $\frac{1}{10}N$ pixels of intensity 0, which sums up to 1. Such a fabric has a very big dark hole – 10% of the image is completely dark.

Now take another rather regular fabric – not uniform, but oscillating and without big holes

$$\mu_{\text{osc}} = \left(\frac{2}{3}\frac{1}{N}, \frac{4}{3}\frac{1}{N}, \frac{2}{3}\frac{1}{N}, \frac{4}{3}\frac{1}{N}, \ldots, \frac{2}{3}\frac{1}{N}, \frac{4}{3}\frac{1}{N} \right).$$

We see that $|\mu_i - \mathring{\mu}_i| = \frac{1}{3N}$ and therefore

$$d_2\left(\mu_{\text{osc}}, \mathring{\mu}\right) = \frac{N^{\frac{1}{2}}}{3N} = \frac{1}{3\sqrt{N}}.$$

The distance of μ_{hole} to uniformity is

$$d_2\left(\mu_{hole}, \mathring{\mu}\right) = \left[(N - \frac{N}{10})(\frac{10}{5N} - \frac{1}{N})^2 + \frac{N}{10}(\frac{1}{N})^2 \right]^{\frac{1}{2}} = \frac{1}{3\sqrt{N}}.$$

Both fabrics – the oscillating one and the one with the big dark hole – have the same d_2-distance, but certainly not the same quality.

d_p-distances have the property that they are invariant against index permutations; since $\mathring{\mu}$ does not change under index permutation we have

$$\left(\sum_{i=1}^{N}\left(\mu_{z(i)}-\frac{1}{N}\right)^{p}\right)^{\frac{1}{p}}=\left(\sum_{i=1}^{N}\left(\mu_{i}-\frac{1}{N}\right)^{p}\right)^{\frac{1}{p}}.$$

By index permutation, many small holes may be transformed into one big hole, and both would have the same d_p distance. This contradicts the idea of quality.

2) A physicist has a natural measure for the distance to uniformity: the entropy. Since we compare with $\mathring{\mu}$, we speak of relative entropy – the entropy in comparison to $\mathring{\mu}$. It is defined by

$$d_{\text{ent}}\,(\mu,\mathring{\mu})=\sum_{i,j}\mu_{ij}\ln\frac{\mu_{ij}}{\mathring{\mu}_{ij}}=\sum_{i,j}\mu_{ij}\ln\mu_{ij}+\ln N.$$

But what was true for d_p – and what is an advantage of entropy in statistical physics – is true for d_{ent} and therefore unsuitable for our purpose: invariant against index permutations, a big hole would count as many small ones.

3) We should take the idea of a hole more seriously. A hole is an index domain (a set of neighbouring pixels) Ω, in which $\mu_{ij}-\frac{1}{N}$ has one sign. It is larger or smaller than $\frac{1}{N}$ everywhere in Ω. The size of the hole is given by the size of Ω and by the deviations of the grey values from average. We call it "hole volume" and define it as

$$\left|\sum_{(i,j)\in\Omega}(\mu_{ij}-\frac{1}{N})\right|.$$

We restrict our consideration again to one-dimensional fabrics for the sake of simplicity. Here, Ω is a section $[\alpha,\beta]$ and the hole volume is

$$\left|\sum_{j=\alpha}^{\beta}(\mu_j-\frac{1}{N})\right|,$$

α,β are boundaries of holes – by definition. The largest hole is therefore

$$\widetilde{D}(\mu,\mathring{\mu})=\max\left\{\left|\sum_{j=\alpha}^{\beta}(\mu_j-\frac{1}{N})\right|,\ \alpha,\beta \text{ boundaries of holes}\right\}.$$

If we pass over a boundary, $\mu_j - \frac{1}{N}$ changes its sign and the volume defined as above becomes smaller.

Therefore we keep the idea of the largest hole in distance (but may change the values of the distance a bit), when we omit the restriction, that α, β are boundaries of holes and define

$$D(\mu, \mathring{\mu}) = \max_{1 \le \alpha \le \beta \le N} \left| \sum_{j=\alpha}^{\beta} (\mu_j - \frac{1}{N}) \right| ,$$

or, again, in two dimensions

$$D(\mu, \mathring{\mu}) = \max_{\Omega \text{ connected}} \left| \sum_{(i,j) \in \Omega} (\mu_{ij} - \frac{1}{N}) \right| .$$

We make a last simplification in restricting the set of connected domains to the set of rectangles. We finally get

$$D(\mu, \mathring{\mu}) = \max \left\{ \left| \sum_{(i,j) \in R} (\mu_{ij} - \frac{1}{N}) \right| , R = \{(i,j) : \alpha_1 \le i \le \beta_1, \alpha_2 \le j \le \beta_2\} \right\} .$$

We go on with this definition; it is a distance and it cares for index permutations. But there is another reason for choosing it: it has a long mathematical tradition – in a field of pure mathematics, namely, theory of numbers. And therefore there is much known about it. And moreover, we used the distance already in Problem 1.

As for one-dimensional images, already in 1917, the famous mathematician **Hermann Weyl** defined D in a paper entitled (originally in German) "On the uniform distribution of numbers modulo 1", Weyl [1917]. It is a paper in the field of analytic number theory, more precisely, in the theory of Diophantine approximation (starting with the question, how exactly any real number may be approximated by a rational number); just for scientific curiosity: Do you know **Dirichlet's approximation theorem** first proved by **Kronecker** saying that for every real number μ and any natural number N, there exist integers n, m such that

$$\left| \mu - \frac{m}{n} \right| < \frac{1}{nN}, \, 1 \le n \le N .$$

One may observe that there are always integers m, n, such that the fraction $\frac{m}{n}$ approximates μ with an error less than $\frac{1}{n^2}$. It is a nice game: Try it!

The same distance also occurs in statistics: for densities μ, ν, one defines the **Kolmogorov-Smirnov distance** as

$$D(\mu,\nu) = \max_{1\leq\alpha\leq\beta\leq N} \left| \sum_{j=\alpha}^{\beta} (\mu_j - \nu_j) \right| .$$

Here again are many results available.

It is one of the pleasures of modelling if one discovers that others working in completely different fields find similar notions and concepts. It is not only a pleasure but also an advantage. Others, in general, quite intelligent people have thought carefully about it. As a consequence, a modeller should have a deep mathematical knowledge; this knowledge might be the only valuable contribution a mathematician can make to the work of interdisciplinary problem-solving groups.

As mentioned earlier, D is a metric in $\mathbb{R}^N$; for example, there is a triangle inequality. In each metric D, there are unit circles, defined as points, whose distance to zero is less or equal one

$$E = \{\mu / D(\mu, 0) \leq 1\} ,$$

(of course, 0 is not a distribution in our sense; but we want to play with the concept to get some feeling).

If $N = 2$, we can draw a picture of E – please verify it.

It is what we are used to when thinking of unit balls, but it is reasonable (the fact that it is invariant to permutation is due to the very small N) and it may work if we are able to find fast algorithms for large N.

4) But before we try more mathematical ideas: μ are probability densities and in spaces of probability measures, we may find more concepts of distances. For example, there is a Lipschitz distance which, if μ is interpreted as a measure, is defined as

$$\rho(\mu,\nu) := \sup\left\{ \left| \int \varphi d\mu - \int \varphi d\nu \right|, \varphi \in C^b \text{ and } |\varphi(x) - \varphi(y)| \leq \|x - y\| \right\} .$$

Here, C^b is the space of bounded continuous functions on a domain in R^N – the concept has to be transferred to our situation with finite domain

$$\rho(\mu,\nu) := \sup\left\{ \left| \sum_{i=1}^{N} (\mu_i - \nu_i)\varphi_i \right|, (\varphi_1, \ldots, \varphi_N) \in \mathbb{R}^N \quad \text{and} \right.$$
$$\left. |\varphi_i - \varphi_{i+1}| \leq \frac{1}{N}, i = 1, \ldots, N, \varphi_{N+1} = 0 \right\} .$$

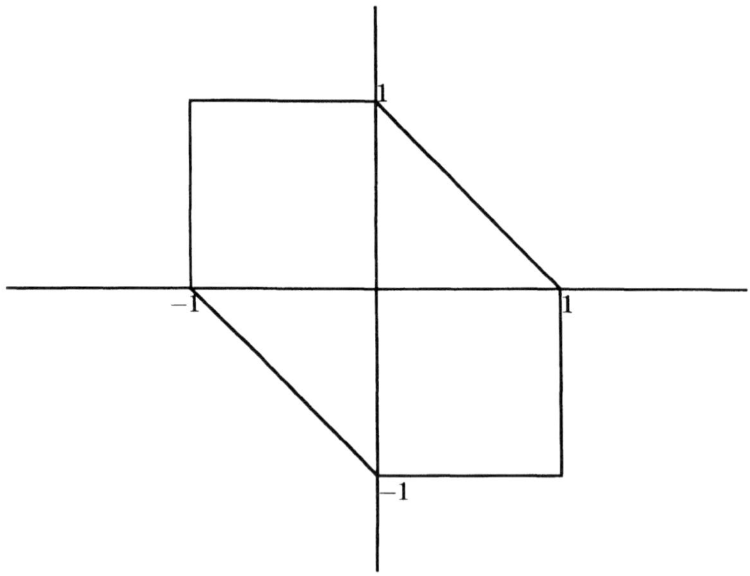

Figure 1.4.2

This seems to be very complicated. We have to maximize with respect to the set

$$\phi = \left\{ (\varphi_1, ..., \varphi_N), |\varphi_1 - \varphi_2| \leq \frac{1}{N}, \cdots |\varphi_{N-1} - \varphi_N| \leq \frac{1}{N}, |\varphi_N| \leq \frac{1}{N} \right\}.$$

This is a problem of linear programming. But, working a bit more, we find an equivalent expression for ρ, which is much simpler to handle and which we formulate as a theorem (We got it from a private communication from J. Wick in 1990).

Theorem 1 : *The* **discrete Lipschitz distance** *is also given by*

$$\rho(\mu, \nu) := \frac{1}{N} \sum_{i=1}^{N} \left| \sum_{j=1}^{i} (\mu_j - \nu_j) \right| .$$

Proof. Let $y_j := \mu_j - \nu_j$. Now

$$\sum_{j=1}^{N} y_j \varphi_j = \sum_{j=1}^{N} \left(\sum_{i=1}^{j} y_i \right) (\varphi_j - \varphi_{j+1}) ,$$

(we get it by Abel's "partial summation", the discrete analogue of partial integration formula).

Putting $y_0 = \varphi_{N+1} = 0$ and denoting $\sum_{i=0}^{j} y_i = Y_j$, we get

$$\sum_{j=1}^{N} y_j\varphi_j = \sum_{j=1}^{N} Y_j\left(\varphi_j - \varphi_{j+1}\right).$$

Since $\left|\varphi_j - \varphi_{j+1}\right| \le \frac{1}{N}$, we get $\left|\sum_{j=1}^{N} y_j\varphi_j\right| \le \frac{1}{N}\sum_{j=1}^{N}|Y_j|$. The last expression is therefore an upper bound for $\rho(\mu,\nu)$. But this bound is even reached for some special φ, which we construct: We denote $\beta_j = \text{sign } Y_j$, such that $|Y_j| = \beta_j Y_j$ and

$$\sum_{j=1}^{N}|Y_j| = \sum_{j=1}^{N}\beta_j Y_j = \sum_{j=1}^{N}\beta_j\sum_{i=1}^{j} y_i \text{ (by definition).}$$

Changing the order of summation, we get

$$\sum_{j=1}^{N}|Y_j| = \sum_{i=1}^{N} y_i \sum_{j=i}^{N}\beta_j.$$

Putting $\mathring{\varphi}_i = \frac{1}{N}\sum_{j=i}^{N}\beta_j$, we see that

$$\left|\mathring{\varphi}_i - \mathring{\varphi}_{i+1}\right| = \frac{\left|\beta_{i+1}\right|}{N} = \frac{1}{N},$$

and

$$\frac{1}{N}\sum_{j=1}^{N}|Y_j| = \sum_{i=1}^{N}\mathring{\varphi}_i y_i.$$

The bound is attained by $\mathring{\varphi}$. Hence,

$$\max_{\varphi}\left|\sum_{i=1}^{N} y_i\varphi_i\right| = \sum_{i=1}^{N}\mathring{\varphi}_i y_i = \frac{1}{N}\sum_{j=1}^{N}|Y_j|,$$

which is the statement of the theorem.

ρ may be interpreted in a "hole" philosophy: $|Y_j|$ is the volume of holes, which begin at the left boundary (at $j = 1$). ρ is therefore not the volume of the largest hole but the average of certain hole volumes.

The unit sphere in R^2 with respect to ρ looks like a parallelepiped.

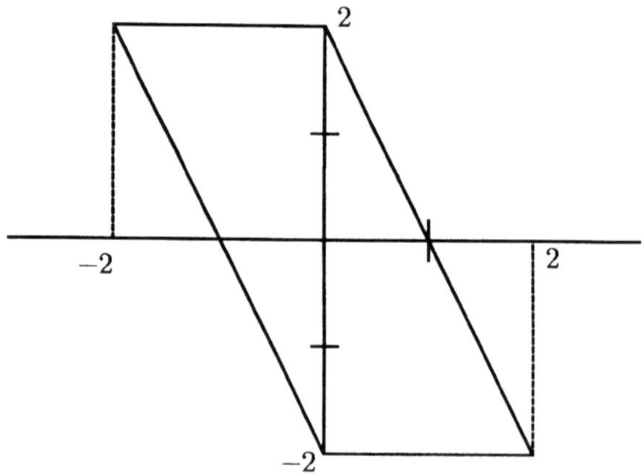

Figure 1.4.3

It is not any more symmetric in (x_1, x_2), not even in 2 dimensions. In two dimensions, the definition (or better representation) of ρ is: Let

$$Y_{ij} = \sum_{k=1}^{i} \sum_{l=1}^{j} (\mu_{kl} - \nu_{kl}) \text{ for } 1 \le i, j \le N,$$

then

$$\rho(\mu, \nu) = \frac{1}{N^2} \sum_{i,j} |Y_{ij}|\,.$$

ρ is, as a model for quality as good as D. Now, we need fast algorithms to compute D and ρ.

1.4.3 Evaluation of our first model

It is important to get fast algorithms to compute D or ρ; realize that N is rather big, up to 6000. Moreover, we need online algorithms – they have to accompany the production process.

For ρ, that is easy: compute the partial sums Y_i or Y_{jk} and take the (arithmetic) average. Our theorem delivers a fast and easy algorithm. What about D? In one dimension, we have

$$
\begin{aligned}
D(\mu, \nu) &= \max_{1 \le \alpha \le \beta \le N} \left| \sum_{j=\alpha}^{\beta} (\mu_j - \nu_j) \right| \\
&= \max_{1 \le \alpha \le \beta \le N} |Y_\beta - Y_{\alpha-1}| = \max_{0 \le \alpha, \beta \le N} |Y_\beta - Y_\alpha| \\
&\quad \text{(please realize that we have substituted)} \\
\alpha &\le \beta \text{ by } \alpha, \beta) \\
&= \max_{0 \le \beta \le N} Y_\beta - \min_{0 \le \alpha \le N} Y_\alpha.
\end{aligned}
$$

This is also very easy: one has to compute the partial sums, their maxima and their minima and the difference. But there is no such formula in two dimensions. It is easier to use an equivalent formulation (equivalent with respect to the modelling quality)

$$D^*(\mu, \nu) = \max_{1 \le i,j \le N} |Y_{ij}| .$$

where D is related to D^* – see Hack [1990] for more details.

The results with D were initially very satisfying, but not completely: sometimes the quality estimated by D (or ρ) and the judgment of the experts disagree.

Fine, says the company, but not fine enough. And the game begins again.

1.4.4 The second model

Clouds and ships are features which have typical sizes. **Clouds** are isotropic irregularities of a size which is much larger than the diameter of the fibres and much smaller than the size of the fabric. **Ships** are anisotropic defects of a length and thickness which have different orders of magnitude.

It is a typical feature of modern applied mathematics to apply a "**multiscale analysis**" to signals, images and functions. This means practically to look at an image first from a very short distance: one sees many small details but loses information about the "whole". Increasing the distance to the image (you may see people in galleries, studying first tiny details of a painting – for example the title! – and then stepping back to get an overall impression of it), one realizes that the details disappear and the features on larger scales come into view. The mathematical model of this process of zooming of magnification and minification is called "**multiscale analysis**". It appears in different forms and connected with different other theories; for example, as wavelets, but also in numerical analysis as multigrid or hierarchical bases (see, for example, Louis, Maaß and Rieder [1997] or Hackbusch [1992]). The general idea is to construct bases which separate the scales, i.e., different elements are representing features at different scales. In that way, it is to filter out features of a certain size. Modern mathematics provides constructions of these bases which allow easy decomposition of images. "Easy" means computationally fast; one way is to care for the orthonormality of the basis. Decomposition is then Fourier – or now: wavelet analysis, since wavelets form orthonormal bases with scale separation.

It should be clear that we first tried to describe clouds and ships with the help of a wavelet analysis; see Stark [1990]. In fact, we started to apply wavelets to industrial problems in 1987 as one of the first outside of France.

The results were not bad but not so impressive either to justify the application of these advanced tools. In 1995, J. Weickert proposed a simpler but similar concept, the so-called "**techniques of pyramids**" known since long in image processing, which is somehow a prototype of multiscale analysis.

We shall present this approach here – it is, as we said, an easy introduction to a modern mathematical concept. It will be applied to the problem of cloudiness and only for 1d images (but here, the extension to 2d is straight-forward).

The process of moving away from an image is modelled by "melting together" any pair of adjacent pixels: magnification, i.e., moving towards the image means the splitting of one pixel into two. How do we do this splitting and pasting? It is clear from the above that dimensions N, which are powers of 2, play a special role. Hence, we start with

$$\mu = (\mu_0, \mu_1, ..., \mu_{2^k})^T \in \mathbb{R}^{2^k+1}.$$

We should see μ as a representation of the complete image, containing all details even on the finest scale. Larger scales mean less information and the image at larger scales will be represented by elements in $\mathbb{R}^{2^j+1}$ with $1 \leq j \leq k-1$. Processes of magnification, the transition from larger to finer scales, must consist in putting new information into it and are mappings from $\mathbb{R}^{2^j+1}$ to $\mathbb{R}^{2^j+1}+1$; we call them interpolation or prolongation operators. Since we restrict our considerations to linear transformations, these operators may be represented by $(2^{j+1}+1) \times (2^j+1)$ matrices. One idea to prolongate is simply to insert the average of two neighbouring grey values:

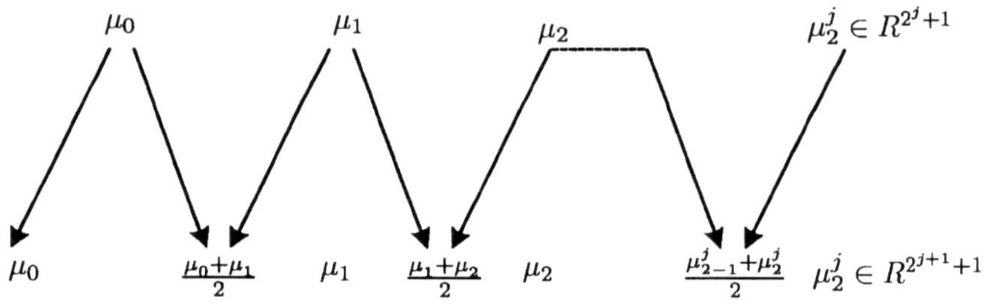

Figure 1.4.4.

The corresponding matrix is

$$I_j = \begin{pmatrix} 1 & 0 & \cdots & \cdots & \cdots & \cdots & \cdots \\ \frac{1}{2} & \frac{1}{2} & 0 & \cdots & \cdots & \cdots & \cdots \\ 0 & 1 & 0 & \cdots & \cdots & \cdots & \cdots \\ 0 & 0 & \frac{1}{2} & \frac{1}{2} & 0 & \cdots & \cdots \\ & & 0 & 1 & 0 & & \\ & & & \ddots & \ddots & \ddots & \\ & & & & 0 & \frac{1}{2} & \frac{1}{2} \end{pmatrix}.$$

For $j = 0$, we put

$$I_0 = \begin{pmatrix} 1 \\ 1 \\ 1 \end{pmatrix}.$$

We see that the row sums are always 1.

The opposite operation of "moving away", a reduction, which removes information, projects larger spaces to smaller ones; reduction operators are represented by $(2^j + 1) \times (2^{j+1} + 1)$ matrices.

A simple pastry process is just to take average values.

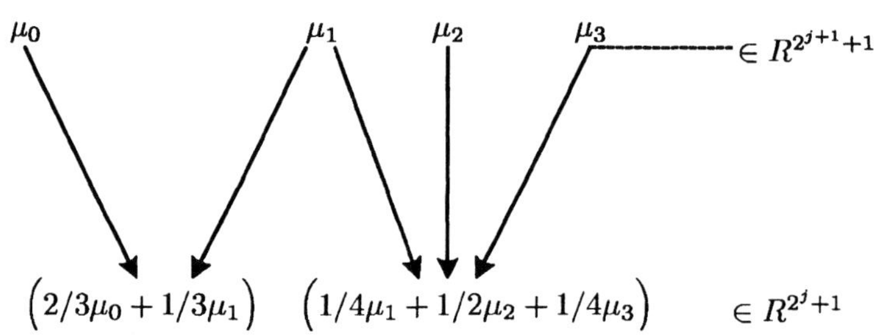

Figure 1.4.5.

We see that at the boundary, we melt μ_m, μ_{m+1} and μ_{m+2} together to one new element $\frac{1}{4}\mu_m + \frac{1}{2}\mu_{m+1} + \frac{1}{4}\mu_{m+2}$; i.e., we "convolute" μ with a mask $\left(\frac{1}{4}, \frac{1}{2}, \frac{1}{4}\right)$[4]. But we "sample" only every second element: $\frac{1}{4}\mu_1 + \frac{1}{2}\mu_2 + \frac{1}{4}\mu_3, \frac{1}{4}\mu_3 + \frac{1}{2}\mu_4 + \frac{1}{4}\mu_5, \ldots$ At the left boundary, we convolute with $\left(\frac{2}{3}, \frac{1}{3}\right)$ at the right correspondingly with $\left(\frac{1}{3}, \frac{2}{3}\right)$.

We get the following matrices:

$$R_j = \begin{pmatrix} \frac{2}{3} & \frac{1}{3} & 0 & \cdots & \cdots & \cdots & \cdots & \cdots \\ 0 & \frac{1}{4} & \frac{1}{2} & \frac{1}{4} & 0 & \cdots & \cdots & \cdots \\ 0 & 0 & 0 & \frac{1}{4} & \frac{1}{2} & \frac{1}{4} & 0 & \cdots \\ \cdots & \cdots & \cdots & \cdots & \cdots & \cdots & \cdots & \cdots \\ \cdots & \cdots & \cdots & \cdots & \cdots & 0 & \frac{1}{3} & \frac{2}{3} \end{pmatrix}.$$

Moreover, we put

$$R_o = \left(\tfrac{1}{3}\ \tfrac{1}{3}\ \tfrac{1}{3} \right) .$$

Again, the sums of row elements are one. We want to make a side remark which connects our consideration with a different approach. To convolute with a sequence $\left(\frac{1}{4}\ \frac{1}{2}\ \frac{1}{4} \right)$ looks like a very rough discretization of the convolution with a Gaussian; i.e., of

$$\widetilde{\mu}(\sigma, x) := \int_{-\infty}^{+\infty} G_\sigma(x-y)\mu(y)dy,$$

where

$$G_\sigma(z) = \frac{1}{\sqrt{4\pi\sigma}} e^{-\frac{z^2}{4\sigma}} ,$$

with a variance σ to be determined. Now we know that $\widetilde{\mu}(t,x)$ solves the **diffusion equation** $\widetilde{\mu}_t = \widetilde{\mu}_{xx}$ with initial value $\widetilde{\mu}(0,x) = \mu(x)$. If our reduction is a discrete approximation of a continuous convolution, it should be a discrete approximation of the diffusion equation. In fact, if we discretize x with a step size Δx such that $\mu_k(t) = \widetilde{\mu}_k(t, k\Delta x)$, $k \in \mathbb{Z}$ and if we discretize t with Δt to get

$$\mu_k(j) = \widetilde{\mu}_k(j\Delta t, k\Delta x),$$

by using a central difference approximation for the second derivative, we end up with

$$\begin{aligned} \mu_k(j+1) &= \mu_k(j) + \frac{\Delta t}{\Delta x^2}\left(\mu_{k+1}(j) - 2\mu_k(j) + \mu_{k-1}(j)\right) \\ &= \frac{\Delta t}{\Delta x^2}\mu_{k+1}(j) + (1 - 2\frac{\Delta t}{\Delta x^2})\mu_k(j) + \frac{\Delta t}{\Delta x^2}\mu_{k-1}(j) ; \end{aligned}$$

with $\frac{\Delta t}{\Delta x^2} = \frac{1}{4}$, we get the above mentioned mask. One step in magnification means therefore one timestep of diffusion. In diffusion processes, perturbations of high frequency disappear quickly; meanwhile, the "long waves" remain. That is exactly what our reduction does, too. Why do we mention this diffusion aspect? The theory of diffusion offers plenty of additional knowledge like maximum principle, monotony, decease of variation. It is rewarding to study diffusion filters in image processing as quite some people have done.

Now we return to our reduction operators for an iterative magnification process. The basis is the complete image $\mu = v^k \in \mathbb{R}^{2^k+1}$ and we partition it step by step.

We get

$$v^j = R_j v^{j+1} \in \mathbb{R}^{2^j+1} \text{ for } j \geq 1, \qquad v^0 \in \mathbb{R}^1 .$$

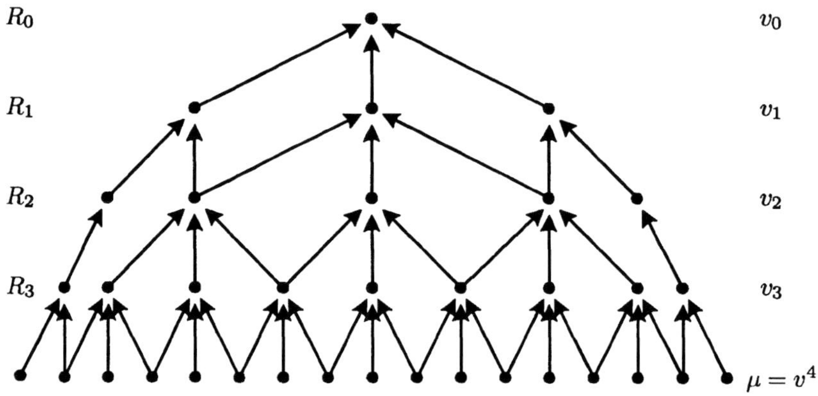

Figure 1.4.6.

In each of these steps, the high-frequency effects are reduced: each step is a low-pass filter and this pyramid, called "**Gaussian pyramid**", may be interpreted as a sequence of low-pass filtered versions of μ.

In each step, information is thrown away. What are these losses? If we want to compare v^j with v^{j+1}, we face the fact that these vectors belong to different spaces. First, we have to "lift" v^j to $\mathbb{R}^{2^{j+1}+1}$ in order to be able to compare. This lifted version should have the same information as v^j. Therefore the lifting is done by our interpolation. $I_j v^j$ is in $\mathbb{R}^{2^{j+1}+1}$ and we are now able to compare

$$v^{j+1} - I_j v^j = w^{j+1}.$$

w^{j+1} contains the information lost by the reduction $v^{j+1} \to v^j$; i.e. by R_j. To give a simple example: if $v^1 = (v_1, v_2, v_3)$, then $v^0 = R^0 v^1 = \frac{1}{3}(v_1 + v_2 + v_3) = \overline{v}$ and $I_0 v^0 = (\overline{v}, \overline{v}, \overline{v})$. Therefore, $w^1 = (v_1 - \overline{v}, v_2 - \overline{v}, v_3 - \overline{v})$ contains the deviation from the average. In general, w^j contains all features belonging to the jth scale. The computation of w^j is comparable with a band pass filter, where the width of the band corresponds to one step in our size scaling. If we write formally $w^0 = v^0$, we may write $w^k, ..., w^0$ in the form of a pyramid; and we call this pyramid the "**Laplace pyramid**".

Given the last reduction $v^0 = w^0$ and the losses $w^k, ..., w^1$, it is easy to reconstruct the whole image recursively

$$\begin{aligned} v^1 &= w^1 + I_0 v^0 \\ v^2 &= w^2 + I_1 v^1 \\ &\vdots \\ \mu = v^k &= w^k + I_{k-1} v^{k-1}. \end{aligned}$$

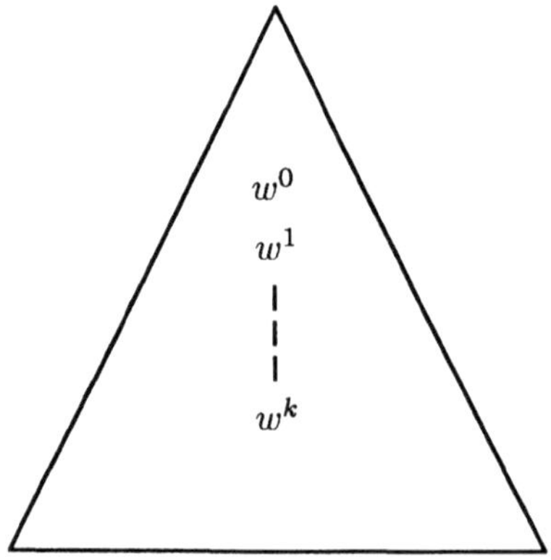

Figure 1.4.7

Storing $w^k, ..., w^0$ means storing the whole image. Of course, $w^k, ..., w^0$ contains more bits than μ–it contains redundant information, but it is still very useful. For example, we may reduce the amount of data by reduction of the kind of information we are not interested in. If we are not interested in details on scales up to order $j < k$, we are content with v^j instead of μ and need only $w^0, ..., w^j$. This set contains $1 + 3 + ... + (2^j + 1) = (j + 1) + 2^{j+1}$ data which is less than $2^{k+1} + 1$ data in the original image.

Remarks.

a) Since w^j should contain only "**fluctuations**" of a certain order of magnitude, their average value should be zero. Whether this condition is fulfilled depends on the reduction at the boundary–our masks $(\frac{2}{3}, \frac{1}{3})$ and $(\frac{1}{3}, \frac{1}{3})$ are properly chosen.

b) We considered only 1d problems. Real 2d images are reduced to the 1d case by treating the rows separately one after the other.

We will now use our **Laplace pyramid** to judge the cloudiness. We look at fluctuations at the jth scale; the larger the L_2-norm of w^j, the more this scale contributes to the nonuniformity. We normalize this L_2-norm and consider

$$\sigma_j(\mu) = \left(\frac{1}{2^j + 1} \left\|w^j\right\|^2\right)^{\frac{1}{2}},$$

which is (since the average of w^j is zero) the "variance" of w^j. Now, what is a cloud? First, not an objective concept: it is defined by the judgment of an expert who looks at the fabric. Weickert has done experiments with 18 experts and found out that all of them weighted irregularities on a middle scale more. If we try to weigh the scales by introducing a weighted measure as quality measure

$$q(\mu) := \sum_{j=1}^{k} c_j \sigma_j^2(\mu),$$

the expert opinion reached maximally the scale $j = 4$ (which corresponds to a cloud size of ca. 7 cm) and it included all scales between 3 and 20 cm. This expert evaluation defined appropriate weights c_j and in computing $q(\mu)$ with these weights, an instrument for an objective and acknowledged quality measurement was developed.

What remains to be done. Strips!

Final remarks.

Pyramids have many features in common with wavelets (which correspond to w^j). They are not as elegant, orthogonal and $w^0, ..., w^k$ therefore contain redundant information. But the price wavelets pay for orthonormality is some inflexibility which often creates problems, especially at the boundary. Pyramids contain a bit of concepts in wavelet and multigrid in a quite simple way. They are able to teach more than just how to handle a problem of quality control.

1.5 Fatigue lifetime

1.5.1 Introduction

Steel may seem to be a very ordinary material, but actually behaves in a complicated way. One of the issues which is still not understood well enough is the slow degradation and eventual destruction of steel-made machinery and equipment. For example, consider the rotor blades of a wind power generator which can be relatively large. The larger they are, the heavier they will be, and the stronger will be the downward pull of gravity upon them. But since the blades rotate, this force will change all the time. If the blade points upward, it will be pushed towards the shaft; if it points downward, it will be pulled away from it. If the blade points side wards, gravity will bend it down−just imagine a bridge over a valley which is supported on only one end. Now, these rotors turn several millions times over the years. If the construction has not been done properly, some degradation occurs during every single turn which, even if it is very slight, will eventually destroy the rotor because of the large number of turns. This phenomenon, called **fatigue**, is actually very common. A railway track experiences a heavy force each time a railway wheel passes over it. **Fatigue cracks** in structural parts of airplanes have reportedly caused crashes and near crashes in the past and, even if detected in time, are a recurrent cause of trouble. It would be very helpful if one could predict them in a reliable manner before they start to appear. The fatigue issue is also very important for the construction of trucks, buses and cars. On one hand you may want to have the truck weighing as little as possible in order to reduce fuel consumption and price of production; on the other hand its parts have to be strong enough to withstand the bumps when travelling over the roads for years. Here the analysis is particularly difficult since, as everybody knows, these bumps are rather irregular.

Although, of course, engineers have done most of the research and development in this area, it has turned out that mathematics contributes new insights and new algorithms for computing estimates of the amount of damage. This is not just speculation; for example, near Kaiserslautern, Germany, a company run by mathematicians, computer scientists and engineers alike, which has had its origin in the group of industrial mathematics headed by Neunzert of Kaiserslautern University, provides software used by the majority of German car producers.

The research group at Kaiserslautern started a systematic study of fatigue analysis since 1984 and have made significant contributions (see, e.g., Dressler and Hack[1996], Dressler, Hack and Krüger [1997] and Brokate and Sprekels [1996]) by establishing a relationship between the phenomenon of hysteresis and fatigue lifetime and developing methods particularly useful for data of long drives or for extrapolated data.

The estimation of the lifetime of a component in the automotive industry like car, aircraft, railways must be carried out to verify the durability of the machine as a whole or a part of it prior to its actual manufacturing. A specific part of a car is usually not destroyed by one large load, but by the growth of very small cracks during many, typical several million, hysteresis loops. At ambient temperatures, the frequency of the oscillation is not taken into account. The damage induced by the individual loops is accumulated according to the famous Palmgren-Miner-rule, according to which experimental results for a given work-piece are usually condensed into an S-N-diagram depicting the Wöhler line, a plot of the (scalar) stress amplitude S versus the number N of cycles (oscillations between two amplitudes; for example, O and S) until destruction occurs. For a sequence of cycles of varying amplitude, the **Palmgren-Miner-rule** of linear damage accumulation evaluates the total accumulated damage as the sum of the contributions $1/N$ from the individual oscillations. The damage induced by these loops is used to estimate the lifetime. The process of finding duration in which a specific part or the whole machine will be destroyed is called the **fatigue analysis**. The duration is called **fatigue life**. Components under consideration may be of different size and shape; so we require a local evaluation of the damage depending on material parameters only. Very often no information about the local value is available. Hence, to get real local values, one has to build a specimen, put into the vehicle and drive it on the test track.

It may be observed that the motivation for data reduction schemes in fatigue analysis is not just the reduction of data (in the sense of storage saving) as suggested by the direct meaning of the word. The main point is to focus attention on the relevant information by intelligent filtering, that is, by omitting the immense mass of data having no effect on damage accumulation. This permits both an effective modular use of modern numerical damage evaluation techniques and the reorganization of test-drive data for test stand experiments. From the point of view of damage analysis, it is well accepted that the **rainflow method** is the optimal rate-independent data reduction scheme for one-dimension load histories. There are stronger points in favour of rainflow reconstruction algorithm from practical point of view and it has no disadvantage against the Markov simulation, e.g., performance and reliability. It is mathematically exact in the sense that there are no approximative or heuristic arguments used. The main goal of this section is to introduce briefly a very effective on-line counting, the so-called "4-point counting" by Krüger [1985], and present **fatigue lifetime** estimate based on rainflow counted data using the local strain approach.

1.5.2 Physical situation, modelling, rate independence, and rainflow counting

Our goal is to estimate the lifetime of a structural component experiencing repetitive loading. These components can be of different size and shape; hence we need a local evaluation of the damage depending on material parameters only. The basic assumption here is that the damage to the components is fatigue-based and damage is not due to heavy loads and loading is one-dimensional. One can model this situation by assuming that the component has one crack, perpendicular to this loading, that is larger than all others. Due to this, the global stress and strain in the component can be thought to be higher at this point (Figure 1.5.1).

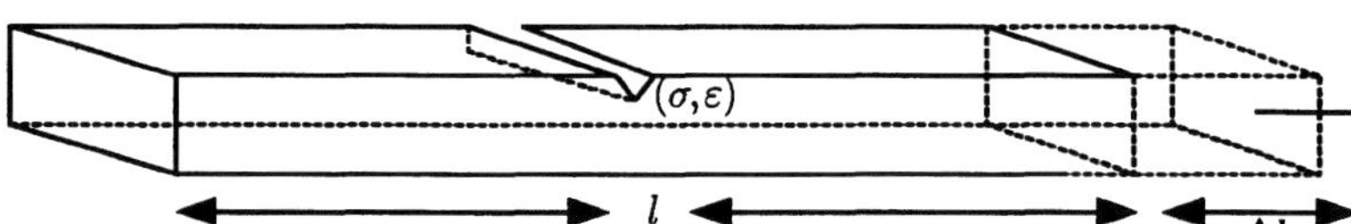

Figure 1.5.1. Schematic drawing of the basic situation.

Thus the global stress and strain are one-dimensional and take their local values at the crack. The breaking will eventually occur at a point when the total accumulated damage exceeds a certain value.

$$\text{Stress} = \sigma = \frac{L}{A} = \frac{\text{Force}}{\text{Cross section Area}}$$
$$\text{Strain} = \epsilon = \frac{\Delta l}{l} = \frac{\text{Extension}}{\text{Original Length}},$$

(σ_1, ϵ_1) are their local values.

Let us consider in more detail the compression of the component at a single expansion component. The expansion will cause an elastic deformation up to a certain point. If this point is passed, then the material undergoes a plastic deformation which causes a permanent damage. Similarly, under compression, the rod will deform first elastically and then plastically. So, in the stress-strain plane, the trajectory $(\sigma(t), \epsilon(t))$ of the history of the sample will follow a so-called hysteresis loop as in Figure 1.5.2.

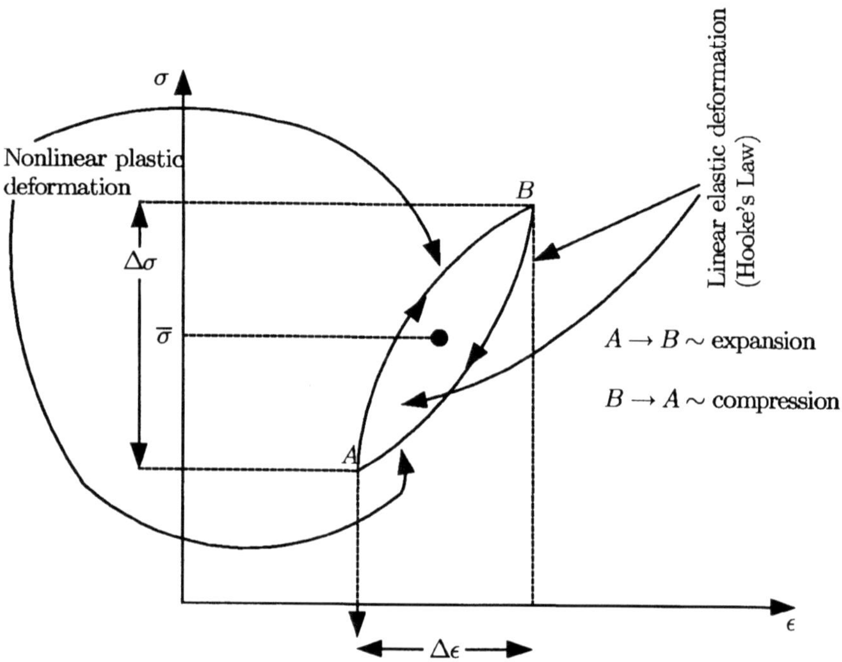

Figure 1.5.2. A single hysteresis loop.

Now, we note that the units of area in the stress-strain plane are energy per unit volume. Thus, each time the trajectory completes a hysteresis loop energy is dissipated, a function of the size of the loop, which it is assumed to permanently damage the structure.

There are two types of curves that make up the time history of the straining of the structure, and they are related by a simple formula. These are called cyclic curve and doubled curve. Initially, the trajectory follows the cyclic curve until its first turning point (corresponding to a change of direction of the force applied). At this point, the trajectory changes its direction to follow the doubled curve because it first must undo what has previously been done before creating strain in the other direction. The formula for the **doubled curve**, when the **cyclic curve** is given by $\epsilon(t) = g(\sigma(t))$, is

$$\tilde{\epsilon}(t) = \pm 2g(\frac{\tilde{\sigma}(t)}{2}),$$

where $\tilde{\epsilon}(t) = \epsilon(t) - \epsilon(t_n)$, $\tilde{\sigma}(t) = \sigma(t) - \sigma(t_n)$ and t_n is the last time at which there is a turning point. This is known as the **Masing law** and, in order to describe the trajectory completely, we must add to it the so-called Memory laws. These refer to Figure 1.5.3.

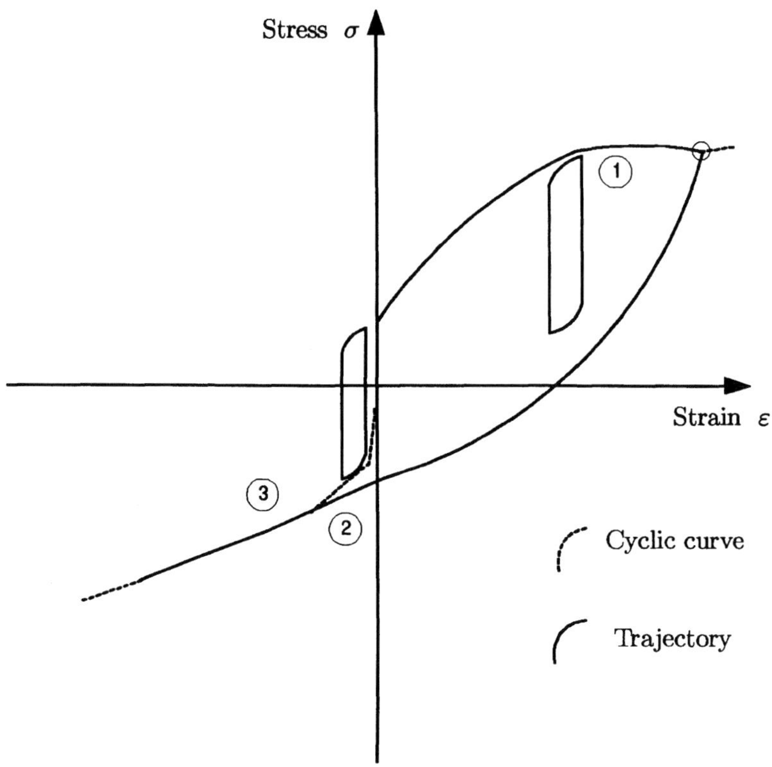

Figure 1.5.3. Memory laws.

(i) After closing a hysteresis loop on the cyclic curve, the trajectory follows the cyclic curve again.

(ii) After closing a hysteresis loop on a basis branch, the trajectory follows this branch again.

(iii) If a branch of the trajectory that started on the cyclic curve ever meets it again, then the trajectory slope changes and it follows the cyclic curve away from the origin.

The first two of these laws can be explained by noting that completing a hysteresis loop wipes this loop from the memory of the trajectory, so it continues along the original track. The third law describes a similar loss of memory of the trajectory when it intersects with the cyclic curve. This is due to the fact that this point could also have been reached by simply traversing the cyclic curve from the origin (the only difference being the accumulated damage).

Load-time graph for a short duration of an experimentally observed data is given in Figure 1.5.4.

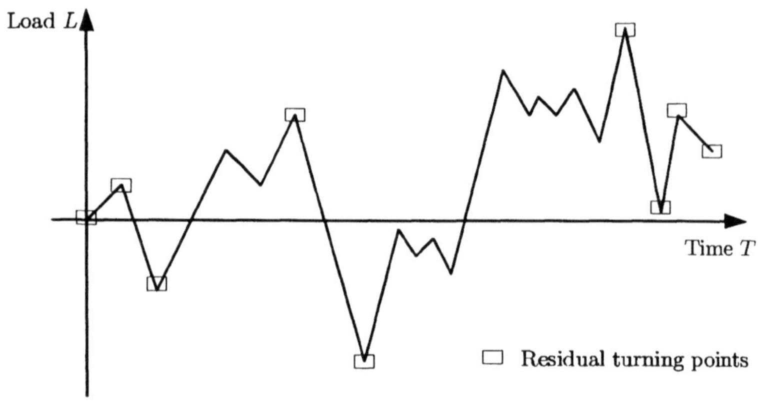

Figure 1.5.4. Time series.

Corresponding to the load-time graph in Figure 1.5.4, the schematic stress-strain trajectory is given in Figure 1.5.5.

The rainflow matrix and the residual graph. We start with a short time series $L(t)$. First, we discretize the load into n intervals of equal length

$$\Delta L = \frac{\max|L|}{n},$$

where n is a given integer, and we denote by L_i the load corresponding to interval i counting in the positive direction. The **rainflow matrix** A is then defined by a_{ij}, being the number of hysteresis cycles whose closing parts start at L_j and end at L_i. So, if $i < j$, the loop is "sitting" and if $i > j$, the loop is "hanging". Furthermore, the loops contained in only one load interval are not counted, i.e., $a_{ii} = 0$ for $i = 1, 2, \cdots, n$. If we delete all hysteresis loops from the time series, then whatever is left is called the **residual graph** or **residue**. Let t_1, t_2, t_3, t_4 be 4 successive residual turning points in Figure 1.5.4 and

$$[L(t_{i+1}), L(t_{i+2})] \subset [L(t_i), L(t_{i+3})].$$

Then we delete t_{i+1} and t_{i+2} from the time series, which is called the **Madelung deletion**. Since there is no hysteresis loop in the residual graph, the load values are

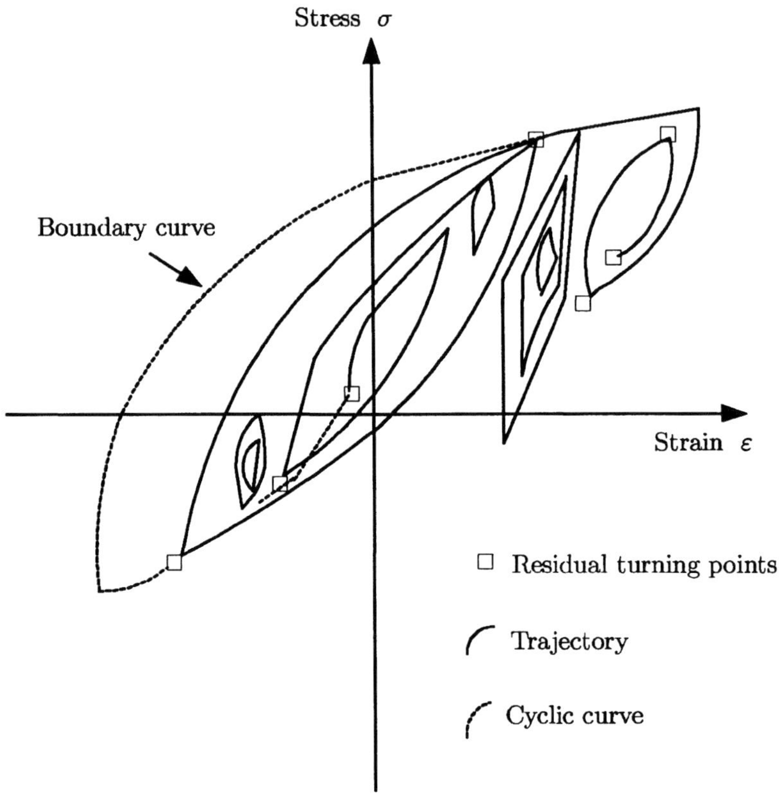

Figure 1.5.5. Trajectory in the stress-strain plane.

of increasing modulus until a possible tailing at the end which may be considered as a nesting of unfinished hysteresis loops. The concept of residual graph is illustrated in Figure 1.5.6.

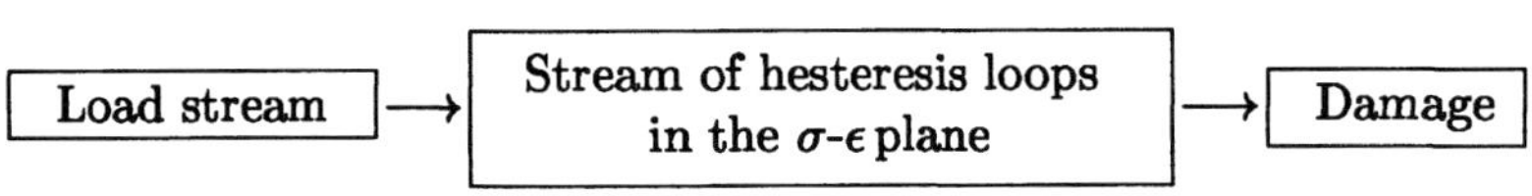

Rainflow counting simply means that, through the repeated Madelung deletions, one eliminates all inner cycles, while noting them down or counting them in the process. Damage can be considered as a functional on the space of loading functions which are piecewise monotone functions v defined on $[0, T]$ into R, where T is a fixed number. This means that there exists a subdivision $0 = t_0 < t_1 < \cdots t_n = T$ such that the restriction of v on any interval $[t_i, t_{i+1}]$ is monotone. In this case, we call $(t_0, t_1, \cdots, t_n)$ a monotonicity division of $[0, T]$. Let $M_{pm}(0, T)$ be the space of all piecewise monotone functions on $[0, T]$. For $v \in M_{pm}(0, T)$ and for a mini-

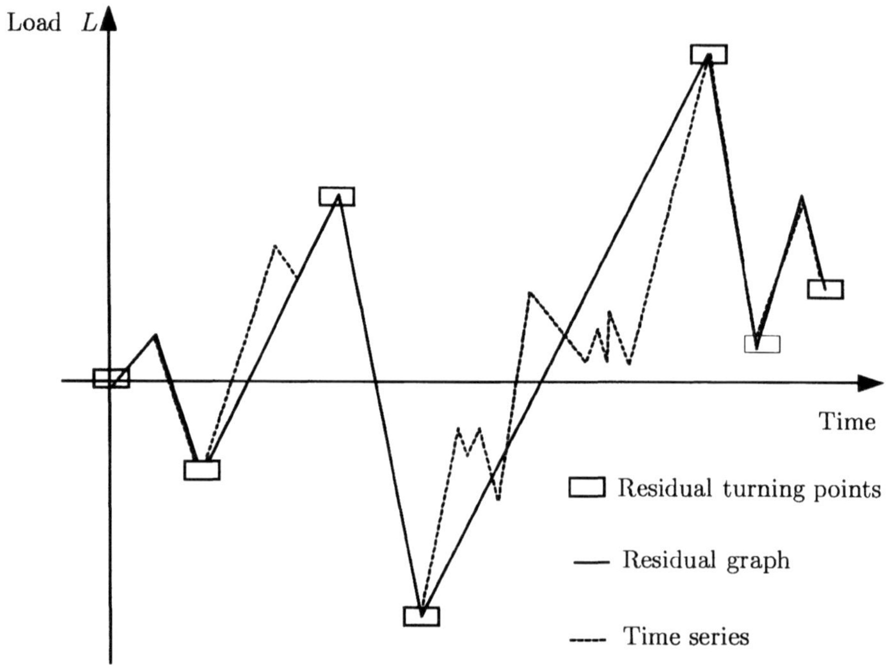

Figure 1.5.6.

mal monotonicity division $(t_0, t_1, \cdots, t_n)$ of $[0, T]$, $(v(t_0), v(t_1), \cdots, v(t_n))$ is called the string of turning points of v. A transformation ϕ on $[0, T]$ into itself is called monotonicity-preserving if ϕ is increasing, $\phi(0) = 0$, and $\phi(T) = T$. A functional $\mathcal{D}$ on $M_{pm}(0, T)$ is called **rate-independent** if for any $v \in M_{pm}(0, T)$ and any monotonicity-preserving ϕ, $\mathcal{D}(v) = \mathcal{D}(v \circ \phi)$ for $v \in M_{pm}(0, T)$, and matching monotonicity division of $[0, T]$. It can be verified that $\mathcal{D}(v) = \mathcal{D}(\tilde{v})$ for any $\tilde{v} \in M_{pm}(0, T)$ with $\tilde{v}(\tilde{t_i}) = v(t_i)$ for all $i = 0, 1, 2, \cdots, n$ and $(\tilde{t_0}, \tilde{t_1}, \cdots, \tilde{t_n})$ a monotonicity division of $\tilde{v}$. This implies that we can define **any rate-independent functional as a functional on the set of turning points**. This concept gives us the first part of data reduction. We just need the string of turning points.

From an algorithmic point of view, a very effective on-line counting is the so-called "4-point" counting given by Krüger et al. [1985] whose refinement and didactical reorganizations are presented in Dressler, Hack and Krüger [1997]. A mathematical description can be seen in Brokate and Sprekels [p.76, 1996]. Having the loop stream, damage is evaluated using material properties, experimental data and a damage parameter. This damage parameter induces a functional $\mathcal{D}$, that maps the stream of hysteresis loop into its damage value indicating how much the stream of loops has damaged the part. Usually a value of zero means not damaged at all, and the value one indicates that the part is broken. Schematically, we get:

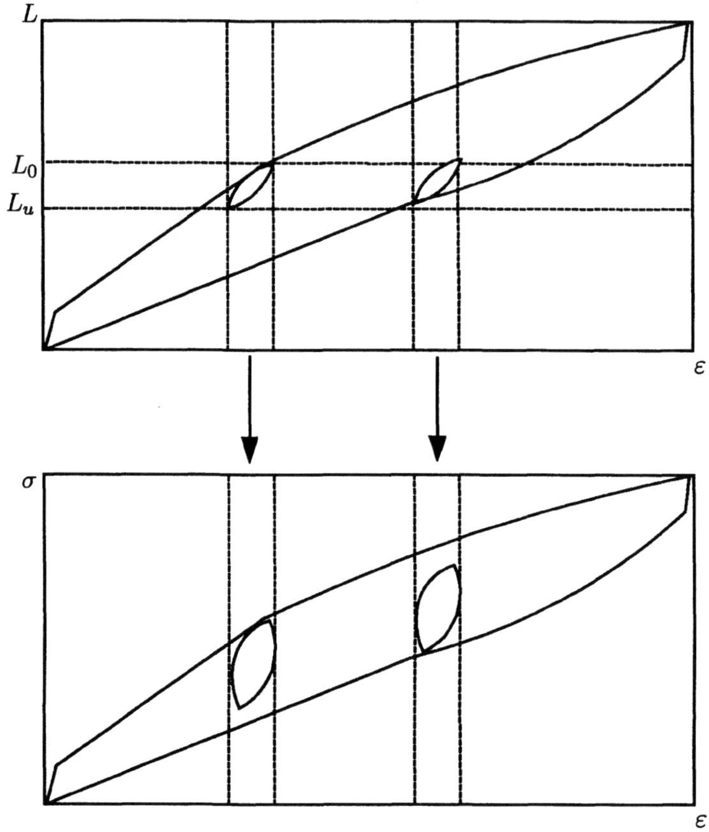

Figure 1.5.7.

1.5.3 Damage estimation using rainflow counted data

In view of the following facts, one looks for the damage of a rainflow counted loading:

(i) The load can be rainflow counted on-line, that is, during the drive on the test track to reduce the data such that much longer drives and much finer measurement can be applied.

(ii) There exist mathematical tools to manipulate the rainflow count, that is, to extrapolate the data in such a way that the resulting rainflow count has the same distribution of hysteresis loop. This gives a much better extrapolation of track data than simply merging several load hysteresis.

(iii) One can manipulate the rainflow matrix to simulate special situations.

(iv) One can generate artificial rainflow matrices.

It may be observed that all information on the original order with respect to time of the hysteresis loop (in the plan load v is load strain) is lost when the data is rainflow counted. We get a set of loops instead of a stream. The problem is that the counting is done on the loading stream and therefore we can only get information on the location of the loops with respect to the loading. We can not read the exact location of the local strains from the rainflow count. For each class of hysteresis loops, we can get the upper and lower loading value L_u, L_0. By using the geometrical and material properties, one can calculate the amplitude of strain ϵ_a (for detail we refer to Neuber [1961]). By the transformation from hysteresis loops in the loading-strain-plane into such in the stress-strain-plane, one also loses the information on the exact location in the stress-strain-plane and one only has the amplitude (σ_a and ϵ_a).

But most of the damage parameters also depend on the location of the hysteresis loops in the **stress-strain-plane**. The reason is that the damage induced by a hysteresis loop of a given stress and strain amplitude is larger under tension than under compression. The crucial value hereby is the mean stress (σ_m). Hence one can not speak of the damage of a rainflow count with respect to such a mean stress-dependent damage parameter. However, it is possible to proceed from a rainflow count to a stream of turning points.

A rainflow count consists of the tuple (RFM, RES), where RFM denotes the rainflow matrix with entries $RFM(i,j)$ giving the number of hysteresis loops from the level i to the level j and RES is the residual of the count.

We consider the set of streams of loading turning points which load to the given rainflow count (RFM, RES). If RC denotes the rainflow count operator, we write formally

$$\mathcal{L} = RC^{-}(RFM, RES) .$$

For each stream we can calculate the damage $\mathcal{D}(H(L))$. Hence we can canonically define the damage of $\mathcal{L}$ and for (RFM, RES) by the expectation value of the damage of the stream of turning points in $\mathcal{L}$, where we assume that every stream has the same probability. This assumption makes sense if we do not have any further information on the loading. Thus

$$\mathcal{D}(RFM, RES) = \frac{1}{|\mathcal{L}|} \sum_{L \in \mathcal{L}} \mathcal{D}(H(L)) .$$

If the stream of turning point consists of the hysteresis loops $\{h_\tau | \tau = 1, \cdots, T\}$, the total damage is evaluated as

$$\mathcal{D}_{tot} = \sum_{\tau=1}^{T} d(h_\tau),$$

where $d(h_\tau)$ denotes the change induced by the hysteresis loop h_τ. For details and implementation of this technique, we refer to Dressler and Hack [1996] where they have also shown the application of this method included in the Fatigue Analysis Tool $FALANCS^{TM}$ to real data.

It may be remarked that the contents of this section as well as those of Chapter 6 are going to be the essential tools for computational work in material sciences such as in the study of elasticity, plasticity, elasto-plasticity with strain hardening and elasto-viscoplasticity.

Chapter 2

Algorithms for Optimization

2.1 Introduction

The traces of optimization can be found in the development of calculus. As far back as 1629, Pierre de Fermat showed that the necessary condition for an extremum (minima or maxima) for a real-valued function of one variable is that the derivative must be zero. A class of extremum problem, that has been the favourite of giants like Jean Bernoulli, Leonhar Euler, Andrien Legendre and Carl Gustav Jacobi since the end of the eighteenth century and the beginning of nineteenth century, is known as the calculus of variation.

The calculus of variation is an infinite-dimensional problem of unconstrained optimization in which the functional to be minimized is defined by an integral. In the second half of the nineteenth century, Karl Weierstrass posed the crucial question of existence of a solution of minima and maxima and answered it for a fairly general situation. In 1939, Nobel Laureate Leonid Vitalevich Kantorovich formulated many problems of Economics in the form of optimization problem for linear functionals known as the **mathematical method of production of planning and optimization** and made extensive study of such problems. Basically, these results constituted the main ideas of the linear programming (See, for example, Kantorovich [1975]). Without having any knowledge of this work, in 1942, George B. Dantzig encountered similar problems of minimizing linear functionals under linear constraints while studying applications of mathematics to industrial problems. Around 1950, David Gale, Harold Willion Kuhn and Albert William Tucker contributed much to this field and also extended this theory for non-linear functionals under constraints. Around 1957, L.S. Pontryagin, V.G. Boltyanskij, R.V. Gamkrelidze and Richard E. Bellman studied the optimal control problem, a special type of optimization problem and a generalization of the calculus of variations where one looks for the solution of a system of differential equations which minimizes or maximizes (extremizes) a functional. In the 1960s, A.Ya-Dubovsky and A. Milyutin, as well as B.N. Pshenichnyj, Lucien W. Neustadt, Hubert Halkin, Jack Warga et al. developed general techniques for obtaining extremum conditions for abstract optimization problems with constraints so as to include the Kuhn-Tucker theorem and the maximum principle obtained earlier. Convex analysis has played a vital role in the study of abstract extremum problem. Richard Terry Rockafellar and his co-workers have made valuable contributions in this area. A fairly general theorem for existence of a unique solution of optimization problem in the setting of reflexive Banach spaces, in particular for Hilbert spaces, has been proved by I. Ekeland and R. Temam in 1974. As a corollary, one obtains from this theorem that the solution of optimization problem for the energy functional has a unique solution under fairly general conditions. Generalization of variational problem (variational equation) in the form of variational inequalities, was introduced by Guido Stampacchia, Jacques-Louis Lions and G. Fichera in 1970s and it has been extensively studied by Alain Bensoussan, C.Baiocchi, G. Duvaut, R. Glowinski, F. Giannessi, G. Isac, N. Kikuchi, D. Kinderlehrer, U. Mosco, P.D. Panagiotopoulos et al. An interesting part of this theory is that a large class of optimization problems including the programming problems (linear and non-linear) are special cases of the variational inequality problems. Shape optimization problems are of vital importance in the design and construction of industrial structures like cars, aircraft, space crafts (Sokolowski and Zolesio [1992]). For a comprehensive account of the above development, we refer to Polyak [1987] and Siddiqi [1986,1994]. A comprehensive algorithm for unconstrained optimization is presented by Goldfarb [1994].

Since the main objective of this chapter is to present those results of optimization which are frequently used in industrial applications, we confine ourselves to optimization algorithms like Newton, Gradient, Conjugate gradient, and quasi-Newton, especially DFP (David-Fletcher-Powell) and BFGS (Broyden-Fletcher-Goldfarb-Shanno).

It may be recalled that the iterative methods, where one constructs sequences which converge to a solution of optimization problems, are known as the optimization algorithms. It may be also observed that a field of more recent developments known as the automatic (or computational) differentiation could influence the developments of optimization algorithms. The algorithmic differentiation is a process for evaluating derivatives which depends only on an algorithmic specification of the function to be differentiated. In actual practice the specification of the function is all or part of a computer programme, and the derivative values are produced by the execution of the program derived from the program of the original function, hence the term "automatic" (Raff). We refer to Griewank and Corliss [1991] and Berz, Bishof, Corliss and Griewank [1996] for a detailed account of this elegant study.

A recent book by Outrata, Kočvara and Zowe [1998] contains valuable contributions in the field of algorithmic non-smooth optimization.

2.2 General results about optimization

Let X be a space with appropriate structure (topological or algebraic or both) and let K be a subset of X. Furthermore, let F be a function on X into R. The general problem of optimization, say (P), is to find an element u of K such that $F(u) \leq F(v)$ for all $v \in K$. If this happens, we write

$$F(u) = \inf_{v \in K} F(v) \qquad \text{(P)},$$

where u is called a **minima** and the value $F(u)$ is called the minimum value. Finding the maxima of F is equivalent to finding the minima of $-F$. Sometimes this problem is called the **constraint optimization problem.** If $K = X$, then (P) is often known as the **unconstraint optimization problem.** If K is a neighbourhood of u, that is, minima exists in a neighbourhood of u, then it is called a **relative minima** or **local minima**. If u is a minima or maxima, then it is called an **extrema**. A minimum or maximum value of F is called **extremum value.** If $X = R$, the set of real number, and $K = [a, b] \subseteq R$ and F is a continuous function on R into itself, then F has a minima on $[a, b]$.

For a topological space X, K, a compact subset of X and F continuous, (P) has a solution. The optimization problem for a Banach space X where K is not compact, for example, K is a unit ball of the Banach space namely $K = \{x \in X / \mid x \mid \leq 1\}$ and F a continuous functional had been an open problem for a long time. This was resolved in 1974 in the context of a reflexive Banach space where K is closed, convex and non-empty subject of X and F bounded continuous and convex functional (see, for example, Siddiqi [1986, p. 233]). The solution is unique if F is also strictly convex. If X is a normed linear space and the derivative of a functional F exists at a point $u \in X$ and u is an extrema of F, then $F'(u) = 0$. This is an extension of a famous result of Fermat that if u is a minima of $F(x)$ on R^n and F is differentiable at u, then $\nabla F(u) = 0$. These results can be extended for an open subset of X in the following manner.

Theorem 2.1. *Let U be an open subset of a normed linear space X and $F : U \subset X \to R$ be a functional. If F is differentiable at an element u of X and F has extremum value at this element, then*

$$F'(u) = 0. \qquad (2.1)$$

Equation (2.1) is often called **Euler's equation.**

Remark 2.1.

(i) The condition U to be open subset is essential as the theorem cannot hold for $X = R, U = [0, 1]$ and $F : [0, 1] \to R$ defined by $F(x) = x$.

(ii) For $X = R^n$, (2.1) is equivalent to the equations

$$\begin{cases} \frac{\partial}{\partial x_1} F(x_1, x_2, \cdots, x_n) = 0; \\ \frac{\partial}{\partial x_2} F(x_1, x_2, \cdots, x_n) = 0; \\ \vdots \\ \vdots \\ \frac{\partial}{\partial x_n} F(x_1, x_2, \cdots, x_n) = 0. \end{cases} \tag{2.2}$$

Theorem 2.2. *Let U be an open subset of R^n, let $\phi_i : U \to R, 1 \le i \le m$ be a sequence of continuous functions whose first derivatives exist and is continuous, and let u be a point of the set M,*

$$M = \{y \in U / \phi_i(y) = 0, 1 \le i \le m\} \subset U, \tag{2.3}$$

at which the derivatives $\phi_i'(u), 1 \le i \le m$, are linearly independent elements of the vector space of all bounded linear functionals on R^n into R.

Suppose that $F : U \to R$ is differentiable at u and F has an extremum value at u over U, then there exist m numbers $\mu_i(u), 1 \le i \le m$, uniquely defined such that

$$F'(u) + \mu_1(u)\phi_1'(u) + \cdots + \mu_m(u)\phi_m'(u) = 0. \tag{2.4}$$

The numbers $\mu_i, 1 \le i \le m$, obtained in this theorem are called the **Lagrange multipliers** associated with the extremum u under the constraints (2.3).

Example 2.1. Let $X = R^2, F(x_1, x_2) = -x_2$, $M = \{(x_1, x_2) \in R^2 \mid \phi(x_1, x_2) = x_1^1 + x_2^2 - 1 = 0\}$. Then equation (2.4) takes the form

$$\frac{\partial}{\partial x_1} F(x_1, x_2) + \mu \frac{\partial}{\partial x_1} \phi(u) = 0$$
$$\frac{\partial}{\partial x_2} F(x_1, x_2) + \mu \frac{\partial}{\partial x_2} \phi(u) = 0.$$

Remark 2.2. In order to solve a problem posed in the form of Theorem 2.2, we need to find the $m+n$ unknown $u_i, 1 \le i \le n$, and $\mu_j, 1 \le j \le m$, which are solutions of the system of $m + n$ equations ($u = (u_1, u_2, \cdots u_n)$)

$$\begin{cases} \frac{\partial F(u)}{\partial u_1} + \mu_1 \frac{\partial \phi_1(u)}{\partial u_1} + \cdots + \mu_m \frac{\partial \phi_m(u)}{\partial u_1} &= 0; \\ \frac{\partial F(u)}{\partial u_2} + \mu_1 \frac{\partial \phi_1(u)}{\partial u_2} + \cdots + \mu_m \frac{\partial \phi_m(u)}{\partial u_2} &= 0; \\ \quad \cdots\cdots \\ \quad \cdots\cdots \\ \frac{\partial F(u)}{\partial u_n} + \mu_1 \frac{\partial \phi_1(u)}{\partial u_n} + \cdots + \mu_m \frac{\partial \phi_m(u)}{\partial u_n} &= 0; \\ \phi_1(u) = 0, \cdots\cdots, \phi_m(u) &= 0. \end{cases} \tag{2.5}$$

The first n equations may also be written in the matrix form

$$\begin{pmatrix} \frac{\partial_u}{\partial u_1} F(u) \\ \vdots \\ \vdots \\ \vdots \\ \frac{\partial}{\partial u_n} F(u) \end{pmatrix} + \begin{pmatrix} \frac{\partial}{\partial u_1}\phi_1(u), & \cdots, & \frac{\partial}{\partial u_1}\phi_m(u) \\ & \vdots & \\ & \vdots & \\ & \vdots & \\ \frac{\partial}{\partial u_n}\phi_1(u), & \cdots, & \frac{\partial}{\partial u_n}\phi_m(u) \end{pmatrix} \begin{pmatrix} \mu_1 \\ \vdots \\ \vdots \\ \vdots \\ \mu_m \end{pmatrix}$$

$$= \nabla F(u) + \sum_{i=1}^{m} \mu_i \nabla \phi_i(u) = 0. \tag{2.6}$$

The following theorem is a generalization of a well-known result of the classical calculus.

Theorem 2.3. *Let U be an open subset of a normed linear space X and $F : U \subset X \to R$ be differentiable on U and twice differentiable at the point $u \in U$. Then*

$$F''(u)(w, w) \geq 0 \text{ for every } w \in X, \tag{2.7}$$

provided F has a minimum value at u.

Extrema and convexity. Theorem 2.1 takes the following form:

Theorem 2.4. *If, in addition to the hypotheses of Theorem 2.1, K is a convex subset of U, then*

$$F'(u)(v - u) \leq 0 \quad \text{for every } v \text{ in a } K \text{ and } u \in K. \tag{2.8}$$

It may be observed that if K is a subspace of X, then (2.8) takes the form

$$F'(u)v = 0 \quad \text{for every } v \in K,$$

and in particular $F'(u) = 0$, the condition (2.1) if $K = X$.

Theorem 2.5. *Let K be a convex subject of a normed linear space*

(i) If a convex functional $F : K \subset X \longleftarrow R$ has a minimum value in a neighbourhood of a point $u \in K$, then u is a minimum value over the entire set K.

(ii) Every strictly convex functional $F : K \subset X \longleftarrow R$ has at most one minima, and that minima say, u is strict; that is,

$$J(u) < J(v) \text{ for all } v \in K.$$

(iii) Let $F : U \subset X \longleftarrow R$ be a convex function defined on an open subset U of X containing U and differentiable at a point $u \in K$. Then the functional has a minima at u belonging to U if and only if

$$F'(u)(v-u) \geq 0 \text{ for every } v \in K. \tag{2.9}$$

If the set K is open, the preceding condition is equivalent to (2.1).

By the Weierstrass theorem of analysis, the optimization problem for a continuous function defined on R^n into R has at least one solution over a closed bounded subset of R^n. The following theorem provides the solution of such a problem for the non-bounded case.

Theorem 2.6. *Let K be a non-empty closed subset of R^n and $F : R^n \to R$ a continuous function which is coercive on U in the sense that $F(x) \to \infty$ as $\|x\| \to \infty$. Then (P) has at least one solution.*

This result has been extended in the infinite-dimensional case as follows.

Theorem 2.7 *Let K be a non-empty, convex, closed subset of a Hilbert space X and $F : K \subset X \to R$ be convex, continuous and coercive in the sense of the above theorem. Then (P) has at least one solution. This solution is unique if, in addition, F is strictly convex.*

It may be remarked that this theorem holds in the context of reflexive Banach spaces while continuity can be replaced by a weaker condition, namely, lower semi-continuity. However, this cannot be extended beyond reflexive Banach spaces as this will break down for L_1 space (space of Lebesgue integrable functions). For proof of the general form of this theorem, we refer to Siddiqi [1986]. Let $a(\cdot,\cdot)$ be a bilinear, continuous and symmetric form on a Hilbert space X and $F \in X^\star$ (space of all bounded linear functionals on X known as the dual space). Then

$$J(v) = \frac{1}{2}a(v,v) - F(v) \tag{2.10}$$

is often called the **energy functional**. It is called **a quadratic functional** if A is a matrix in (2.11). In view of some well known results of functional analysis (See, for example, Siddiqi, 1986), (2.10) can be written as $J(v) = \frac{1}{2}\langle Au, v\rangle -$

$\langle y, v\rangle$, for every $v \in X$ and an element $y \in X$ (associated with F by the Riesz representation theorem), $A : X \to X$, bounded linear. A relation of the type

$$a(u,v) = F(v) \text{ or} \langle Au, v\rangle = \langle y, v\rangle, \tag{2.11}$$

for every $v \in U$, a subspace of X and u an element of U, is known as the **variational equation.** Let K be a non-empty, closed and convex subset of a Hilbert space X, then the problem of finding $u \in K$ such that

$$a(u, v-u) \geq F(v-u) \text{ for every } v \in K\,, \tag{2.12}$$

or equivalently,

$$\langle Au, v-u\rangle \geq \langle y, v-u\rangle \text{ for every } v \in K$$

is known as the **variational inequality problem.** Inequality (2.12), is called a **variational inequality** (See Siddiqi, 1986). It can be proved as a corollary of Theorem 2.7 (see Siddiqi, 1986) that the optimization problem (P) for the energy functional J has a unique solution provided $a(\cdot,\cdot)$ is coercive, that is, there exists a number α such that $\alpha > 0$ and $a(v,u) \geq \alpha\|v\|^2$ for every $v \in X$. It can also be proved that finding the solution of the problem (P) for the energy functional over a non-empty, convex and closed subset K of X is equivalent to finding the solution of the variational inequality (2.12) where $a(\cdot,\cdot)$ is bilinear, bounded, coercive and symmetric. If K is a subspace of X and, in particular, X itself, then finding the solution of the variational equation (2.11)is equivalent to finding the optimization problem for $J(v)$ under the conditions mentioned above on $a(\cdot,\cdot)$, that is, there exists a unique $u \in K$ such that

$$J(u) \leq J(v) \text{ for all } v \in K \text{ if and only if } a(u,v) = F(v) \text{ for all } v \in K\,,$$

or

$$\langle Au, v\rangle = \langle y, v\rangle \quad \text{or } Au = y\,.$$

Similarly, if K is a non-empty, convex and closed subset, then there exists a unique $u \in K$ such that

$$J(u) \leq J(v) \text{ for all } v \in K \text{ if and only if } a(u, v-u) \geq F(v-u) \text{ for all } v \in K,$$

or

$$\langle Au, v-u\rangle \geq \langle y; v-u\rangle.$$

For details of the above mentioned results, we refer to Ciarlet [1989] and Siddiqi [1986].

2.3 Special classes of optimization problem

2.3.1 Programming problem

Unconstrained optimization problems of the following type are called programming problems:

(i) Let $U = \left\{v \in X/\phi_i(v) \leq 0, 1 \leq i \leq m', \phi_i(v) = 0, m' + 1 \leq i \leq m\right\}$ where $\phi_i : X \to R, 1 \leq i \leq m$. For $F : U \subset X \to R, X$ a normed linear space, (P) is called **a non-linear programming problem with inequality constraints** if $m' = m$ and with **equality constraints** if $m' = 0$.

(ii) If $U = \{v \in X/\phi_i(v) \leq 0, 1 \leq i \leq m\}$ and F and ϕ_i are convex functionals, then the optimization problem (P) is called **a convex programming problem.**

(iii) If $X = R^n, J(v) = \frac{1}{2}\langle Av, v\rangle - \langle y, v\rangle, A = A^T \in A_n(R), y \in R^n$, the matrix A is assumed to be positive definite $U = \left\{v \in R^n / \sum_{j=1}^n \phi_i(v_j) \leq d_i,\right.$ $\left.1 \leq i \leq m\right\}$, where $\phi_i, i = 1, 2 \cdots m$ are affine and hence convex then (P) is called **a quadratic programming problem.** Here $\phi_i = Ax + d, x \in R^n$, and $d \in R^n$.

(iv) If $X = R^n, J(v) = \sum_{i=1}^n \alpha_i v_i, \alpha_i \in R, \quad v = (v_1, v_2, \cdots, v_n) \in R^n$ and

$$U = \left\{ v \in R^n / \sum_{j=1}^n a_{ij} v_j \leq d_i, \quad 1 \leq i \leq m \right\},$$

where $A = (a_{ij})$, then the problem (P) is called **a linear programming problem.**

2.3.2 Calculus of variation

Let $X = C^1[a, b]$ be a space of all continuous functions whose first derivatives exist and are continuous such that

$$F(x(t)) = \int_a^b l(x(t), x'(t), t)dt,$$

where $l(\cdot, \cdot)$ is continuous in all arguments and is continuously differentiable in x and x'. Then (P) is called the **calculus of variation problem.** The function $x(t)$ which yields extremum values of F must satisfy the Euler-Lagrange equation; namely,

$$\frac{\partial l}{\partial x} - \frac{d}{dt}\left(\frac{\partial l}{\partial x'}\right) = 0. \tag{2.13}$$

2.3.3 Minimum norm problem and projection

Let M be a closed convex subset of a Hilbert space X and let $\varphi = \inf_{x\in M} ||x||$. Then there exists an element u of M such that $\varphi = ||u||$. This result provides an answer to the problem of existence of an element whose norm is less than or equal to the norms of all elements over a subset. For a given element w of X, there exists an element $P_M(w) \in M$ such that $||w - P_M(w)|| = \inf_{v\in M} ||w - v||$. The element $P_M(w)$ satisfies the condition

$$\langle P_M(w) - w, v - P_M(w)\rangle \geq 0 \text{ for all } v \in M, \tag{2.14}$$

and, conversely, if an element $u \in M$ satisfies the condition

$$\langle u - w, v - u\rangle \geq 0 \text{ for all } v \in M, \tag{2.15}$$

then $P_M(w) = u$.

The element $P_M(w)$ is called the **projection** of an element $w \in X$ on M and $P_M : X \to M$, the operator P_M on X onto the set M is called the **projector operator**. The projector operator is non-expansive, that is, it satisfies the condition

$$||P_M(w_1) - P_M(w_2)|| \leq ||w_1 - w_2|| \text{ for all } w_1, w_2, \in X .$$

P_M is linear if and only if M is a subspace and (2.14) takes the form

$$\langle P_M(w) - w, v\rangle = 0 \text{ for every } v \in M.$$

2.3.4 Optimal control problem for a system represented by differential equations

We consider a system described by nonlinear differential equations

$$x'(t) = f(x(t), u(t)) \text{ on } [0,T], \text{ a real finite-time interval,} \tag{2.16}$$

$$\text{with } x(0) = x_0, \tag{2.16(a)}$$

where $x(t) \in R^n$ is the state vector, $u(t) \in R^m$ is the control vector and $f : R^n \times R^m \to R^m$ is assumed to be continuously differentiable in its arguments. For the class of admissible controls, we take

$$U = C^1(0,T;R^n) = \left\{f : [0,T] \leftarrow R^n/f' \text{ exists and continuous}\right\},$$

and suppose that for all $u \in U$, equation (2.16) has a unique solution $x \in C'(0,T;R^n)$. We call $x(t)$ the trajectory corresponding to the control $u(t)$. We suppose that the terminal time T is fixed and $x(t)$ satisfies

$$G(x(T)) = c, \tag{2.17}$$

where $G : R^n \longrightarrow R^m, c \in R^n$. The functional to be extremized is

$$J(u) = \int_0^T l(x,u)dt, \tag{2.18}$$

where $l : R^{m+n} \longrightarrow R$, and we suppose l and G to be continuously differentiable. The **optimal control problem** is to find an admissible pair (x,u) which extremises (minimizes or maximizes) $J(\cdot)$ under the conditions of equations (2.16) and (2.17). For more details, see Siddiqi [1986].

2.4 Newton algorithm and its generalization

The **Newton method** deals with the search of zeros of the equation $F(x) = 0$, $F : U \subset X \longrightarrow Y$, X and Y are normed spaces, in particular, for $X = Y = R, F : R \longrightarrow R$ or $X = R^n$ and $Y = R^n, F : R^n \to R^n$ and U an open subset of X (open interval of R or open ball of R^n). Once we have this method, the functional F can be replaced by F' or ∇F to obtain the algorithm for finding the extrema of F, that is, zeros of F' or ∇F which are extremum points of F. One can easily check that if $F : [a,b] \longrightarrow R$ and $|F'(x)| < 1$, then $F(x) = 0$ has a unique solution, that is, F has a unique zero. For the function $F : U \subset R \longrightarrow R, U$ the open subset of R, the Newton method is defined by the sequence

$$u_{k+1} = u_k - \frac{F(u_k)}{F'(u_k)}, \quad k \geq 0. \tag{2.19}$$

u_0 is an arbitrary point of open set U. The geometric meaning of (2.19) is that each point u_{k+1} is the intersection of the axis with the tangent at the point u_k. This particular case suggests the following generalization for the functional $F : U \subset X \longrightarrow Y$: For an arbitrary point $u_0 \in U$, the sequence $\{u_k\}$ is defined by

$$u_{k+1} = u_k - \left\{F'(u_k)\right\}^{-1} F(u_k), \tag{2.20}$$

under the assumption that all the points u_k lie in U. If $X = R^n, Y = R^n, F(u) = 0$ is equivalent to

$$\begin{aligned} F_1(u) &= 0, \quad u = (u_1, u_2, \cdots u_n) \in R^n \\ F_2(u) &= 0 \\ F_3(u) &= 0 \\ &\vdots \\ F_n(u) &= 0\,, \end{aligned}$$

where $F_i : R^n \to R, i = 1, 2, \cdots n$. A single iteration of the Newton method consists in solving the linear system

$$\begin{cases} F'(u_k)\,\Delta u_k = -F(u_k) \text{ with matrices} \\ F'(u_k) = \left(\dfrac{\partial F_i(u_k)}{\partial x_j}\right)_{i,j}, \end{cases} \tag{2.21}$$

and then setting

$$u_{k+1} = u_k + \triangle u_k \, .$$

It may be observed that if F is an affine function, that is, $F(x) = A(x)+b$, $A = (a_{ij})$ is a square matrix of size n, that is, $A \in \mathcal{A}_n(R)$ and $b \in R^n$, then the iteration described above reduces to the solution of the linear system $Au_k = b$. In this case, the method converges in a single iteration.

We now look for:

(i) sufficient conditions which guarantee the existence of a zero of the function F, and

(ii) an algorithm for approximating such an element u, that is, for constructing a sequence $\{u_k\}$ of points of U such that

$$\lim_{k\to\infty} u_k = u.$$

We state below two theorems concerning the existence of a unique zero of F and state their corollaries for the existence of the unique zero of ∇F. The extrema of F will exist at the zero of ∇F.

Theorem 2.8. *Let X be a Banach space, U be an open subset of X, Y a normed linear space and $F : U \subset X \to Y$ be differentiable over U. Suppose that there exist three constants α, β and γ such that $\alpha > 0$ and $S_\alpha(u_0) = \{u \in X / \|u - u_0\| \leq \alpha\} \subseteq U$*
(i)

$$\sup_{k\geq 0} \sup_{u\in S_\alpha(u_0)} \|A_k^{-1}(u)\|_{BL(X,Y)} \leq \beta, A_k(u) = A_k \in BL(X,Y)$$

is bijective.
(ii)

$$\sup_{k\geq 0} \sup_{x'\in S_\alpha(u_0)} \| F'(x) - A_k(x') \|_{BL(X,Y)} \leq \frac{\gamma}{\beta}, \text{ and } \gamma < 1 \, .$$

(iii)

$$\|F(u_0)\| \leq \frac{\alpha}{\beta}(1-\gamma) \, .$$

Then the sequence defined by

$$u_{k+1} = u_k - A_k^{-1}(u_{k'})F(u_k), k \geq k' \geq 0 \tag{2.22}$$

is entirely contained within the ball and converges to a zero u of F in $S_\alpha(u_0)$ which is unique. Furthermore

$$\|u_k - u\| \le \frac{\|u_1 - u_0\|}{1-\gamma}\gamma^k. \tag{2.23}$$

Theorem 2.9. *Let X be a Banach space, U an open subset of X, $F : U \subset X \to Y$, and Y be a normed linear space. Furthermore, let F be continuously differentiable over U. Suppose that u is a point of U such that*

$$\left\{\begin{array}{ll} F(u) = 0, A = F'(u) : X \longrightarrow Y, & \text{bounded linear and bijective} \\ \sup\limits_{k\ge 0} \|A_k - A\|_{BL(X,Y)} \le \dfrac{\lambda}{\|A^{-1}\|_{BL(Y,X)}}, & \text{and } \lambda < 1/2 \end{array}\right\}.$$

Then there exists a closed ball, $S_r(u)$, with centre u and radius r such that for every point $u_0 \in S_r(u)$, the sequence $\{u_k\}$ defined by

$$u_{k+1} = u_k - A_k^{-1}F(u_k), \quad k \ge 0, \tag{2.24}$$

is contained in $S_r(u)$, and converges to a point u, which is the only zero of F in the ball $S_r(u)$. Furthermore, there exists a number γ such that

$$\gamma < 1 \text{ and } \|u_k - u\| \le \gamma^k \|u_0 - u\|, k \ge 0. \tag{2.25}$$

As a consequence of Theorem 2.8, we get the following result:

Corollary 2.1. *Let U be an open subset of a Banach space X and let $F : U \subset X \to R$ which is twice differentiable in the open set U. Suppose that there are three constants α, β, γ such that $\alpha > 0$ and $S_\alpha(u_0) = \{v \in X \mid \|v - u_0\| \le \alpha\} \subset U$, $A_k(v) \in BL(X, X^\star)$ and bijective for every $v \in S_\alpha(u_0)$ and*

$$\left\{\begin{array}{l} \sup\limits_{k\ge 0}\ \sup\limits_{v\in S_\alpha(u_0)} \|A_k^{-1}(v)\|_{BL(X^\star,X)} \le \beta \\ \sup\limits_{k\ge 0}\ \sup\limits_{v,v'\in S_\alpha(u_0)} \|F''(v) - A_k(v')\|_{BL(X,X^\star)} \le \dfrac{\gamma}{\beta}, \text{ and} \\ \gamma < 1, \|J'(u_0)\|_{X^\star} \le \dfrac{\alpha}{\beta}(1-\gamma). \end{array}\right.$$

Then the sequence $\{u_k\}$ defined by

$$u_{k+1} = u_k - A_k^{-1}(u_{k'})F'(u_k), k \ge k' \ge 0$$

is contained in the ball $S_\alpha(u_0)$ and converges to a zero of F', say u, which is the only zero in this ball. Furthermore,

$$\|u_k - u\| \le \frac{\|u - u_0\|}{1-\gamma}\gamma^k.$$

As a consequence of Theorem 2.9, we get the following result.

Corollary 2.2 *Let U be an open subset of a Banach space X and let $F : \Omega \subset X \to R$ be a function which is twice differentiable in U. Moreover, let u be a point of U such that*

$$F'(u) = 0, F''(u) \in BL(X, X^\star)$$
$$\sup_k \|A_k - F''(u)\|_{BL(X,X^\star)} \leq \frac{\lambda}{\|(F''(u))^{-1}\|_{BL(X^\star,X)}} \text{ and } \lambda < \frac{1}{2}.$$

Then there exists a closed ball $S_r(u)$ with centre u and radius $r > 0$ such that, for every point $u_0 \in S_r(u)$, the sequence $\{u_k\}$ defined by $u_{k+1} = u_k - A_k^{-1}F'(u_k)$, is contained in $S_r(u)$ and converges to the point u, which is the only zero of F' in the ball. Furthermore, $u_{k+1} = u_k - A_k^{-1}(u_k)F'(u_k)$ converges geometrically, namely, there exists a γ such that $\gamma < 1$ and $\|u_k - u\| \leq \gamma^k \|u_0 - u\|, k \geq 0$.

Remark 2.3.

(i) Let $X = R^n$, the generalized Newton method of Corollary 2.1 take the form

$$u_{k+1} = u_k - A_k^{-1}(u_{k'})\nabla F(u_k), k \geq k' \geq 0, \tag{2.26}$$

where $A_k(u_k)$ are invertible matrices of order n, $\nabla F(u_k)$ denotes the gradient vector of the function F at the point u_k; $\left((R^n)^\star \text{ is identified with } R^n\right)$. In particular, the original Newton method corresponds to

$$u_{k+1} = u_k - \left\{\nabla^2 F(u_k)\right\}^{-1} \nabla F(u_k), k \geq 0, \tag{2.26a}$$

where the matrix $\nabla^2 F(u_k)$ is Hessian of the function F at the point u.

(ii) The special case, $A_k(u_{k'}) = \varphi^{-1}I$, is known as the **gradient method with fixed parameter.**

(iii) The special case, $A_k(u_{k'}) = -\varphi_k^{-1}I$, is called the **gradient method with variable parameter.**

(iv) The special case, $A_k(u_{k'}) = -(\varphi(u_k))^{-1}I$, is called the **gradient method with optimal** parameter, where the number $\varphi(u_k)$ (provided it exists) is determined from the condition

$$F(u_k - \varphi(u_k))\nabla F(u_k) = \inf_{\varphi \in R} F(u_k - \varphi \nabla F(u_k)). \tag{2.27}$$

General definition of gradient method. Every iterative method for which the point u_{k+1} is of the form

$$u_{k+1} = u_k - \varphi_k \nabla F(u_k), \quad \varphi_k > 0,$$

is called a **gradient method**. If φ_k is fixed, it is called a **gradient method with fixed parameter**, while it is called a **gradient method with variable parameter** provided φ_k is variable.

Theorem 2.10. *Let $X = R^n$ and the functional $F : X \to R$ be elliptic, that is, there is a positive constant α such that $F(x) \geq \alpha||x||^2$ for all $x \in X$. Then the gradient method with optimal parameter converges.*

Remark 2.4.

(i) The following properties of elliptic functionals are quite useful (For details we refer to Ciarlet [1989]):

(a) Let $F : X \longrightarrow R$ (X is a Hilbert space, in particular $X = R^n$) be strictly convex and coercive, then it satisfies the inequality

$$F(v) - F(u) \geq \langle \nabla J(u), v - u\rangle + \frac{\alpha}{2}||v - u||^2 \text{ for every } u, v \in X. \quad (2.28)$$

(b) If F is twice differentiable, then it is elliptic if and only if

$$\langle \nabla^2(u)w, w\rangle \geq \alpha||w||^2 \text{ for every } w \in X. \quad (2.29)$$

(c) A quadratic functional over R^n

$$\begin{cases} F(v) = \frac{1}{2}\langle Av, v\rangle - \langle y, v\rangle, A \text{ is the } n \times n \text{ matrix and} \\ A = A^T \text{ is elliptic if and only if} \\ \langle \nabla^2 F(u)w, w\rangle = \langle Aw, w\rangle \geq \lambda_1||w||^2 \text{ for all } u, w \in R^n, \end{cases} \quad (2.30)$$

where λ_1 denote the smallest eigenvalue of A.

(ii) Let $J(v) = \frac{1}{2}\langle Av, v\rangle - \langle y, v\rangle, A : R^n \to (R^n)^\star = R^n$.

Since $\nabla J(u_k)$ and $\nabla J(u_{k+1})$ are orthogonal and $\nabla J(v) = Av - y$, we have

$$\langle \nabla J(u_{k+1}), \nabla J(u_k)\rangle = \langle A(u_k - \varphi(u_k)(Au_k - y)) - y, Au_k - y\rangle = 0.$$

This implies that $\varphi(u_k) = \frac{||w_k||^2}{\langle Aw_k, w_k\rangle}$ where $w_k = Au_k - y = \nabla J(u_k)$.
A single iteration of the method then takes the following form:

(i) Calculate the vector $w_k = Au_k - y$.

(ii) Calculate the number $\varphi(u_k) = \frac{||w_k||^2}{\langle Aw_k, w_k\rangle}$.

(iii) Calculate the vector

$$u_{k+1} = u_k - \varphi(u_k)w_k .$$

Theorem 2.11. *Let $F : R^n \to R$ be a functional which is differentiable and satisfies the following properties. There are two positive constants α and β such that*

(i) $\langle \nabla F(v) - \nabla F(u), v - u\rangle \geq \alpha ||v - u||^2$ *for all* $v, u \in R^n$ *and* $\alpha > 0$,

(ii) $||\nabla F(v) - \nabla F(u)|| \leq \beta ||v - u||$ *for every* $u, v \in R^n$.

Furthermore, let there exist two numbers a and b such that

$$0 < a \leq \varphi_k \leq b < \frac{2\alpha}{\beta^2} \text{ for every } k.$$

Then the gradient method with variable parameter converges and the convergence is geometric in the sense that there exists a constant γ depending on α, β, a, b such that $\gamma < 1$ and $||u_k - u|| \leq \gamma^k ||u_0 - u||$.

Remark 2.5.

(i) If F is twice differentiable, then condition (i) can also be written in the form $\sup ||\nabla^2 F(u)|| \leq \beta$.

(ii) In the case of an elliptic quadratic functional $F(v) = \frac{1}{2}\langle Av, v\rangle - \langle y, v\rangle$, one iteration of the method takes the form

$$u_{k+1} = u_k \varphi_k (Au_k - y), \quad k \geq 0,$$

and it follows from Theorem 2.11 that the method is convergent if $0 < a \leq \varphi_k \leq b \leq 2\lambda_1/\lambda_n^2$, where λ_1 and λ_n are the least and the largest eigenvalues of the symmetric positive definite matrix A.

Proof of Theorem 2.8. First of all, we prove that for dvery integer $k \geq 1$,

$$\begin{aligned} ||u_k - u_{k-1}|| &\leq \beta ||F(u_{k-1})||, \\ ||u_k - u_0|| &\leq \alpha \text{ equivalently } u_k \in S_\alpha(u_0) \\ ||F(u_k)|| &\leq \frac{\gamma}{\beta} ||u_k - u_{k-1}||. \end{aligned}$$

We apply the method of the finite induction for the proof. Let us show that the results are true for $k = 1$; that is,

$$\begin{aligned} &||u_1 - u_0|| \leq \beta ||F(u_0)|| \\ &||u - u_0|| \leq \alpha, \quad ||F(u_1)|| \leq \frac{\gamma}{\beta} ||u_1 - u_0||. \end{aligned}$$

Putting $k = 1$ in relation (2.22), we get

$$u_1 - u_0 = -A_0^{-1}(u_0) F(u_0), \tag{2.31}$$

which implies that $||u_1 - u_0|| \le \beta||F(u_0)|| \le \alpha(1-\gamma) \le \alpha$ by the hypotheses of the theorem. Further, from (2.31), we can write

$$F(u_1) = F(u_1) - F(u_0) - A_0(u_0)(u_1 - u_0).$$

By the Mean Value Theorem applied to the function $u \to F(u) - A_0(u_0)u$, we have

$$||F(u_1)|| \le \sup_{u \in S_\alpha(u_0)} ||F'(u) - A_0(u_0)|| \, ||u_1 - u_0|| \le \frac{\gamma}{\beta}||u_0 - u_1||,$$

by condition (ii) of the theorem.

Let us assume that the desired results are true for the integer $k = n-1$. Since $u_n - u_{n-1} = -A_{n-1}^{-1}(u_{(n-1)'})F(u_{n-1})$, it follows that $||u_n - u_{n-1}|| \le \beta||F(u_{n-1})||$ which gives the first relation for $k = n$. Then we have

$$\begin{aligned} ||u_n - u_{n-1}|| = ||A_{n-1}^{-1}(u_{n-1})F(u_{n-1})|| &\le \beta||F(u_{n-1})|| \\ &\le \beta\tfrac{\gamma}{\beta}||u_{n-1} - u_{n-2}|| \\ &\cdots \\ &\cdots \\ &\le \gamma^{n-1}||u_1 - u_0|| \,. \end{aligned}$$

This implies that

$$\begin{aligned} ||u_n - u_0|| &\le \sum_{i=1}^{n} ||u_i - u_{i-1}|| \le \left\{\sum_{i=1}^{n} \gamma^{i-1}\right\} ||u_1 - u_0|| \\ &\le \frac{||u_1 - u_0||}{1-\gamma} \le \frac{\beta}{1-\gamma}||F(u_0)|| \le \frac{\beta\alpha}{\beta}(1-\gamma) = \alpha, \end{aligned}$$

which means that $u_n \in S_\alpha(u_0)$.

For the proof of the last relation, we write

$$F(u_n) = F(u_n) - F(u_{n-1}) - A_{n-1}(u_{(n-1)}{}')(u_n - u_{n-1}).$$

By applying the Mean Value Theorem to the function $u \to F(u) - A_{(n-1)}(u_{(n-1)'})u$, we get

$$\begin{aligned} ||F(u_k)|| &\le \sup_{u \in S_\alpha(u_0)} ||F'(u) - A_{n-1}(u_{(n-1)'})|| ||u_n - u_{n-1}|| \\ &\le \frac{\gamma}{\beta}||u_n - u_{n-1}||, \end{aligned}$$

and the last relation is established for n. Hence these three relations are true for all integral values of k.

We now prove the existence of a zero of the functional F in the ball $S_\alpha(u_0)$. Since

$$\left\{ \begin{aligned} ||u_{k+m} - u_k|| &\le \sum_{i=1}^{m-1} ||u_{k+i+1} - u_{k+i}|| \\ &\le \gamma^k \sum_{i=0}^{m-1} \gamma^i ||u_i - u_0|| \le \frac{\gamma^k}{1-\gamma}||u_1 - u_0|| \to 0 \text{ as } k \to \infty, \end{aligned} \right. \tag{2.32}$$

where $\{u_k\}$ is a Cauchy sequence of points in the ball $S_\alpha(u_0)$ which is a closed subspace of a complete metric space X (X, a Banach space). This implies that there exists a point $u \in S_\alpha(u_0)$ such that

$$\lim_{k\to\infty} u_k = u.$$

Since F is differentiable and therefore continuous, we get

$$||F(u)|| = \lim_{k\to\infty} ||F(u_k)|| \leq \frac{\gamma}{\beta} \lim_{k\to\infty} ||u_k - u_{k-1}|| = 0\ ,$$

which, in turn, implies $F(u) = 0$ by the first axiom of the norm. By taking the limit $m \to \infty$ in (2.32), we find that $||u_k - u|| \leq \frac{\gamma^k}{1-\gamma}||u_1 - u_0||$ is the desired result concerning geometric convergence.

Finally, we show that u is unique. Let v be another zero of F, that is, $F(v) = 0$. Since $F(u) = F(v) = 0$,

$$v - u = -A_0^{-1}(F(u) - F(v) - A_0(u_0)(v-u)),$$

from which it follows that

$$||v - u|| \leq ||A_0^{-1}(u_0)|| \sup_{u\in S_\alpha(u_0)} ||F'(v) - A_0(u_0)|| ||v-u|| \leq \gamma||v-u||,$$

which implies that $u = v$ as $\gamma < 1$.

Proof of Theorem 2.9.

(i) First of all, we show the existence of constants α and β such that

$$\alpha > 0, S_\alpha(u) = \{x \in X / ||x - u|| \leq \alpha\} \subset U, \tag{2.33}$$

and

$$\sup_{k\geq 0} \sup_{x\in S_\alpha(u)} ||I - A_k^{-1}F'(x)|| \leq \beta < 1. \tag{2.34}$$

For every integer k, we can write $A_k = A(I + A^{-1}(A_k - A))$ with $||A^{-1}(A_k - A)|| \leq \lambda < 1$ in view of a condition of the theorem. Thus, A_k are isomorphisms from X onto Y and moreover,

$$\begin{aligned} ||A_k^{-1}|| &= ||(A(I + A^{-1}(A_k - A))^{-1}|| \\ &\leq ||(I + A^{-1}(A_k - A)^{-1}|| ||A^{-1}|| \leq \frac{||A^{-1}||}{1-\lambda}. \end{aligned}$$

This implies that

$$\begin{aligned} ||I - A_k^{-1}A|| &= ||A_k^{-1}A_k - A_k^{-1}A|| \leq ||A_k^{-1}|| \quad ||A_k - A|| \\ &\leq ||A_k^{-1}|| \frac{\lambda}{||A^{-1}||} \qquad \text{for } \lambda < \frac{1}{2} \\ &\leq \frac{||A^{-1}||}{1-\lambda} \frac{\lambda}{||A^{-1}||}, \end{aligned}$$

or

$$||I - A_k^{-1}A|| \leq \frac{\lambda}{1-\lambda} = \beta^{'} < 1.$$

Let δ be such that $\beta' < \beta^{'} + \delta = \beta < 1$. This implies that

$$||I - A_k^{-1}F^{'}(x)|| \leq ||I - A_k^{-1}A|| + ||A_k^{-1}(A - F^{'}(u))||.$$

From here, (2.33) and (2.34) follow immediately keeping in mind the continuity of the derivative F' and the fact that $A = F'(u)$.

(ii) Let u_0 be any point of the ball $S_\alpha(u)$ and $\{u_k\}$ be the sequence defined by $u_{k+1} = u_k - A_k^{-1}F(u_k)$; each of these elements lies in $S_\alpha(u)$. This implies that $\{u_k\}$ is well defined. Since $F(u) = 0$, we have

$$u_{k+1} - u = u_k - A_k^{-1}F(u_k) - \left(u - A_k^{-1}F(u)\right).$$

By the Mean Value Theorem applied to the function, $x \rightarrow x - A_k^{-1}F(x)$ shows that

$$||u_{k+1} - u|| \leq \sup_{x \in S_\alpha(u)} ||I - A_k^{-1}F'(u)||||u_k - u|| \leq \gamma||u_k - u||.$$

By (2.34) and continuing in this, way we get

$$||u_{k+1} - u|| \leq \gamma^{k-1}||u_1 - u||,$$

which is the geometric convergence. This relation also implies that $u_k \longrightarrow u$ as $k \longrightarrow \infty$ as $\beta < 1$.

(iii) The zero of F, point u is unique. For this let v be another point such that $F(v) = 0$. The sequence $\{u_k\}$ corresponding to $u_0 = v$ is a stationary sequence, since $u_1 = u_0 - A_0^{-1}F(u_0) = u_0$ and; on the other hand, it converges to the point u by the above discussion. This implies $u = v$.

For the proof of Theorem 2.10 we refer to Ciarlet [1989, *pp*.300 – 301] and we prove here Theorem 2.11.

Proof of Theorem 2.11. In a gradient method with variable parameter, we have $u_{k+1} = u_k - \varphi_k \nabla F(u_k)$. Since $\nabla F(u) = 0$ for a minima at u, we can write $u_{k+1} - u = (u_k - u) - \varphi_k \{\nabla F(u_k - \nabla F(u)\}$. This implies that $||u_{k+1} - u||^2 = ||u_k - u|| - 2\varphi_k < \nabla F(u_k) - \nabla F(u), u_k - u > + \varphi_k^2 ||\nabla F(u_k) - \nabla F(u)||^2 \leq \{1 - 2\alpha\varphi_k + \beta^2\varphi_k^2\} ||u_k - u||$, under the condition that $\varphi_k > 0$. If

$$0 < a \leq \varphi_k \leq b < \frac{2\alpha}{\beta^2},$$

then

$$1 - 2\alpha\,\varphi_k + \beta^2\varphi_k^2 < 1,$$

and so

$$\|u_{k+1} - u\| \leq \gamma\|u_k - u\| \leq \gamma^{k+1}\|u - u_0\|,$$

where $\gamma < 1$ which depends on α, a, b and β. This also implies the congence of $\{u_k\}$.

Newton Algorithm for finding $\inf_{x \in R^n} F(x), F : R^n \longrightarrow R^n$.

Algorithm 2.1:

Data. $u_0 \in R^n$
Step 0. Set $k = 0$
Step 1. Compute the Newton search direction
$h_k = -H(u_k)^{-1}\nabla F(u_k)$
Here $H(u)$ is the Hessian Matrix of F
Step 2. Set $u_{k+1} = u_k + h_k$
replace k by $k+1$ and go to step 1.

Newton-Armijo Algorithm 2.2:

Parameters $\alpha \in (0, 1/2), \beta \in (0, 1)$
Data . $u_0 \in R^n$
Step 0. Set $i = 0$
Step 1. Compute the Newton search direction

$$h_i = -H(u_i)^{-1}\nabla F(u_i)$$

Stop if $h_i = 0$.
Step 2. Compute the Armijo step size
$\lambda_i = \max_{k \in N}\{\beta^k \mid F(u_i + \beta^k h_i) - F(u_i) \leq \alpha\beta^k < h_i, \nabla F(u_i) >\}$
Step 3. Set $u_{i+1} = u_i + \lambda_i + h_i$,
replace i by $i+1$ and go to Step 1.

2.5 Conjugate gradient method

The **conjugate gradient** method deals with the minimization of the quadratic functional on $X = R^n$; that is,

$$J : v \in R^n \longrightarrow \frac{1}{2}\langle Av, v\rangle - \langle b, v\rangle,$$

or

$$J(v) = \frac{1}{2}\langle Av, v\rangle - \langle b, v\rangle,$$

where A is the $n \times n$ matrix. Starting with an initial arbitrary vector u_0, we set $d_0 = \nabla J(u_0)$. If $\nabla J(u_0) = 0$, the algorithm terminates. Otherwise, we define the number

$$r_0 = \frac{\langle \nabla J(u_0), d_0 \rangle}{\langle Ad_0, d_0 \rangle},$$

then the vector u_1 is given by

$$u_1 = u_0 - r_0 d_0.$$

Assuming that the vectors $u_1, d_1, \cdots, u_{k-1}, d_{k-1}, u_k$ have been constructed which assumes that the gradient vectors $\nabla J(u_l), 0 \leq l \leq k-1$ are all non-zero, one of two situations will prevail: either $\nabla J(u_k) = 0$ and the process terminates, or $\nabla J(u_k) \neq 0$, in which case we define the vector

$$d_k = \nabla J(u_k) + \frac{||\nabla J(u_k)||^2}{||\nabla J(u_{k-1})||^2 d_{k-1}},$$

then the number r_k and u_{k+1} are given by

$$r_k = \frac{\langle \nabla J(u_k), d_k \rangle}{\langle Ad_k, d_k \rangle},$$

and

$$u_{k+1} = u_k - r_k d_k,$$

respectively.

This elegant algorithm was developed by Hestenes and Stiefel in 1952. This method converges in at most n iterations (For proof, see Polyak [1987, p. 69]). For a computer programme for implementation of these algorithms, we refer to Press et al. [1992].

The study of the conjugate gradient method for non-quadratic function on R^n into R began in sixties. Details of these methods and their comparative merits in different cases can be found in Powel[1986], Gilbert and Nocedal [1992] and Polak [1997]. However, we present here the essential ingredients of two of these best methods, namely, Fletcher-Reeves(FR) and Polak-Ribiére (PR).

Let $F : R^n \longrightarrow R$, we look for $\inf_{v \in R^n} F(v)$ where F is twice differentiable. The point at which $\inf_{v \in R^n} F(v)$ is attained will be denoted by $\arg\inf(x)$. Starting with an arbitrary vector u_0, one assumes the vectors $u_1, u_2, \cdots u_k$ to have been constructed, which means that the gradient vectors $\nabla F(u_i), 0 \leq i \leq n-1$, are non-zero. In such situations, either $\nabla F(u_n) = 0$ and the algorithm terminates, or $\nabla F(u_n) \neq 0$, in which case, vectors u_{n+1} is defined (if it exists and is unique) by the relations

$$u_{n+1} = u_n - r_n d_n \text{ and } F(u_{n+1}) = \inf_{\rho \in R} F(u_n - \rho d_n),$$

the successive descent directions d_i being defined by the recurrence relation:

$$\begin{aligned} d_0 &= \nabla J(u_0) \\ d_i &= \nabla F(u_i) + \frac{\langle \nabla F(u_i), \nabla F(u_i) - \nabla F(u_{i-1})\rangle}{\|\nabla F(u_{i-1})\|^2} d_{i-1}, \ 1 \le i \le n \\ r_i &= \frac{\langle \nabla F(u_i), \nabla F(u_i) - \nabla F(u_{i-1})\rangle}{\|\nabla F(u_{i-1})\|^2}, \end{aligned}$$

is called the Polak-Ribiére formula and in this case the conjugate gradient method is called the **Polak-Ribiére conjugate gradient method** and one denotes r_i by r_i^{PR}.

The case

$$r_i = \frac{\|\nabla F(u_i)\|^2}{\|\nabla F(u_{i-1})\|^2}$$

is called the **Fletcher-Reeves formula** and the corresponding method is called the **Fletcher-Reeves conjugate gradient method**. Such r_i is denoted by r_i^{FR}. It may be observed that the Polak-Ribiére conjugate gradient method is more efficient in practice.

Polak-Ribiére conjugate gradient algorithm:

Data. $u_0 \in R^n$
Step 0. Set $i = 0$, $d_0 = \nabla F(u_0)$, and $h_0 = -d_0$
Step 1. Compute the step size
$\lambda_i = \arg\min_{\lambda \ge 0} F(u_i + \lambda h_i)$
Step 2. update: Set $u_{i+1} = u_i + \lambda_i h_i$

$d_{i+1} = \nabla F(u_{i+1})$

$r_i^{PR} = \frac{\langle d_{i+1}, d_{i+1} - d_i\rangle}{\|d_i\|^2}$

$h_{i+1} = -d_{i+1} + r_i^{PR} h_i$

Step 3. Replace i by $i+1$ and go to Step1.

Fletcher-Reeves conjugate gradient Algorithm:

Data. $u_0 \in R^n$
Step 0. set $i = 0, d_0 = \nabla F(u_0,$ and $h_0 = -d_0$
Step 1. Compute the step size

$\lambda_i = \arg\ \min_{\lambda \ge 0} F(u_i + \lambda h_i)$

Step 2. Update: Set $u_{i+1} = u_i + \lambda_i h_i$

$d_{i+1} = \nabla F(u_{i+1})$

$r_i^{FR} = \frac{\|d_{i+1}\|^2}{\|d_i\|^2}$

$h_{i+1} = -d_{i+1} + r_i^{FR}$

Step 3. Replace i by $i+1$ and go to Step 1.

2.6 Variable metric methods (DFP and BFGS methods)

Variable metric methods are used to compute the gradient of functions at arbitrary points. Variable metric methods come in two main flavours: one is the Davidon-Fletcher-Powell (DFP) algorithms and the other is the Broyden-Fletcher-Goldfarb-Shanno (BFGS). However, the BFGS scheme is superior in the sense that it has a low round-off error and converges faster in comparison to other methods. Consider a function $F : R^n \to R$ whose value and the gradient ($\nabla F(x)$) can be calculated at a given n-dimensional point u as the origin of the coordinate system. Our goal is to find $u \in R^n$ at which F has minimum value, that is, $F(u) \leq F(x)$ for all $x \in R^n$. In the first place, we approximate $F(\cdot)$ by a quadratic functional by Talylor's expansion. Then we look for a point of R^n at which the quadratic functional attains minimum and subsequently find the minimum value. The function F can be approximated by its Taylor series

$$\begin{cases} F(x) = F(u) + \sum_i \frac{\partial F}{\partial x_i} x_i + \frac{1}{2} \sum_{i,j} \frac{\partial^2 F}{\partial x_i \partial x_j} x_i x_j + \cdots \\ \qquad \simeq C - b \cdot x + \frac{1}{2} x \cdot A \cdot x \\ \qquad = F(u) + \frac{1}{2} \langle Ax, x \rangle - \langle b, x \rangle, \end{cases} \tag{2.2}$$

where

$$C = F(u), b = -\nabla F(x), A = (a_{i,j}) = \frac{\partial^2 F}{\partial x_i \partial x_j} = \nabla^2 F(x),$$

$$x = (x_1, x_2 \cdots x_n) \in R^n.$$

In approximation, the gradient of F at u, that is, $\nabla F(u) = \langle Au, u \rangle - \langle b, u \rangle, b \in R^n, x \in R^n$ where $\langle \cdot, \cdot \rangle$ denotes the inner product. u is an extrema of F if $\nabla F(u) = 0$ or$\langle Au, u \rangle = \langle b, u \rangle$or $Au = b$. Thus, finding the extrema of F is approximately equivalent to solving the matrix equation $Au = b$, that is, equivalent to finding A^{-1}. Newton's method can be employed for this task.

The matrix A whose components are the second partial derivatives of the function is called the **Hessian matrix** of the function at the given point. The basic idea of the variable metric method is to build up, iteratively, a good approximation to the inverse Hessian matrix A^{-1}, that is, to construct a sequence of matrices H_i with the property

$$\lim_{i \to \infty} H_i = A^{-1}. \tag{2.36}$$

It is much better if the limit is achieved after n iterations instead of ∞. Consider finding a minimum by using Newton's method to search for a zero of the gradient of the function. Near the current point x_i, we have the second order approximation $F(x) = F(x_i) + \langle (x - x_i), \nabla F(x_i) \rangle + \frac{1}{2} \langle A(x - x_i), (x - x_i) \rangle$; so

$$\nabla F(x) = \nabla F(x_i) + A(x - x_i). \tag{2.37}$$

In Newton's method, we set $\nabla F(x) = 0$ to determine the next iteration point

$$x - x_i = -A^{-1}\nabla F(x_i). \tag{2.38}$$

The left hand side in (2.38) is the finite step which is required for exact minima; the right hand side is known if we have an accurate $H \simeq A^{-1}$.

The "quasi" in quasi-Newton is because we do not use the actual Hessian metric of F but instead use an approximation of it. Now, consider the descent direction of F at x_i. This is the direction p along which F decreases: $\langle \nabla F(u).,p\rangle < 0$.

For the Newton direction (2.38) to be a descent direction, we must have

$$\langle \nabla F(x_i),(x - x_i)\rangle = -\langle A(x - x_i),(x - x_i)\rangle < 0\,; \tag{2.39}$$

that is, A must be positive definite.

When we are not close enough to the minimum, taking the full Newton step with p even with positive definite A need not decrease the function. Here we can use the line search and the backtracking strategy to choose a step along the direction of Newton step p.

These properties are valid for DFG algorithm, Press et al. [1992] which update H_{i+1}. Subtracting equation (2.38) at x_{i+1} from that same equation at x_i gives

$$x_{i+1} - x_i = A^{-1}(\nabla F_{i+1} - \nabla F_i), \tag{2.40}$$

where $\nabla F_i \equiv \nabla F(x_i)$. We require that the new approximation H_{i+1} satisfies equation (2.40); that is,

$$x_{i+1} - x_i = H_{i+1}(\nabla F_{i+1} - \nabla F_i)\,;$$

i.e., the updating formula should be of the form $H_{i+1} = H_i +$ correction. The DFP (Davidon-Fletcher-Powell) updating formula is

$$H_{i+1} = H_i + \frac{y_i y_i^T}{y_i^T s_i} - \frac{H_i s_i s_i^T H_i}{s_i^T H_i s_i}\,, \tag{2.41}$$

where

$$\begin{aligned} s_i &= (\nabla F_{i+1} - \nabla F_i) \\ y_i &= (x_{i+1} - x_i). \end{aligned}$$

The BFGS (Broyden-Fletcher-Goldfarb-Shanno) formula is the same but with one additional term; i.e., $\cdots + [s_i^T H_i s_i]u\; u^T$ where

$$u = \frac{y_i}{y_i^T s_i} - \frac{H_i s_i}{s_i^T H_i s_i}\,. \tag{2.42}$$

All these methods are invariant under linear transformation of the variables and converge to a strict local minimizer.

For a computer programme, see Press et al. [1992]. For details concerning algorithms and convergence of these formulae, we refer to Polak [1997].

2.7 Problems

Problem 2.1. Let $m \geq n > 0$ and $A = (a_{ij})$, $i = 1, 2, \cdots, m, j = 1, 2, \cdots, n$ and $y \in R^m$. Then write down the solution of optimization problem for the function $F(x) = \|Ax - y\|$ over R^n.

Problem 2.2. (a) Let $F : R^n \longrightarrow R$ be continuously differentiable in an open convex set $U \subset R^n$. Then show that $u \in U$ can be an extrema of F on U if $\nabla F(u) = 0$.

(b) Extend (a) for $F : X \longrightarrow Y$ where Y is a normed linear space and X is a vector space and F is a Gâteau differentiable function.

Problem 2.3. (a) Discuss the application of the Newton method for finding a minimizer of $f(x) = \sin x$.

(b) Under what conditions for a real-valued function f defined on $[a, b]$, $f(x) = 0$ has a solution? Justify your answer.

Problem 2.4. Explain the concept of Fibonacci search and the golden section search.

Problem 2.5. Let $Ax = y$, where A is as $n \times n$ matrix and x and y are elements of R^n. Write down a sequence of approximate solutions of this equation and examine its convergence. Under what condition on A this equation has necessarily the unique solution?

Problem 2.6. Explain the concept of the steepest descent and apply it to study the optimization of the functional F defined on a Hilbert space H as follows:

$$F(x) = \langle Ax, x \rangle - 2\langle y, x \rangle ,$$

where A is a self-adjoint positive definite operator on H, $x, y \in H$. For any $x_1 \in H$, construct a sequence $\{x_n\}$ where

$$x_{n+1} = x_n + \frac{\langle z_n, z_n \rangle}{\langle Az_n, z_n \rangle} z_n,$$

where $z_n = y - A$, and show that $\{x_n\}$ converges to x_0 in H which is the unique solution of $Ax = y$. Furthermore, show that by defining

$$F(x) = \langle A(x - x_0), x - x_0 \rangle,$$

the rate of convergence satisfies

$$\langle x_n, x_n \rangle \leq \frac{1}{m} F(x_n) \leq \frac{1}{m} (1 - \frac{m}{M})^{n-1} F(x_1) .$$

Problem 2.7. Show that the Polak-Ribiére conjugate gradient algorithm converges under appropriate conditions.

Problem 2.8. Examine the convergence of the DFP algorithm.

Chapter 3

Maxwell's Equations, Finite and Boundary Element Methods

What are the essential needs of modern life? One can say without any hesitation, electric lighting, heating and cooling, telephones, electric motors and generators, radio, television, X-ray machine, ECG machine, radars and weather forecasting system equipment, etc. In fact, the invention of electromagnetism embracing electricity and magnetism completely revolutionized our way of life. Every day, some new tools, techniques and equipment are being developed based on the concepts of the electromagnetism theory. The backbone of modern electromagnetism is a system or a set of four equations established by the Scottish Scientist James Clark Maxwell, a professor at the University of Cambridge, around 1872. This system is now known as the system of Maxwell's equations. In the last three decades, two powerful methods, namely, the finite element and the boundary element, have been invented to find out solutions of physical phenomena represented by differential equations.

The main object of this chapter is to give a brief introduction to this elegant system of mathematical models along with these two methods. In Section 3.1.1, we give a brief historical note along with the physical laws which led to the derivation of Maxwell's equations. Consequences of Maxwell's equations are presented in Section 3.1.2 while variational (weak) formulation of Maxwell's equations are discussed in Section 3.1.3. Sections 3.2.1 to 3.2.4 are devoted to the various aspects of the finite element method. The boundary element method is introduced in Section 3.3, and problems are presented in Section 3.4.

3.1 Maxwell's equations

3.1.1 Brief historical note and physical laws

Electromagnetic theory encompasses electric and magnetic theory. It deals with the electromagnetic fields, a region in which electric and magnetic forces act. As the story goes, in about 600 B.C., a Greek mathematician, astronomer and philosopher Thales observed that when amber is rubbed with silk it produces sparks which led to the invention of electricity, electron and electronics (the Greek word for amber is electron). He also observed the attractive power between pieces of a natural magnetic rock called lead-stone found at a place called Magnesia, from where the words magnet and magnetism are derived. Around 1600 A.D., William Gilbert of England invented the electroscope for measuring electrostatic effects and pointed out that the earth itself is a huge magnet. Around 1750, American scientist-statesman Benjamin Franklin discovered lightning of rod with electricity, positive and negative charges and established the conservation law of charge. After this invention, Charles Augustine de Coulomb of France invented torsion balance for measuring electric and magnetic forces. During the same period, the famous German mathematician and astronomer Karl Friedrich Gauss formulated his famous divergence theorem relating the volume and surface. In 1800, Alessandro Volta of Italy invented Volta cell and electric battery. In 1819, the Danish physicist Hans Christian Oersted noted that electricity produces magnetism. In the following years, the french Physicist André Marie Ampére observed that the atoms in a magnet are magnetized by tiny electric current circulating in them. At this time, George Simon of Germany pronounced his famous law relating current, voltage and resistance. In 1831, Michael Faraday of London demonstrated that a changing magnetic field produces an electric current. The above mentioned inventions provided motivation to James Clark Maxwell, a professor at the Cambridge University, to develop a unified theory of electricity and magnetism which he published in the form of a classic treatise in 1873. He postulated that light is electromagnetic in nature and that electromagnetic radiation of other wavelengths should be possible. He discovered four basic laws governing the interaction of bodies which are magnetised or electrically charged or both. The bodies may be static or moving. He established four equations known as the Maxwell's equations. The acceptance of Maxwell's theory by the scientific community had to wait for 15 years until it was vindicated by the German physicist Heinrich Hertz through experiments.

Objects like metals which permit easy passage of charges are called **conductors.** On the other hand, the objects in which charges are not free to pass are called **insulators** or **dielectrics**.

The force between two point charges placed at a certain distance r is given by **Coulomb's law**

$$F = \frac{kq_1q_2}{r^2}, \tag{3.1}$$

where

F = Force between two point charges q_1 and q_2;
r = Distance between q_1 and q_2;
k = Proportionality constant.

In the international system $k = \frac{1}{4\pi\epsilon}$, where ϵ is the permittivity of the medium in which the charges are situated, ϵ has the dimension of capacitance per unit length.

The permittivity of vacuum ϵ_0 is

$$\epsilon_0 = \frac{1}{36\pi} nFm^{-1} \cong \text{ permittivity of air.}$$

Force is a vector which can be written as

$$F = \hat{r}\frac{q_1 q_2}{4\pi\epsilon_0 r^2} ,$$

where $\hat{r}$ is the unit vector pointing in the direction of line joining the charges. Electric field intensity E is defined as the force per unit charge

$$E = \frac{F}{q_2} = \hat{r}\frac{q_1}{4\pi\epsilon_0 r^2} , \tag{3.2}$$

where q_2 is the positive test charge.

The **electric field intensity** is a vector having the same direction as the force F but differing in dimension and numerical magnitude. The (average) **electric charge density** φ is equal to the total charge q in a volume V divided by the volume. Thus, $\varphi = \frac{q}{V}$. The SI unit of charge density is Coulomb per cubic meter (cm^{-3}). φ can be defined as

$$\varphi = \lim_{\Delta V \to 0} \frac{\Delta q}{\Delta V} ,$$

and hence φ can be treated as a point function. It is often called the volume charge density to distinguish it from the surface charge density and the linear charge density. The electric potential is a measure of energy for some kind of unit quantity. The work or energy per unit charge required to transport the test charge from a point x_1 to point x_2 is called the difference in the electric potential of the points x_2 and x_1. Potential is a scalar quantity, that is, it has magnitude but no direction. The total electric potential at a point is the algebraic sum of the individual component potentials at the point. The gradient of the potential at a point is defined as the potential rise Δv across an element of length Δl along a field line divided by Δl, with the limit of the ratio taken as Δl approaches zero. Thus, if V is the electric potential, then

$$\text{Gradient of } V = \lim_{\Delta l \to 0} \frac{\Delta V}{\Delta l} = \frac{dV}{dl} .$$

The gradient of $V = -E$, as a potential rise occurs when moving against the electric field, the direction of the gradient is opposite to that of the field. The usual notation for the gradient of V is grad V or ∇V.

$$\text{Gradient of } V = \text{ grad } V = \nabla V = -E.$$

In rectangular coordinates

$$E = -\nabla V = -(\hat{x}\frac{\partial V}{\partial x} + \hat{y}\frac{\partial V}{\partial y}) = \hat{x}E_x + \hat{y}E_y,$$

where

$$E_x = \frac{\partial V}{\partial x}, \quad E_y = \frac{\partial V}{\partial y}.$$

The combination of two point charges q of opposite sign separated by a small distance l is called an **electric dipole**, and the product ql is called the **electric dipole moment**.

A **dimension** defines some physical characteristic like length, mass, time, velocity, force.

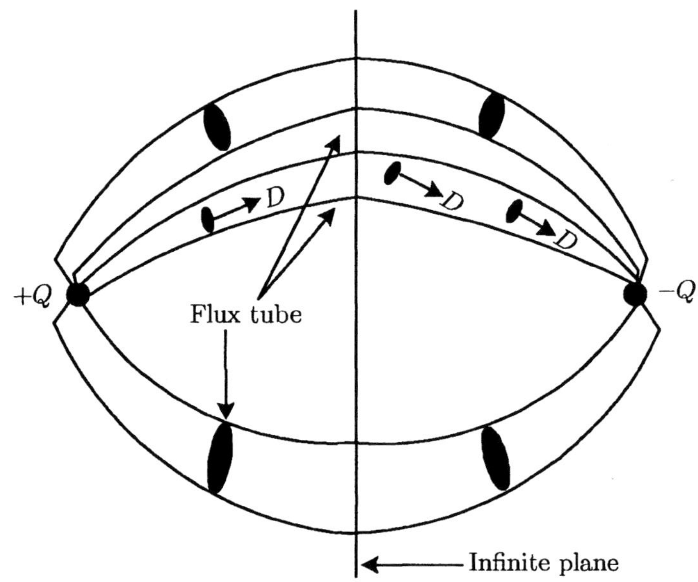

Figure 3.1. Electric field between two point charges of opposite sign with flux tubes joining them. The tube follows the electric field lines. Each tube has a constant amount of flux.

If in Figure 3.1, each tube represents a constant amount of charge or electric flux ψ then, at any point, there is a flux density D, proportional to E, as given by $\frac{\psi}{A}$, where A is the cross-sectional area of the tube. Thus, the electric flux for a tube is given by

$$\psi = DA, \quad D = \text{ average flux density/cm}^{-2}.$$

It can be seen that

$$D = \epsilon_0 E \text{ (See Kraus, [1992]).} \tag{3.3}$$

Gauss' law of electric field. The total electric flux through any closed surface equals the charge enclosed. The equivalent form of this law is

$$\oint_S D \cdot dS = \oint_V \varphi dV = q, \tag{3.4}$$

where $\oint_S$ denotes the double or surface integral over closed surface and $\oint_V$ denotes the triple or volume integral throughout the region enclosed.

If we replace φ by $\nabla \cdot D$, then we get

$$\oint_S D \cdot ds = \oint_V \nabla \cdot DdV.$$

This result is known as the **divergence theorem** or the **Gauss theorem**. This relation holds not only for D but for any vector function. In other words, the divergence theorem states that the integral of the normal component of a vector function over a closed surface S equals the integral of the divergence of that vector throughout the volume V enclosed by the surface S.

A conductor can conduct or convey electric charge. In static situations, a conductor may be treated as a medium in which the electric field is always zero. A medium is **homogeneous** if its physical characteristics like mass, density, and molecular structure do not vary from point to point. If the medium is not homogeneous, it is called **non-homogeneous**, or **inhomogeneous** or **heterogeneous**. A medium is called **linear** with respect to an electrostatic field if the flux density D is proportional to the electric field intensity E. This is the case in free space where $D = \epsilon_0 E$ (ϵ_0 is permittivity constant). In material media, permittivity constant may not be constant. Such material is called non-linear. A material is called **isotropic** if its properties are independent of direction. Crystalline media or certain plasma may have direction characteristics and such materials are called **non-isotropic** or **anisotropic**. Here, we shall confine ourselves to isotropic, linear and homogeneous media.

Electric current J is a vector of R^3 that measures a flux of electric charge. It is also called the current density vector. The flux of electric charges across a surface element dS in the sense of unit normal n to $dS = J \cdot ndS$. The **Ohms law** states that the potential difference or voltage V between the ends of a conductor is equal to the product of its resistance R and the current I. The Ohms law at a point is given by the relation

$$J = \sigma E, \tag{3.5}$$

where

J = current density,
E = Field intensity,
σ = conductivity of the material.

Gauss' law for magnetic fields. A moving charge constitutes an electric current and possesses a **magnetic field**. If a charge q moving with a velocity V experiences a force F, there must be a magnetic field $B = \frac{F}{qV}$, the force being perpendicular to both the field and the direction of motion of the charge. The magnetic field is a vector quantity and it is also regarded as a **magnetic flux density** and is given by $B = \frac{\psi_m}{A}$, $\psi_m = |B|\, A \cos\, \alpha$ where

ψ_m = magnetic flux through area A
$|B|$ = magnitude of magnetic flux density B
α = angle between normal to area A and direction of B .

Dimensionally,

$$\psi_m = \frac{\text{force}}{\text{current moment}} \times \text{ area},$$

or

$$\psi_m = \int\int B \cdot ds\,.$$

Gauss' law for magnetic fields states that the **magnetic flux** through closed Gaussian surface is zero; that is,

$$\oint_S B \cdot dS = 0. \tag{3.6}$$

This implies that the divergence of B equals zero; that is,

$$\nabla \cdot B = 0. \tag{3.7}$$

Amperés law. Let

$$H = \frac{B}{\mu}\,, \tag{3.8}$$

where B = magnetic flux density and μ = permeability of medium (H and B are vectors having the same direction). The amperés law states that the line integral of H around a single closed path is equal to the current enclosed; namely,

$$\oint H \cdot dl = I, \tag{3.9}$$

where

H = magnetic field, A m^{-1}
dl = infinite small element of path length, m
I = current enclosed, A.

Faraday's law. As we know, a current carrying conductor produces a magnetic field. In fact, the reverse effect is also possible, that is, a magnetic field can produce a current in a closed circuit but under the condition that the magnetic flux linking the circuit must be changing. Consider the situation in Figure 3.2(a) of the closed wire loop. A magnetic field with flux density B is normal to the plane of the loop. The induced current flows in such a way as to oppose the original direction. It produces its own magnetic field and the direction of flow of the current is such that the secondary magnetic field acts in the opposite direction to the original magnetic field (see Figures 3.2(b) and 3.2(c)).

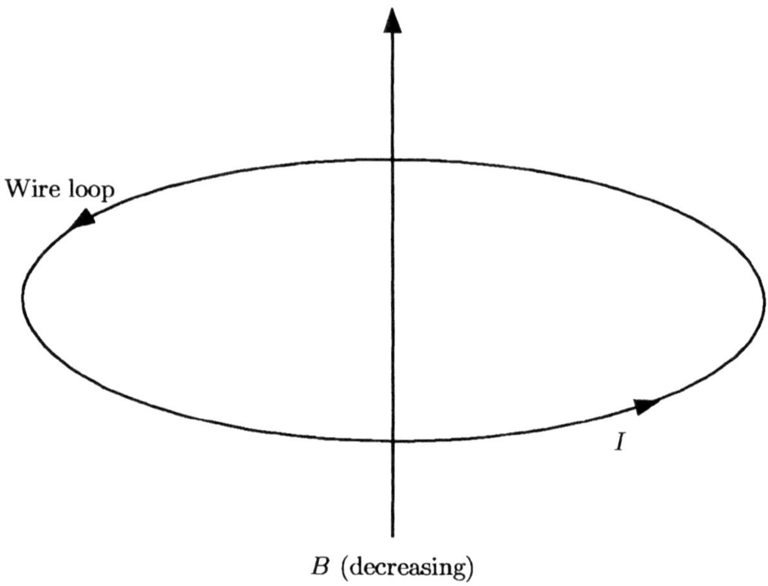

Figure 3.2(a). Relation between decreasing flux density B and induced current I in loop.

According to the Faraday's law, "the total electromotive force (emf) induced in a closed circuit is equal to the time rate of decrease of the total magnetic flux linking the circuit". In mathematical language, it means that

$$V = -\frac{d\psi_m}{dt} ,$$

where

$$\begin{aligned} V &= \text{total emf} \\ \psi_m &= \text{total flux} \\ t &= \text{time} \end{aligned}$$

or

$$V = -\int_S \frac{\partial B}{\partial t} \cdot dS , \tag{3.10}$$

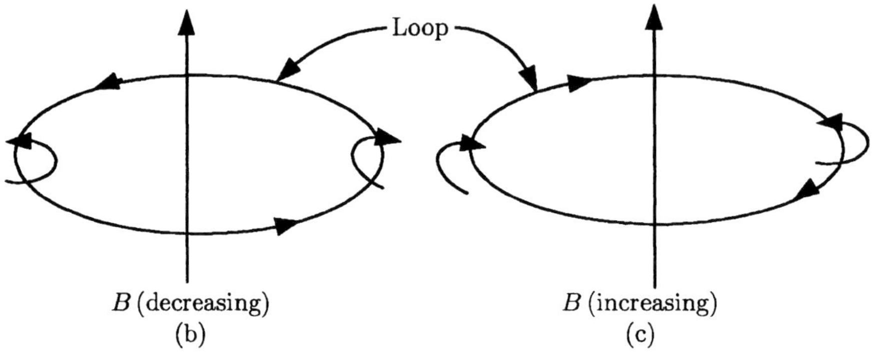

Figure 3.2(b) and 3.2(c). Induced currents for decreasing and increasing flux density B.

where $\int_S$ indicates a double or surface integral ($\int\int$) over a surface S and B is the flux density. For more details, we refer to Kraus [1992].

The notions mentioned in this subsection and the subsequent subsections have wide applications in fields like technology for producing electricity, namely, alternators, transformers, circuit breakers; usage of strong currents, namely, electric circuits and their various components, resistance, inductance, capacities; controlling high frequencies in generators, lines, antennae, receivers, electron beam and their applications in radio, T.V., radar, automatic control devices, telecommunication equipment. Some of the other areas of applications are optical cavity of lasers, quantum electronics, instrumentation, linear and non-linear optics including fibre optics and laser circuits besides electro-chemistry.

3.1.2 Maxwell's equations and their consequences

Maxwell's macroscopic equations (Maxwell's equations in continuous media). Let $E(\cdot,\cdot)$ and $B(\cdot,\cdot)$ be two functions defined on the whole space $R^3_x \times R_t$ with vector values in R^3 which are called the electric field and the magnetic induction. Assume that $E(\cdot,\cdot)$ and $B(\cdot,\cdot)$ are related with $D(\cdot,\cdot)$ and $H(\cdot,\cdot)$ through the relations $D = E + P, -P$ is the electric field generated by $\hat{j}$ where $j = \hat{j} + J$ and $H = B - M$, M is the magnetic field generated by $\hat{\varphi}$ where $\varphi = \hat{\varphi} + \rho$. D is called the **electric induction** or the **electric displacement** or **flux density** while H is called the **magnetic field**. Furthermore, let $\rho(\cdot,\cdot)$ and $J(\cdot,\cdot)$ be defined on $R^3_x \times R_t$ with $\rho(x,t) \in R$ and $J(x,t) \in R^3$. The following set of equations are called

Maxwell's macroscopic equations:

$$-\frac{\partial D}{\partial t} + \text{curl } H = J \qquad (3.11)(\text{i})$$

$$\text{div } D = \rho \qquad (3.11)(\text{ii})$$

$$\frac{\partial B}{\partial t} + \text{curl } E = 0 \qquad (3.11)(\text{iii})$$

$$\text{div } B = 0, \qquad (3.11)(\text{iv})$$

where

$$E = (E_1, E_2, E_3), \quad x = (x_1, x_2, x_3)$$

$$\text{div } E = \sum_{i=1}^{3} \frac{\partial E_i}{\partial x_i}$$

$$\text{curl } E = (\frac{\partial E_3}{\partial x_2} - \frac{\partial E_2}{\partial x_3}, \frac{\partial E_1}{\partial x_3} - \frac{\partial E_3}{\partial x_1}, \frac{\partial E_2}{\partial x_1} - \frac{\partial E_1}{\partial x_2}).$$

E, B, ρ, J are called respectively the **electric field, magnetic induction, charge density** and **current density**. D is called the **electric conduction** or **electrical displacement** or **flux density** while H is known as the **magnetic field.**

Remark 3.1. (i) (a) Since (3.11)(i) is derived from the Ampére's law, it is often called the **Maxwell-Ampére's law**.

(i) (b) (3.11)(ii) is a consequence of the Gauss' electric law and so it is known as the **Maxwell-Gauss electric law.**

(i) (c) (3.11)(iii) is a consequence of Faraday's law and therefore it is called the **Maxwell-Faraday law.**

(i) (d) Similarly, (3.11)(iv) is nothing but the **Gauss' magnetic law.**

(ii)(a) Relation (3.9) can also be written as

$$\oint H \cdot dl = \int\int_A J \cdot dS = I = \int_S J \cdot dS, \qquad (3.12)$$

as the current I enclosed by the path is given by the integral of the normal component of J over the surface A. Or,

$$\oint H \cdot dl = \int_S (J + \frac{\partial D}{\partial t}) \cdot dS. \qquad (3.13)$$

Applying Stoke's theorem to (3.13), we get (3.11)(i).

(ii) (b). By Gauss' law

$$\oint D \cdot dS = \int \varphi \, dV. \qquad (3.14)$$

Applying (3.14) to an infinitesimal volume, we get (3.11)(ii).

(ii) (c) Replacing V in (3.10) by the line integral of E around the circuit, we get

$$\oint E \cdot dl = -\int_S \frac{\partial B}{\partial t} \cdot dS. \tag{3.15}$$

By applying Stokes theorem to (3.15), we get (3.11)(iii).

(ii) (d) In view of (3.7), we get (3.11)(iv), which is also known as **Maxwell's magnetic field equation.**

(ii) (e) For further details of the derivation of Maxwell's equations, the continuity relation for the electricity and equations of conservation laws, we refer to Dautry and Lions [1990], Kraus [1992], Itzykson and Zuber [1980] and Landau and Lipschitz [1959].

Remark 3.2. (i) We have employed here the systems unit called natural in which the speed of light in *vacuo*, $c = 1$ and the permeability of the vacuum, $\mu_0 = 1$.

(ii) If, in equations (3.11), $P = 0$ and $M = 0$, then these take the form

$$-\frac{\partial E}{\partial t} + \text{curl } B = j \qquad (3.16)(i)$$

$$\text{div } E = \varphi \qquad (3.16)(ii)$$

$$\frac{\partial B}{\partial t} + \text{curl } E = 0 \qquad (3.16)(iii)$$

$$\text{div } B = 0. \qquad (3.16)(iv)$$

The system of the four equations of (3.16) is called the system of **Maxwell's microscopic equations** or **Maxwell's equations** in vacuo.

Remark 3.3. *Maxwell's equations in perfect media.* For perfect media $D = \epsilon E$, $H = \frac{1}{\mu}B$, with ϵ and μ being positive constants depending on the characteristic of the medium considered known as the permittivity or dielectric constant and the permeability or the magnetic permeability constant, respectively. Putting these values of D and H in (3.11), we have the following system of equations:

$$-\epsilon\mu\frac{\partial E}{\partial t} + \text{curl } B = \mu J \qquad (3.17)(i)$$

$$\text{div } E = \frac{\varphi}{\epsilon} \qquad (3.17)(ii)$$

$$\frac{\partial B}{\partial t} + \text{curl } E = 0 \qquad (3.17)(iii)$$

$$\text{div } B = 0. \qquad (3.17)(iv)$$

Remark 3.4. Since $D = \epsilon E$ and $E = -\nabla V$, we have $D = -\epsilon \nabla V$ and putting this value in (3.11)(ii), we get

$$\nabla^2 V = -\frac{\varphi}{\epsilon}, \tag{3.18}$$

which is the **Poisson's equation.**

If $\varphi = 0$, then (3.18) is reduced to the **Laplace's equation**:

$$\nabla^2 V = 0. \tag{3.19}$$

In rectangular coordinates,

$$\nabla^2 V = \frac{\partial^2 V}{\partial x^2} + \frac{\partial^2 V}{\partial y^2} + \frac{\partial^2 V}{\partial z^2} = 0\,.$$

The static potential distribution for any conductor configuration can be determined by solving Laplace's equation under appropriate boundary conditions.

Remark 3.5. In static electromagnetism, we find solutions of Maxwell's equations which are independent of time. In this case, (3.11) reduces to

$$\begin{aligned} \text{curl } H &= J && (3.20)(\text{i}) \\ \text{div } D &= \varphi && (3.20)(\text{ii}) \\ \text{curl } E &= 0 && (3.20)(\text{iii}) \\ \text{div } B &= 0. && (3.20)(\text{iv}) \end{aligned}$$

In electrostatics domain, $J = B = H = 0$, we have

$$\begin{aligned} \text{div } D &= \varphi && (3.21)(\text{i}) \\ \text{curl } E &= 0. && (3.21)(\text{ii}) \end{aligned}$$

For magnetostatics, $\varphi = 0,\ E = D = 0$, we get

$$\begin{aligned} \text{curl } H &= J && (3.21)(\text{i}) \\ \text{div } B &= 0. && (3.21)(\text{ii}) \end{aligned}$$

Remark 3.6. Derivation of the wave equation. Let us assume that $\varphi = 0, j = 0$. By differentiating (3.16)(i) with respect to the time and by keeping in view the formula

$$\text{curl curl } E = -\Delta E + \text{grad div } E,$$

we find that E satisfies the wave equation

$$\frac{\partial^2 E}{\partial t^2} = \Delta E = \frac{\partial^2 E}{\partial x^2}\,.$$

In a similar manner, by differentiating (3.16)(iii), we find that B satisfies the wave equation; that is,

$$\frac{\partial^2 B}{\partial t^2} = \frac{\partial^2 B}{\partial x^2}\,.$$

Remark 3.7. Relation between charge density and current density. Differentiating (3.16)(ii) with respect to t and by taking the divergence on both sides of (3.16)(i), we get the equation

$$\frac{\partial \varphi}{\partial t} + \text{div } j = 0, \tag{3.22}$$

which is called the **equation of the conservation of electricity or the continuity relation for electricity.**

Remark 3.8. The vector field P and M are called, respectively, the **polarization** and the **magnetization** vector.

Remark 3.9. Maxwell's equations and the Lorentz condition. In equations (3.16), we consider the functions

$$\begin{cases} (x,t) \longrightarrow A(x,t) \in R^3, & \text{and} \\ (x,t) \longrightarrow \phi(x,t) \in R, \end{cases}$$

which are called the **potential vector** and the **scalar potential**, respectively. $A(\cdot,\cdot)$ and $\phi(\cdot,\cdot)$ are related to E and B by the relation

$$B = \text{curl } A \tag{3.23(i)}$$

$$E = -\text{ grad } \phi - \frac{\partial A}{\partial t}. \tag{3.23(ii)}$$

On substituting the values of B and E in equations (3.16)(i) and (3.16)(ii), we obtain the inhomogeneous linear systems

$$\frac{\partial^2 A}{\partial t^2} - \Delta A + \text{ grad } (\text{div } A + \frac{\partial \phi}{\partial t}) = j \tag{3.24(i)}$$

$$-\Delta\phi - \frac{\partial}{\partial t} (\text{div } A) = \varphi\,. \tag{3.24(ii)}$$

It can be noted that A and ϕ are not defined in a unique manner by (3.23) starting from E and B. If A and ϕ satisfy (3.23), then for each arbitrary function F of x and t, A' and ϕ' defined by

$$A' = A + \text{ grad } F \tag{3.25(i)}$$

$$\phi' = \phi - \frac{\partial F}{\partial t}, \tag{3.25(ii)}$$

also satisfy (3.23).

The transformation $(A, \phi) \to (A', \phi')$ given by (3.25) is called a **gauge transformation.**

By (3.25), we have

$$\text{div } A' + \frac{\partial \phi'}{\partial t} = \text{ div } A + \frac{\partial \phi}{\partial t} + \Delta F - \frac{\partial^2 F}{\partial t^2}. \tag{3.26}$$

Taking for F, a solution of the equation

$$\Delta F - \frac{\partial^2 F}{\partial t^2} = - (\text{div } A + \frac{\partial \phi}{\partial t}), \tag{3.27}$$

and assuming A and ϕ known, we find that a pair (A_L, ϕ_L) can be chosen such that

$$\text{div } A_L + \frac{\partial \phi_L}{\partial t} = 0. \tag{3.28}$$

This relation is called the **Lorentz condition.**

Keeping in view the Lorentz condition, (3.24) can be written as

$$\frac{\partial^2 A_L}{\partial t^2} - \Delta A_L = j \tag{3.29)(i}$$

$$\frac{\partial^2 \phi_L}{\partial t^2} - \Delta \phi_L = \varphi. \tag{3.29)(ii}$$

It may be observed that the pair (A_L, ϕ_L) is not unique when j and φ are known.

Remark 3.10. Maxwell's equations and the wave equation. If φ and A are chosen such that $\frac{\partial \phi}{\partial t} + \text{div } A = 0$, then keeping in view (3.29) and (3.24), Maxwell's equations in vacuum (equation (3.16)) can be written as the wave equation in the absence of charge and current ($\varphi = 0, j = 0$). That is, we have

$$\begin{cases} \frac{\partial^2 A}{\partial t^2} & = \Delta A, x \in R^3, t \geq 0\,; \\ A(x,0) & = A^0(x), x \in R^3\,; \\ \frac{\partial A}{\partial t}(x,0) & = A'(x), x \in R^3\,, \end{cases} \tag{3.30}$$

$$\begin{cases} \frac{\partial^2 \phi}{\partial t^2} & = \Delta \phi, x \in R^3\,; \\ \phi(x,0) & = \phi^0(x), x \in R^3\,; \\ \frac{\partial \phi}{\partial t}(x,0) & = \phi'(x), x \in R^3\,, \end{cases} \tag{3.31}$$

where A and ϕ are, respectively, the vector potential and scalar potential.

Each component of the potential vector is the solution of the wave problem; namely,

$$\begin{cases} \frac{\partial^2 u}{\partial t^2} & = \Delta u, \; x \in R^n, \quad t > 0\,; \\ u(x,0) & = u^0(x)\,; \\ \frac{\partial u}{\partial t}(x,0) & = u'(x)\,, \end{cases} \tag{3.32}$$

where u^0 and u' are given functions or distributions with a velocity of propagation which is the speed of light in the vacuum, taken equal to 1 in the natural system of units.

Let $u = u(x,t) \in R$ be a solution of the wave equation

$$\frac{\partial^2 u}{\partial t^2} - \Delta u = 0 \text{ in } R_x^3 \times R_t.$$

We put $u^1 = u, u^2 = \dfrac{\partial u}{\partial t}$, then

$$FM = (E^1, B^1) = (\text{curl curl}(xu^1),\ \text{curl}(xu^2))$$
$$FE = (E^2, B^2) = (-\text{curl}(xu^2),\ \text{curl curl}(xu^1))$$

can be shown to be the solutions of Maxwell's equations (3.16) where $\varphi = 0, j = 0$. This can be obtained by using the identity curl curl curl $(x\phi) = \ -\text{curl}(x\Delta\phi)$.

FM is called the **transverse magnetic wave** while FE is called the **transverse electric wave.** $U = (u^1.u^2)$ is called the **Debye potential.** For further details, we refer to Dautry and Lions [1990] and Schulenberger [1978].

Remark 3.11. The Coulomb condition. In place of (3.28), we can impose the condition div $A = 0$, then (3.24)(ii) implies that $\Delta\phi = -\varphi$, that is, ϕ satisfies the Poisson equation in R_x^3 where ϕ and φ possibly depend on the time. In general, we can choose for the solution ϕ

$$\phi(x', t) = \int_{R^3} \frac{\varphi(x,t)}{|x' - x|} dx, \quad x' \in R^3, t \in R. \tag{3.33}$$

The right hand side expression in (3.33) has a meaning by convolution if φ is a distribution with compact support. Normally, this ϕ is called a Newtonian potential but in electromagnetism this ϕ is called the Coulomb potential and the condition div $A = 0$ is called the **Coulomb condition.**

3.1.3 Variational formulation of Maxwell's equations

Stable media. The results presented in this section are based on Duvaut-Lions [1972, Chapter VII], and Dautry and Lions [1990, Vol. 3, Chapter IX, pp. 239–264]. Let Ω be a domain in R^3 with a regular bounded boundary Γ. The open set Ω may or may not be bounded. We want to find the vector fields B, D, J that satisfy

$$\frac{\partial D}{\partial t} + J \ - \text{curl }(\hat{\mu}\ B) = G_1 \ \text{ in } \Omega \tag{3.34)(i}$$

$$\frac{\partial B}{\partial t} + \text{curl}(\hat{\epsilon}\ D) = G_2 \ \text{ in } \Omega, \tag{3.34)(ii}$$

where G_1 and G_2 satisfy the conditions

$$\text{div } G_2 = 0, G_2.n = 0, \hat{\mu} = \frac{1}{\mu}, \hat{\epsilon} = \frac{1}{\epsilon}.$$

(μ and ϵ are as in equation (3.17)). $\hat{\epsilon}$ and $\hat{\mu}$ are strictly positive and remain bounded which may depend on x, in particular, may be piecewise constant.

Let

$$B(x,0) = B_0(x),\ D(x,0) = D_0(x).$$

Furthermore, div $D = \varphi$, div $B = 0$.

$$B_0(0) = B_0, D_0(0) = D_0 \quad \text{on } \Omega \tag{3.35(i)}$$

$$J = \sigma\ \hat{\epsilon}\ D \text{ on } \Omega. \qquad \text{(Stable Media)}, \tag{3.35(ii)}$$

and

$$n \wedge D = 0. \tag{3.35(iii)}$$

We define the space $H(\text{curl};\ \Omega)$ as follows:

$$H(\text{curl};\ \Omega) = \left\{v \in (L_2(\Omega))^3 /\ \text{curl } v \in (L_2(\Omega))^3\right\}.$$

It is a Hilbert space with respect to the inner product which induces the norm

$$\|v\|^2 = \|v\|^2_{(L_2(\Omega))^3} + \|\text{ curl } v\|^2_{(L_2(\Omega))^3}. \tag{3.36}$$

Let

$$H_0(\text{curl};\ \Omega) = \left\{ \begin{array}{l} v \in H(\text{curl};\ \Omega) \mid n \wedge v = 0 \text{ on } \Gamma, \text{ where } n \text{ is the normal} \\ \quad \text{to } \Gamma \text{ directed towards the exterior of } \Omega \end{array} \right\}.$$

$(\mathcal{D}(\Omega))^3$ is dense in H_0 (curl; Ω).

Let

$$\mathcal{H} = (L_2(\Omega))^6 = (L_2(\Omega))^3 \times (L_2(\Omega))^3. \tag{3.37}$$

An inner product $\langle , \rangle$ is defined on $\mathcal{H}$ in the following manner:

$$\left\{ \begin{array}{ll} \langle \Phi, \Phi_\star \rangle_{\mathcal{H}} & = \langle \phi, \phi_\star \rangle + \langle \psi, \psi_\star \rangle, \text{ where} \\ \Phi & = \{\phi, \psi\} \text{ and } \Phi_\star = \{\phi_\star, \psi_\star\} \text{ belong to } \mathcal{H}. \end{array} \right. \tag{3.38}$$

An operator $\mathcal{A}$ is defined on $\mathcal{H}$ as follows:

$$\begin{array}{ll} D(\mathcal{A}) & = \left\{ \begin{array}{l} \Phi = \{\phi, \psi\} \in \mathcal{H} \mid \text{ curl } (\hat{\epsilon}\psi) \in (L_2(\Omega))^3 \\ \text{and curl } (\hat{\mu}\psi) \in (L_2(\Omega))^3, n \wedge \phi = 0 \text{ on } \Gamma \end{array} \right\} \\ \mathcal{A}\Phi & = \{-\text{curl } (\hat{\mu}\psi), \text{curl } (\hat{\epsilon}\phi)\} \in \mathcal{H}. \end{array} \tag{3.39}$$

It has been shown [Duvaut-Lions 1972] that $D(\mathcal{A})$ is dense in $\mathcal{H}$ and $\mathcal{A}$ is closed and

$$\mathcal{A}^\star = -\mathcal{A},\ \ D(\mathcal{A}^\star) = \mathcal{D}(\mathcal{A}).$$

The following problem is the variational formulation or weak formulation of Maxwell's equations in the stable media (equations (3.34)).

Let

$$\mathcal{M}\Phi = \{\sigma\, \hat{\varepsilon}\, \phi,\ 0\ \} \text{ if } \Phi = \{\phi, \psi\}, \tag{3.40)(i}$$

which defines

$$\mathcal{M} \in \mathcal{L}(\mathcal{H}, \mathcal{H}) . \tag{3.40)(ii}$$

Find $U = \{D, B\}$ such that

$$U \in L^{\infty}(0, T, \mathcal{H}) \tag{3.40)(iii}$$

$$\int_0^T \left[-\langle U, \frac{\partial \Phi}{\partial t} \rangle \mathcal{H} - \langle U, \mathcal{A}\Phi \rangle \mathcal{H} + \langle \mathcal{M}U, \Phi \rangle \mathcal{H} \right] dt$$

$$= \int_0^T \langle G, \Phi \rangle \mathcal{H}\ dt + \langle \mathsf{U}_0, \Phi(0) \rangle \mathcal{H} \quad \forall \Phi \in \mathcal{H}\,, \tag{3.40)(iv}$$

and

$$\begin{cases} \Phi \in L_2(0, T; \mathcal{D}(\mathcal{A})) \\ \partial\Phi/\partial t \in L_2(0, T, \mathcal{H}), \Phi(T) = 0\,, \end{cases} \tag{3.40)(v}$$

$$\begin{cases} G = \{G_1, G_2\} \in L_2(0, T, \mathcal{H}) \\ \mathsf{U}_0 = \{D_0, B_0\} \in \mathcal{H}\,. \end{cases} \tag{3.40)(vi}$$

For proof, see Lemma 4.5 [Duvaut-Lions, 72, p. 347].

Theorem 4.1 [Duvaut-Lions, 72, p. 347]. The system of equations (3.40)(i) – (vi) has a unique solution.

3.1.4 Variational formulation (weak formulation) of magnetostatics of a surface current

In this section, we look for the solutions B of the following equations:

$$\text{div } B = 0 \text{ in } R^3 \tag{3.41)(i}$$

$$\text{curl } B = 0 \text{ in } R^3 \setminus \Omega = \Omega' \tag{3.41)(ii}$$

such that

$$W = \frac{1}{2\mu} \int_{\Omega} |B|^2 dx + \frac{1}{2\mu'} \int_{\Omega'} |B|^2 dx < \infty,\ B \in (L_2(R^3))^3,\ \Omega \subset R^n. \tag{3.42}$$

We consider the case where the surface current J_Γ on Γ boundary of Ω is given, that is, we find $B \in (L_2(R^3))^3$ satisfying

$$(P_1) \begin{cases} \text{div } B = 0 \text{ in } R^3 & (3.43)(\text{i}) \\ \text{curl } B = 0 \text{ in } \Omega \text{ and } \Omega' & (3.43)(\text{ii}) \\ \left[\dfrac{-B}{\mu} \wedge n \right]_{\Gamma} = J_\Gamma, J_\Gamma \text{ given with div } J_\Gamma = 0 & (3.43)(\text{iii}) \end{cases}$$

where $\left[\frac{-\beta}{\mu} \wedge n\right]_\Gamma$ denotes the jump in the quantity $\frac{-B}{\mu} \wedge n$ with n normal to Γ oriented to the exterior of Ω across Γ, that is, on denoting by $B_{\Omega'}$ and B_Ω, the restriction of B to Ω' and Ω, $\left[\frac{B}{\mu} \wedge n\right]_\Gamma = \left(\frac{B_{\Omega'}/\Gamma}{\mu'} - \frac{B_\Omega/\Gamma}{\mu}\right) \wedge n$.

Let

$$V = H(\text{div } 0, R^3) = \left\{\hat{B} \in (L_2(R^3))^3/\text{div } \hat{B} = 0\right\} \quad \text{and}$$
$$W = \left\{\hat{A} \in (H_1(R^3))^3/\text{div}\hat{A} = 0\right\}.$$

By the Poincaré lemma for each $\hat{B} \in V$, there exists unique $\hat{A} \in W$ such that $\hat{B} = \text{curl } \hat{A}$. By applying the techniques of the Fourier transform, it can be verified that the mapping $\hat{A} \in W \to \hat{B} = \text{curl } \hat{A} \in V$ is an isometry of the Beppo-Levi space on the space V.

Let

$$\begin{cases} a(B, \hat{B}) = \dfrac{1}{2\mu} \displaystyle\int_\Omega B \cdot \hat{B} dx + \dfrac{1}{2\mu'} \displaystyle\int_{\Omega'} B \cdot \hat{B} dx, \quad \forall B, \hat{B} \in V \\ a_0(A, \hat{A}) = a(\text{curl } A, \text{ curl } \hat{A}), \forall A, \hat{A} \in W, \end{cases} \tag{3.44}$$

where $a(\cdot,\cdot)$ and $a_0(\cdot,\cdot)$ are continuous and coercive on V and W, respectively. In fact, for $\mu_m = \inf\left\{\mu, \mu'\right\}$, $\mu_M = \sup\left\{\mu, \mu'\right\}$, we have,

$$\frac{1}{2\mu_M} \int_{R^3} |\hat{B}|^2 \, dx \leq a(\hat{B}, \hat{B}) \leq \frac{1}{2\mu_m} \int_{R^3} |\hat{B}|^2 \, dx \quad \forall \hat{B} \in V. \tag{3.45}$$

It can be observed that each element $\hat{A} \in W$ admits a trace $\hat{A}/\Gamma \in (H^{1/2}(\Gamma))^3$.

Theorem 3.1. *For each given surface current J_Γ such that $J_\Gamma \in (H^{-\frac{1}{2}}(\Gamma))^3$, $J_\Gamma \cdot n = 0$, almost every where on Γ, and div $J_\Gamma = 0$, then the problem (P_1) is equivalent to the variational problem:*

Find $B \in V$ (respectively $A \in W$, $B =$ curl A) such that

$$a(B, B) = a_0(A, \hat{A}) + \frac{1}{2} \int_\Gamma J_\Gamma \cdot \hat{A}_{/\Gamma} d\,\Gamma, \quad \forall \hat{B} = \text{ curl } \hat{A} \in V, \tag{3.46}$$

or

$$\forall \, \hat{A} \in W$$

has a unique solution.

For proof, we refer to Dautry and Lions [1990, p. 241].

Remark 3.12. Finding the solution of the variational problem (3.46) is equivalent to solving the optimization problem: Find $B \in V$ (or $A \in W$, $B = \text{curl } A$) such that

$$F(B) = \inf_{\hat{B} \in V} F(\hat{B}), \tag{3.47}$$

where F is a quadratic functional defined on V as

$$F(\hat{B}) = a(\hat{B}, \hat{B}) - \int_{\Gamma} J_{\Gamma} \hat{A}_{\Gamma} d\Gamma, \quad \forall \hat{B} = \text{ curl } \hat{A} \in V. \tag{3.48}$$

Variational formulations of various other problems of electromagnetism can be found in Dautry-Lions [1990, vol. 3 pp. 243-263].

Bossavit [1993] has presented applications of Maxwell's equations in areas like electrical motors, microwave ovens, resonant cavities. He has published a series of papers on various aspects of Maxwell's equations with boundary conditions, see for example, Bossavit [1995, 1993, 1991]. Hoppe and Wohlmuth [1981] have studied the interior boundary value problem for Maxwell's equations in the time harmonic case. He has obtained qualitative results by a-priori estimates established therein and has also established estimates for the global discretization error in various norms of the underlying spaces of approximating vector fields. Non-conforming finite element methods have been applied. Reissel [1995] has also studied variational formulation and numerical solutions of such problems. Monographs of Colton and Kress [1983], Jin [1993] and Wang [1995] provide a comprehensive account of the variational (weak) formulation and numerical solutions of the Maxwell's equations. These results are of vital importance in areas like fibre communication systems. An automatic mesh generator for the finite element method along with all procedures is incorporated in a software package available with Wang [1995]. A book by Krizek and Neittanmäki [1996] and papers by Alonso and Valli [1997, 1999] present some current aspects of the Maxwell's equations.

3.2 Finite element method

3.2.1 Introduction to numerical methods

As the things stand today, it is a formidable task to find exact solutions of most of the real life models. During the last half of the century, attempts have been made to develop techniques and methods to find approximate solutions of these models and to examine whether the approximate solutions converge to exact solutions or not. The error estimations between the exact and approximate solutions have been studied extensively in the recent past. New endeavours are made to find methods and techniques which will require minimum time in evaluating the models with maximum accuracy. These methods, techniques and tools constitute the subject of 'scientific computation'. Very often, the models are in the form of ordinary and partial differential equations with boundary or initial conditions. By an exact solution we mean a solution of these boundary and initial boundary problems, while the corresponding approximate solution is a solution of an algebraic equation obtained from the given problem by discretizing the analytical model. The following methods for finding approximate solutions of models are well known to mathematicians, engineers and other users of mathematics:

(i) **The Rayleigh-Ritz method,**

(ii) **Galerkin's method,**

(iii) **The weighted residuals method,**

(iv) **The collocation method,**

(v) **The least square method,**

(vi) **The finite difference methods,**

(vii) **Multigrid methods,**

(viii) **Finite volume methods,**

(ix) **Particle methods,**

(x) **The finite element method,**

(xi) **The boundary element Method**, and

(xii) **The wavelet method.**

In this subsection, we briefly introduce some of these methods. A detailed account of the finite element method is presented in the following subsections while a comprehensive account of the boundary element method is given in Section 3.3. The wavelet method is introduced in Chapter 5 along with the updated references. The particle methods are introduced in Chapter 4. The finite difference, the finite element and the boundary element methods have been the main competitor of each other. Lucid accounts of their advantages and disadvantages can be found in Dautry and Lions [1990, vol. 4, pp. 168-170 and 369-370], Hammond [1986], Brebbia et al. [1985], and Reddy [1985]. However, this area is rapidly growing and in recent years methods like the multigrid finite element, wavelet-based adaptive finite element are attracting more attention due to their superior performance in many areas. See, for example, Hackbusch [1985, 1994, 1995], Bramble [1993], Brenner and Scott [1994], Hoppe and Wohlmuth [1997] Canuto and Cravero [1997] and references therein. In cases of complicated geometric regions, the finite element methods and/or their combination with boundary element and multigrid methods have an edge over the other methods in general. Some sort of weak formulation (variational formulation) of given models is required in all methods. The theory of distributions developed by the French mathematician Laurent Schwartz during 1945-1950 and the Sobolev spaces studied by the Russian physicist, S.L.Sobolev in the years 1937-38 have wide applications in the weak-formulation of initial and boundary value problems. The finite element methods can be treated as approximation methods in the Sobolev spaces. However, we shall not follow this approach here and we refer to Ciarlet [1978], and Ciarlet and Lions [1991] for this type of study. This approach is also introduced in Siddiqi [1986] along with updated literature. Mathematical models can be written in the form

$$\begin{cases} Tu = f \text{ on } \Omega \\ Su = g \text{ on } \Gamma \,, \end{cases} \tag{3.49}$$

where $\Omega \subset R^n$ (in particular $n = 1, 2, 3$), Γ is the boundary of Ω, T and S are differential operators, f and g are elements of an appropriate function space, especially a Sobolev space X. For example,

(i) $T = \nabla, \quad S = I, g = 0$

(ii) $T = \nabla, \quad S = \dfrac{\partial}{\partial n}, g = 0$

(iii) $T = \dfrac{d^2}{dx^2}, \quad f = x^2, S = I, g = 0$

$$X = H^1(\Omega) \quad = \left\{ f \in L_2(\Omega) / \frac{\partial f}{\partial x} \in L_2(\Omega) \right\}.$$

In most of the methods, we are required to write (3.49) in the variational form (weak form): Find $u \in X$ such that

$$a(u, v) = F(v), \tag{3.50}$$

where $a(\cdot, \cdot)$ is a bilinear continuous form on the space X and F is a bounded linear functional on X into R. We know that if $a(\cdot, \cdot)$ is also symmetric, then finding the solution u of (3.50) is equivalent to finding $u \in X$ such that

$$\begin{cases} J(u) \leq J(v) \text{ for all } v \in X, \quad \text{where} \\ J(v) = \dfrac{1}{2} a(v, v) - F(v) \,. \end{cases} \tag{3.51}$$

It may be recalled that by an **exact** or **classical solution** of a differential equation, we mean a function that identically satisfies the equation and the specified boundary or initial conditions. **Variational (weak) solution** of a differential equation with or without boundary conditions is the solution of an associated variational problem. The exact solution is sufficiently differentiable as required by the equation while the variational solution is not sufficiently differentiable to satisfy the differential equation but differentiable enough to satisfy a variational equation (problem) equivalent to the differential equation.

(i) Rayleigh-Ritz method. The Rayleigh-Ritz method deals with the approximate solution of (3.50) in the form of a finite series

$$u_m = \sum_{j=1}^{m} c_j \phi_j + \phi_0 \,, \tag{3.52}$$

where the coefficients c_j, called the Rayleigh-Ritz coefficients, are chosen such that equation (3.50) holds for $v = \phi_i, i = 1, 2 \cdots m$; that is,

$$a(\phi_i, \sum_{j=1}^{m} c_j \phi_j + \phi_0) = F(\phi_i), i = 1, 2 \cdots m. \tag{3.53}$$

Since $a(\cdot, \cdot)$ is bilinear, (3.53) takes the form

$$\sum_{j=1}^{m} a(\phi_i, \phi_j) c_i = F(\phi_i) - a(\phi_i, \phi_0), \tag{3.54}$$

or

$$Ac = b, \tag{3.55}$$

where

$$A = \left(a(\phi_i, \phi_j)\right)_{ij}$$

is a matrix

$$b^T = [b_1, b_2, \cdots, b_m] \text{ with } b_i = F(\phi_i),$$

and

$$c = [c_1, c_2, \cdots, c_m],$$

which represents a system of m linear algebraic equations in m unknowns c_i. The columns (and rows) of coefficient matrix A must be linearly independent in order that the coefficient matrix in (3.55) can be inverted. Thus, for symmetric bilinear forms, the Rayleigh-Ritz method can be viewed as one that seeks a solution of the form in equation (3.52) in which the parameters are determined by minimizing the quadratic functional (energy functional) given in (3.51). After substituting u_m of equation (3.52) for u into (3.51) and integrating, the functional $J(u)$ becomes an ordinary function of the parameters $c_1, c_2 \cdots$. The necessary condition for the minimum of $J(c_1, c_2, \cdots c_m)$ is that

$$\frac{\partial J(\cdots)}{\partial c_1} = \frac{\partial J(\cdots)}{\partial c_2} = \cdots\cdots \frac{\partial J(\cdots)}{\partial c_m} = 0. \tag{3.56}$$

This gives m linear algebraic equations in m unknowns, $c_j, j = 1, 2 \cdots m$. It may be observed that (3.54) and (3.56) are the same in the symmetric case while they differ in the non-symmetric case. In other words, we get the same c_i's by solving (3.54) and (3.56) separately. In the non-symmetric case, we determine the m unknowns by solving the linear algebraic equations (matrix equations) (3.55). The choice of $\{\phi_j\}$, $j = 1, 2 \cdots m$ is crucial and this should be the basis of the **Hilbert space**.

(ii) The Galerkin method. Let $R = T(u) - b \neq 0$ where $b = T(u_0)$ in Ω such that

$$\langle R, w \rangle = \int_{\Omega} \text{Rwd}\Omega = 0. \tag{3.57}$$

where $w = \sum_{i=1}^{m} \beta_i \phi_i$.

For linear T, (3.57) gives a linear system of algebraic equations from which β_i's can be determined. In the linear case, the Rayleigh-Ritz and Galerkin methods are identical. It is a common practice to choose w as a variation of u; that is,

$$w = \delta u = \delta\alpha_1\phi_1 + \delta\alpha_2\phi_2 + \delta\alpha_3\phi_3 + \dots, \tag{3.58}$$

where $\delta\alpha_i = \beta_i$ for all i. Thus the Galerkin method seeks an approximate solution to (3.50) in the form of $w = \sum_{i=1}^{m} \beta_i \phi_i$, and determines the coefficients β_i from the condition that the residual R is orthogonal to w.

(iii) The weighted residual method. T in (3.49) can be chosen as any one of the following operators:

$$Tu = -\frac{d}{dx}(a\frac{du}{dx}) \qquad (3.59)(i)$$

$$Tw = \frac{d^2}{dx^2}(b\frac{d^2w}{dx^2}) \qquad (3.59)(ii)$$

$$Tu = -\frac{d}{dx}(u\frac{du}{dx}) \qquad (3.59)(iii)$$

$$Tu = -\left[\frac{\partial}{\partial x}(k_x\frac{\partial u}{\partial x}) + \frac{\partial}{\partial y}(k_y\frac{\partial u}{\partial y})\right] \qquad (3.59)(iv)$$

$$T(u,v) = u\frac{\partial u}{\partial x} + v\frac{\partial u}{\partial y} + \frac{\partial^2 u}{\partial x^2} + \frac{\partial}{\partial y}(\frac{\partial u}{\partial y} + \frac{\partial v}{\partial x}).$$

In this method also, the solution u of (3.49) is approximated by expression of the form

$$u_m = \phi_0 + \sum_{j=1}^{m} c_j\phi_j\ , \tag{3.60}$$

where ϕ_0 must satisfy all boundary conditions, say $\phi_0 = 0$, if all the specified boundary conditions are homogeneous and ϕ_j must satisfy all conditions as mentioned in the Rayleigh-Ritz method as well as continuity. However, continuity can be relaxed if weak formulation is possible for the given problem.

$$E = T(u_m) - f \neq 0, \tag{3.61}$$

in general, is called a **residual** or **error** in the equation. Once ϕ_0 and ϕ_1 are selected, E is simply a function of the independent variables and the parameters c_j. In the **weighted residual method**, the parameters are determined by setting the integral of a weighted residual of the approximation to zero, that is, setting the condition

$$\int_\Omega \psi_i(x,y)E(x,y,c_j)dxdy = 0 \qquad i = 1, 2\cdots m\ , \tag{3.62}$$

where ψ_i are **weight functions** which are linearly independent. It may be remarked that if $\psi_i = \phi_i$ for all i, then we get the Galerkin method as a special case of the weighted residual method. The case $\psi_i \neq \phi_i$ is sometimes referred to as the **Petrov-Galerkin method.**

For linear T, (3.62) reduces to

$$\sum_{j=1}^{m} \left(\int_\Omega \psi_i T(\phi_j) dxdy \right) c_j = \int_\Omega \psi_i \left[f - T(\phi_0) \right] dxdy \,,$$

or

$$\sum_{j=1}^{m} T_{ij} c_j = f_i \,, \tag{3.63}$$

where

$$T_{ij} = \int_\Omega \phi_i T(\phi_j) dxdy \,. \tag{3.64}$$

It is clear that the matrix $[T] = (T_{ij})$ is not symmetric as $T_{ij} \neq T_{ji}$. The details of this method can be found in Finlayson [1972].

(iv) The collocation method. In this method we look for an approximate solution u_m of (3.49) of the form (3.60) by requiring the residual in the equation to be identically zero at m selected points $x^i = (x_i, y_i), i = 1, 2, 3 \cdots m$ in the domain Ω; that is,

$$E(x^i, y^i, c_j) \;=\; 0, \quad i = 1, 2 \;\cdots\; m. \tag{3.65}$$

The selection of the points x^i is vital for obtaining an accurate solution. The collocation method is a special case of the residual method when $\psi_i = \delta(x - x^i)$, where $\delta(x)$ is the Dirac delta function characterized by the equation

$$\int_\Omega f(x)\delta(x - \xi) \; dx = f(\xi) \,. \tag{3.66}$$

For more details, we refer to Douglas and Dupont [1973], Prenter and Russel [1976] and wheeler [1978].

(v) The least squares method. This is a special case of the weighted residual method where $\psi_i = \dfrac{\partial E}{\partial c_i}$, when we are looking for a solution, as in (3.62). We shall determine the parameter c_i from the condition

$$\frac{\partial}{\partial c_i} \int_\Omega E^2(x, y, c_j) dxdy = 0 \,, \tag{3.67}$$

which is the necessary condition for the integral of the square of the residual (3.63) to be minimum. On differentiation, (3.67) takes the form

$$\int_\Omega \frac{\partial E}{\partial c_i} E \, dxdy = 0 \, . \tag{3.68}$$

For linear T, (3.68) takes the form

$$\sum_{j=1}^{m} \left(\int_\Omega T(\phi_i)T(\phi_j)dxdy \right) c_j = \int_\Omega T(\phi_i)\left[f - T(\phi_0)\right] dxdy \, . \tag{3.69}$$

We refer to Bramble-Nitsche [1973], Prenter and Russel [1976], and Locker and Prenter [1978] for more details.

(vi) Finite difference method. In the finite difference method, we replace differential operators in (3.49) by a **difference quotient**. Suppose $Tu = -\Delta u$ and $Su = u_{/\Gamma}, g = 0$, $\Omega = \{(x,y)/0 < x < 1, 0 < y < 1\}$ in (3.49), that is, we want to solve the Dirichlet problem

$$\begin{cases} -\Delta u = f \text{ in } \Omega \\ u_{/\Gamma} \quad = 0 \, . \end{cases} \tag{3.70}$$

We explain the general procedure through the following example. Let N be an integer ≥ 1 and $h = \frac{1}{N+1}$. A mesh on the square Ω is the set of points $(x_i = ih, y_j = jh), i = 0, 1, \cdots N+1, j = 0, 1, 2 \cdots N+1$, while these points are called the nodes of the mesh. The finite difference method comprises obtaining an approximate solution, that is, an approximation of u satisfying (3.70) at the points $(x_i, y_j), i, j = 0, 1, 2 \cdots N+1$ and it is based on the Taylor formula; namely,

$$\begin{cases} -\Delta u(x_i, y_j) = \dfrac{1}{h^2}\left[4\, u_{i,j} - u_{i+1,j} - u_{i-1,j} - u_{i,j+1} - u_{i,j-1}\right] \\ \qquad + \frac{h^2}{12}\left[\frac{\partial^4 u}{\partial x^4}(x_i + \theta_i h, y_i) + \frac{\partial^4 u}{\partial y^4}(x_i, y_i + \theta_j h)\right] \, , \end{cases} \tag{3.71}$$

with $u_{ij} = u(x_i, y_j)$ and $|\theta_i| \leq 1 \quad i, j = 1, 2, 3 \cdots N$ where the 4-th order derivative of u exists and is continuous.

The two main points of this method are:

(1) neglecting the rest of the expansion in which the coefficient $\dfrac{h^2}{12}$ is small; and

(2) requiring that equation (3.70) is satisfied at all the points (x_i, y_j), $j = 1, 2, 3, \ldots, N$ of the mesh, the quantity $-\Delta u(x_i, y_j)$ being approximated, in conformity with (1) by the difference quotient

$$\frac{1}{h^2}\left[4u_{ij} - u_{i+1,j} - u_{i-1,j} - u_{i,j+1} - u_{i,j-1}\right] \, .$$

Putting $f_{ij} = f(x_i, y_j)$, we get in this way N^2 equations in N^2 unknowns $u_{ij}, i, j = 1, 2 \cdots N$; that is,

$$\frac{1}{h^2}\left[4u_{ij} - u_{i+1,j} - u_{i-1,j} - u_{i,j+1} - u_{i,j-1}\right] \;= f_{ij}, \; i, j = 1, 2 \cdots N \, . \tag{3.72}$$

The boundary condition $u_{/\Gamma} = 0$ is taken into account in equation (3.71) by requiring that

$$u_{0,j} = u_{N+1,j} = u_{i,0} = u_{i,N+1} = 0, \quad i,j = 1,2,3\cdots N . \tag{3.73}$$

The system of equations (3.72) and (3.73) is then written in the form of matrix equation as

$$AU = F. \tag{3.74}$$

The solution of (3.74) is the approximate solution of (3.70). There exists a fairly good literature on this method published in sixties, for example, one may see references in Dautry and Lions [1990, vol. 4]. The problem of convergence of the approximate solution to the exact solution has also been investigated. Weak formulation of (3.49) and applying the finite difference method to it is known as the **variational approximation method for finite difference.** This has been studied by Felippa [1973]. For further general results, we refer to Céa [1964] and Aubin [1972].

(vii) Multi-grid method. We shall see in the next two subsections that applications of the finite and the boundary element methods reduce the continuous problems, namely, ordinary and partial differential equations to matrix equations. In recent years, several robust and adaptive algorithms have been developed to solve these equations known as the multi-grid methods which we do not present here due to the limitation of space. In full multi-grid methods for elliptic partial differential equations, one works on a sequence of meshes where a number of pre- and/or post-smoothing steps are performed on each level. Interested readers in acquiring a good knowledge of this new development may go through Hackbusch [1985, 1989, 1994], Bramble [1993], Brenner and Scott [1994] and references therein.

(viii) Finite volume method. In the finite volume method, integral form of the equations representing laws of fluid dynamics are discretized. The flow field domain is subdivided into a set of non-overlapping cells that cover the whole domain on which the conservation laws are applied. In the finite volume method, the term cell is used for element. On each cell conservation laws which are the basic laws of fluid dynamics are applied to determine the flow field variables in some discrete points of the cells, called nodes. Cells can be triangular, quadrilateral, etc. They can be elements of a structured grid or non-structured grid. In this method a function space for the solution need not be defined and nodes can be chosen in a way that does not imply an interpolation structure. Since the computational grid is not necessarily orthogonal and equally spaced, the definition of derivative by Taylor's expansion is impossible. Furthermore, there is no mechanism like a weak formulation and therefore this method is best suited for flow problems in primitive variables where the viscous terms are absent (Euler equations) or not very important (high Reynolds number Navier Stokes equations). For a lucid description of this method, we refer to Kroner [1997] and Wendt [1991].

3.2.2 Finite element method

The finite element method is a variational-method-based technique to solve differential equations. In this method continuous problems described by a differential equation are written into an equivalent variational form (weak form) and the approximate solution of this variational problem is assumed to be a linear combination, $\sum_i c_i\psi_j$, of approximation functions ψ_i. The constants c_i's are determined by the associated variation form. The finite element method provides a systematic technique for deriving the approximation functions for simple subregions which constitute a geometrically complex region. In this method, the approximation functions are piecewise polynomials, that is, polynomials that are defined only on a subregion, called an element. As mentioned earlier, phenomena in nature or real-life problems can be expressed in the form of mathematical models employing the laws of physics and with the aid of the given physical conditions. As we see these models are nothing but algebraic, differential, integral or operators equations. Finding the exact solutions of these equations is a formidable task. In the case where analytic solution is not feasible, we look for an approximate solution. The finite difference method and variational methods such as the Ritz and Galerkin methods were well known until the beginning of the sixties for finding approximate solutions. A systematic study of variational formulation of the boundary value problems and their discretization began in the early seventies. Since early 1950, engineer Argyris started the study of certain techniques for structural analysis which are now known as the **primitive finite element method.** The work representing the beginning of finite element was contained in a paper of Turner, Clough, Martin and Topp [1956], where endeavour was made for a local approximation of the partial differential equations of linear elasticity by the usage of assembly strategies, an essential ingredient of finite element method. In 1960, Clough termed these techniques as "finite element method". Between 1960 and 1980, several conferences were organized in different parts of the world, mainly by engineers, to understand the intricacies of the method. A paper by Zlamal [1968] is considered as the first most significant mathematical contribution in which analysis of interpolation properties of a class of triangular elements and their application to the second- and fourth-order linear elliptic boundary value problems is carried out. Valuable contributions of Ciarlet, Strang, Fix, Schultz, Birkhoff, Bramble and Zlamal, Babuska, Aziz, Varga, Raviart, Lions, Glowinski, Nitsche, Brezzi have enriched the field. Proceedings of the conferences edited by Whiteman [1973-79] and the book by Zienkiewicz and Cheung [1967] have popularized the method among engineers and mathematicians alike. The Finite Element Handbook edited by Kardestuncer and Norrie [1987] and the Handbook of Numerical Analysis edited by Ciarlet and Lions [1991] provide updated literature. Wahblin [1995] contains some of the current research work in this field. In short, one can say that there is no other approximation method which has had such a vast impact on the theory and applications of numerical methods. It has been practically applied in every conceivable area of engineering: structural analysis, semiconductor devices, meteorology, flow through porous media, heat conduction, wave propagation, electromagnetism, environmental studies, safing sensors, geomechanics, biomechanics, aeromechanics and acoustics; to name a few.

The **finite element method** is popular and attractive due to the following reasons: The method is based on weak formulation (variational formulation) of boundary and initial value problems. This is a critical property because it provides a proper setting for the existence of even discontinuous solution to differential equations, for example, distributions, and also because the solution appears in the integral of a quantity over a domain. The fact that the integral of a measurable function over an arbitrary domain can be expressed as the sum of integrals over an arbitrary collection of almost disjoint subdomains whose union is the original domain, is a very important point in this method. Due to this fact, the analysis of a problem can be carried out locally over a subdomain, and by making the subdomain sufficiently small, polynomial functions of various degrees are sufficient for representing the local behaviour of the solution. This property can be exploited in every finite element program which allows us to focus the attention on a typical finite element domain and to find an approximation independent of the ultimate location of that element in the final mesh. The property stated above has important implications in physics and continuum mechanics and consequently, the physical laws will hold for every finite portion of the material.

Some important features of the **finite element methods** are

(1) arbitrary geometries,

(2) unstructured meshes,

(3) robustness, and

(4) sound mathematical foundation.

Arbitrary geometries means that, in principle, the method can be applied to domains of arbitrary shapes and with arbitrary boundary conditions. By unstructured meshes, we mean that, in principle, one can place finite elements anywhere ranging from the complex cross-sections of biological tissues to the exterior of aircraft to internal flows in turbo machinery, without strong use of a globally fixed coordinate frame. Robustness means that the scheme developed for assemblage after local approximation over individual elements is stable in appropriate norms and insensitive to singularities or distortions of the meshes (This property is not available in classical difference methods).

The method has a sound mathematical basis as convergence of an approximate solution of the abstract variational problem (a more general form of variational inequality problem) and error estimation of the abstract form in a fairly general situation and their particular cases have been systematically studied in the last two decades. These studies make it possible to lift the analysis of important engineering and physical problems above the traditional empiricism prevalent in many numerical and experimental studies.

In Subsection 3.2.3, we shall present error estimation of an approximate solution of an abstract variational problem in the context of a Hilbert space and its exact solution along with the general concept of **finite element method, finite element** and main steps for solving a boundary value problem. In Subsection 3.2.4, detailed procedures will be indicated for solving concrete problems.

3.2.3 Abstract finite element method

Let H be a Hilbert space and $a(\cdot,\cdot)$ be a bounded bi-linear form on $H \times H$ into R. For each $F \in H^\star$, the dual space of H (the space of all bounded linear functionals on H) the variational equation (problem):
There exists $u \in H$ such that

$$a(u,v) = F(v) \text{ for all } v \in H \tag{3.75}$$

has a unique solution provided $a(\cdot,\cdot)$ satisfies the **coercivity** or **ellipticity** condition, namely $\exists\, \alpha > 0$ such that

$$a(u,u) \geq \alpha \|u\|^2 \text{ for all } u \in H.$$

This result is known as the Lax-Milgram Lemma (for proof, one may see Siddiqi [1986] or Ciarlet [1978] which mainly depends on the Riesz representation theorem concerning representation of the elements of $H^\star$ in terms of the inner product on H).

Conformal finite element method. The procedure for finding a finite-dimensional subspace H_h of H such that there exists $u_h \in H_h$ satisfying the equation

$$a(u_h, v_h) = F(v_h) \text{ for all } v_h \in H_h \tag{3.76}$$

is called the **conformal finite element method** or simply the **finite element method.** Equation (3.75) is known as the **abstract variational problem** and equation (3.76) is called the **approximate problem.** If H_h is not a subspace of H, the above method is called the **non-conformal finite element method**.

Equation (3.76) is nothing but a matrix equation of the form

$$AU = b, \tag{3.77}$$

where $U = (\alpha_1, \alpha_2, \alpha_3 \cdots \alpha_{N(h)})$, $N(h)$ = dimension of H_h

$$A^T = (a(w_i, w_j))i,j \qquad (3.77)(i)$$

$$b = (F(w_1), F(w_2) \cdots F(w_{N(h)})) \qquad (3.77)(ii)$$

$$u_h = \sum_{i=1}^{N(h)} \alpha_i w_i \qquad (3.77)(iii)$$

$$v_h = \sum_{j=1}^{N(h)} \beta_j w_j, \qquad (3.77)(iv)$$

where α_i and β_j are real numbers $i, j = 1, 2, 3 \cdots, N(h)$. The choice of the basis $\{w_i\}_i$ of H_h, $i = 1, 2 \cdots N(h)$, is of vital importance, namely, choose a basis of H_h which makes A a sparse matrix so that the computing time is reasonably small. In the terminology of structural engineers, A and $L(w_j)$ are called the stiffness matrix and the load vector, respectively.

If $a(\cdot,\cdot)$ is symmetric, then finding the solution of (3.75) is equivalent to finding the optimization problem

$$J(u) = \inf_{v \in H} J(v), \tag{3.78}$$

where $J(v) = \frac{1}{2}a(v, v) - F(v)$ is called the energy functional.

In this case, the **finite element method** is nothing but the Rayleigh-Ritz-Galerkin method. It is quite clear that (3.76) and the approximate problem for (3.78), namely,

$$J(u_h) = \inf_{v_h \in H_h} J(v_h), \tag{3.79}$$

where $J(v_h) = \frac{1}{2}a(v_h, u_h) - f(v_h)$ have unique solutions.

Finding $||u - u_h||$, where u and u_h are the solutions of (3.75) and (3.76), respectively, is known as the **error estimation**. The problem $u_h \to u$ as $h \to 0$, that is, $||u_h - u|| \to 0$ as $h \to 0$ or $N = \frac{1}{h} \to \infty$ is known as the **convergence problem**.

Error estimation.

Theorem 3.2 [Céa's Lemma]. *There exists a constant C independent of the subspace H_h such that*

$$||u - u_h||_H \leq C \inf_{v_h \in H_h} ||u - v_h||_{H_h} = Cd(u, H_h), \tag{3.80}$$

where $C = \frac{M}{\alpha}$, independent of H_h, M is the constant associated with the continuity (boundedness) of $a(\cdot,\cdot)$ and α is the coercivity constant. If $a(\cdot,\cdot)$ is symmetric, then the degree of approximation is improved, that is, we get $C = \sqrt{\frac{M}{\alpha}}$ which is less than the constant in the non-symmetric case.

Theorem 3.3 [First Strang Lemma]. *Let H be a Hilbert space and $H_h(\cdot)$ be its finite-dimensional subspace. Further, let $a(\cdot,\cdot)$ be a bilinear bounded and elliptic form on H and $F \in H^\star$. Assume that u_h is the solution of the following approximate problem. Find $u_h \in H_h$ such that*

$$a_h(u_h, v_h) = F_h(v_h) \text{ for all } v_h \in H_h\,, \tag{3.81}$$

where $a_h(\cdot,\cdot)$ is a bilinear form defined on H_h and $F_h(\cdot)$ is a linear functional defined on H_h. Then

$$||u - u_h|| \leq C \left(\inf_{v_h \in H_h} \left\{ ||u - v_h|| + \sup_{w_h \in H_h} \frac{|a(v_h, w_h) - a_h(v_h, w_h)|}{||w_h||} \right\} \right.$$
$$\left. + \sup_{w_h \in H_h} \frac{|F(w_h) - F_h(w_h)|}{||w_h||} \right),$$

provided $a_h(\cdot,\cdot)$ is uniformly H_h-elliptic, that is, $\exists\ \beta > 0$ such that $a_h(v_h, v_h) \geq \beta\|v_h\|^2$ for all $v_h \in H_h$ and all h.

Note: It may be observed that (i) $a_h(\cdot,\cdot)$ and $F_h(\cdot)$ are not defined for all the elements of H, and (ii) equation (3.81) has a unique solution under the given conditions.

Theorem 3.4 [Second Strang Lemma]. *Let u_h be a solution of the following approximate problem (discrete problem): Find $u_h \in H_h$ such that*

$$a_h(u_h, v_h) = F(v_h) \quad \text{for all } v_h \in H_h\,, \tag{3.82}$$

where $a_h(\cdot,\cdot)$ is as in Theorem 3.3 and $F \in H^\star$. Then there exists a constant C independent of the subspace H_h such that

$$\|u - u_h\|_H \leq C\left(\inf_{v_h \in H_h} \|u - v_h\| + \sup \frac{\|a_h(u, w_h) - F(w_h)\|}{\|w_h\|_h}\right),$$

where $H = H_0^1(\Omega)$.

We prove here Theorems 3.2 and 3.3 while the proof of Theorem 3.4 is along similar lines (See Ciarlet, 1991, pp. 212–213, for proof and more information about results in this direction).

Proof of Theorem 3.2. By (3.75) and (3.76), we get $a(u,v) - a(u_h, v_h) = F(v) - F(v_h)$ and this gives $a(u, v_h) - a(u_h, v_h) = 0$, for $v = v_h$. By bilinearity of $a(\cdot,\cdot)$, we get

$$a(u - u_h, v_h) = 0 \ \ \forall\ v_h \in H_h \Rightarrow a(u - u_h, v_h - u_h) = 0 \tag{3.83}$$

by replacing v_h by $v_h - u_h$.

Since $a(\cdot,\cdot)$ is elliptic,

$$a(u - u_h, u - u_h) \geq \alpha\|u - u_h\|^2\,,$$

or

$$\frac{1}{\alpha} a(u - u_h, u - v_h + v_h - u_h) \ \geq\ \|u - u_h\|^2\,,$$

or

$$\frac{1}{\alpha}\ [a(u - u_h, u - v_h) + a(u - u_h, v_h - u_h)] \geq \|u - u_h\|^2.$$

Using (3.83), this becomes

$$\frac{1}{\alpha}\,[a(u - u_h, u - v_h)] \geq \|u - u_h\|^2\,,$$

or

$$\frac{1}{\alpha}\ M\|u - u_h\|\ \|u - v_h\|\ \geq \|u - u_h\|^2,$$

using boundedness of $a(\cdot,\cdot)$; namely,

$$|a(u,v)| \leq M||u||||v|| \,.$$

This gives us

$$||u-u_h|| \leq \frac{M}{\alpha}||u-v_h|| \quad \forall v_h \in H_h \,,$$

or

$$||u-u_h|| \leq C \inf_{v_h \in H_h} ||u-v_h||.$$

When the bilinear form $a(\cdot,\cdot)$ is symmetric, there is a remarkable interpretation of the approximate solution, namely, the approximate solution u_h is the projection of the exact solution u over the subspace H_h with respect to the inner product $a(\cdot,\cdot)$ as $a(u-u_h,v_h)=0$ for all $v_h \in H_h$. Thus, we get

$$a(u-u_h,u-u_h) = \inf_{v_h \in H_h} a(u-v_h,u-v_h).$$

By the properties of ellipticity and boundedness of $a(\cdot,\cdot)$, we get

$$\alpha||u-u_h||^2 \leq M||u-v_h||||u-v_h|| \,,$$

or

$$||u-u_h|| \leq \sqrt{\frac{M}{\alpha}}||u-v_h|| \text{ for all } v_h \in H_h \,.$$

Thus

$$||u-u_h|| \leq C \inf_{v_h \in H_h} ||u-v_h||, \text{ where } C = \sqrt{\frac{M}{\alpha}}.$$

Proof of Theorem 3.3. We have

$$||u-u_h|| \leq ||u-v_h|| + ||u_h-v_h||$$

by the triangular inequality of the norm and

$$\beta||u_h-v_h||^2 \leq a_h(u_h-v_h,u_h-v_h) \text{ by coercivity.} \tag{3.84}$$

By continuity of the bilinear form $a(\cdot,\cdot)$, (3.84) takes the form

$$\begin{aligned}\beta||u_h-v_h||^2 \leq{}& a(u-v_h,u_h-v_h) + \{a(v_h,u_h-v_h) - a_h(v_h,u_h-v_h)\} \\ &+ \{F_h(u_h-v_h) - F(u_h-v_h)\} \,,\end{aligned}$$

or

$$\begin{aligned}\beta\|u_h - v_h\| &\leq M\|u - v_h\| \\ &+ \frac{|a(v_h, u_h - v_h) - a_h(v_h, u_h - v_h)|}{\|u_h - v_h\|} \\ &+ \frac{|F_h(u_h - v_h) - F(u_h - v_h)|}{\|u_h - v_h\|} \\ &\leq M\|u - v_h\| + \sup_{w_h \in H_h} \frac{|a(v_h, w_h) - a_h(v_h, w_h)|}{\|w_h\|} \\ &+ \sup_{w_h \in H_h} \frac{|F_h(w_h) - F(w_h)|}{\|w_h\|}\,.\end{aligned}$$

By putting the value of $\|u_h - v_h\|$ in the first inequality and taking the infimum over H_h, we get the desired result.

Remark 3.13. (i) Theorem 3.3 is a generalization of Cea's lemma as $a_h(\cdot,\cdot) = a(\cdot,\cdot)$ and $F_h(\cdot) = F(\cdot)$ in the case of conformal finite element method (the case when $H_h \subset H$).

(ii) The problem (3.75) can be written in the form

$$Au = f, \tag{3.75$'$}$$

where $A : H \to H$ bounded and linear.

By a well-known result (see, for example, Theorem 2.31 in Siddiqi [1986]), there exists a bounded linear operator A on H into itself ($H^\star = H$) such that $\langle Au, v\rangle = a(u, v)$. By the Riesz theorem for each v, there exists a unique $f \in H$ such that $F(v) = \langle f, v\rangle$. Thus (3.75) takes the form $\langle Au, v\rangle = \langle f, v\rangle$ which gives (3.75)$'$.

Convergence results. As a consequence of Theorem 3.2, we find that $\|u_h - u\| \to 0$ as $h \to 0$, that is, the approximate solution u_h converges to the exact solution of (3.75) if there exists a family $\{H_h\}$ of subspaces of the space H such that for each $u \in H$,

$$\lim_{h\to 0} \inf_{v_h \in H_h} \|u - v_h\| = 0\,. \tag{3.85}$$

If $\|u - u_h\| \leq Ch^\alpha$ for $\alpha > 0$ where C is a positive constant independent of u and u_h and h is the characteristic length of an element, then α is called the rate of convergence. It may be observed that the convergence is related to the norm under consideration, say, L_1 norm, energy norm or L_∞ norm (or sup norm).

There is vast literature concerning convergence for special types of norms and spaces and special types of boundary and initial value problems. For interested readers we refer to Wahlbin [1995], Zeinisek [1990], Ciarlet-Lions [1991], Kardestuncer and Norrie [1987] and Siddiqi [1994]. We present here a convergence theorem for a more general model namely variational inequality including variational formulation as a special case.

Many physical phenomena are modelled by the following problem:

Let K be a non-empty closed convex subset of a Hilbert space H. Find $u \in K$ such that

$$a(u, v-u) \geq F(v-u) \text{ for all } v \in K, \tag{3.86}$$

where $a(\cdot,\cdot)$ is a bounded bilinear form over H and $F \in H^{\star}$. It is clear that if K is a subspace of H, then (3.86) takes the form (3.75). Inequality (3.86) is known as a **variational inequality** whose study was initiated around 1960 by Fichera, Stampacchia and Lions. There exists a vast literature on this theme; for comprehensive and updated references we refer to Glowinski [1984], Siddiqi [1994, 1994], and Kikuchi and Oden [1988]. Inequality (3.86) and its generalizations are natural models for contact problems, flow through porous media and many economic theories; see, for example, Siddiqi [1997] for an updated literature and new areas of applications of the variational inequalities and their finite element analysis.

Let H_h be a closed subspace of H and K_h be a subset of H_h for all h such that $\{K_h\}_h$ satisfies the following conditions:

(i) If $\{v_h\}_h$ is such that $v_h \in K_h$ for all h and $\{v_h\}_h$ is bounded in H, then the weak cluster points of $\{v_h\}_h$ belong to K.

(ii) There exists a subset U of H with $\overline{U} = K$ and $r_h : U \to K_h$ such that $\lim_{h\to 0} r_h\, v = v$ strongly in H, for all $v \in U$.

The problem. Find $u_h \in K_h$ such that

$$a(u_h, v_h - u_h) \geq F(v_h - u_h) \text{ for all } v_h \in K_h \tag{3.87}$$

is called **the approximate** (discrete) **problem.**

It is known that (3.87) has a unique solution if $a(\cdot,\cdot)$ is elliptic. Under the above conditions, the solution u_h of the discrete problem (3.87) converges to the exact solution u of (3.86). For proof of this result, we refer to Glowinski [1984].

Finite elements. Let Ω be a polygonal domain in R^2, $(\Omega \subset R^2)$. A finite collection of triangles T_h satisfying the following conditions is called a **triangulation:**

(i) The union of the triangles with boundaries is equal to the polygonal domain, that is

$$\overline{\Omega} = \cup_{K\in T_h}\overline{K}.$$

(ii) Two members, say, K_1 and K_2 of T_h are either equal or disjoint, that is, for any two distinct members K_1 and K_2 of T_h, their intersection is either empty, or a vertex common to T_1 and T_2, or an edge common to them.

Let $P(K), K \in T_h$, be a function space defined on K such that $P(K) \subset H^1(K)$ (Sobolev space of order 1). Generally, $P(K)$ is taken as a space of polynomials of some degree, and is contained in $H^1(\Omega)$. This is essential for proving the convergence of the method under consideration and has significant advantages in solving (3.76). For proof, see Siddiqi [1986, p. 255] or Ciarlet [1978].

For $h = \max_{K \in T_h}$ (diameter of K), $N(h) =$ the number of nodes of the triangulation, $P(K) = P_1(K) =$ space of polynomials of degree less than or equal to 1 in x and y, $H_h = \{v_h \mid v_h/K \in P_1(K), K \in T_h\}$ is contained in $C^0(\Omega)$, space of real-valued continuous functions on $\overline{\Omega}$, and the functions $w_i, i = 1, 2 \cdots N(h)$, defined by

$$\begin{aligned} w_i &= 1 \text{ at } i\text{-th node} \\ &= 0 \text{ at other nodes,} \end{aligned}$$

form a basis of H_h and so H_h is an $N(h)$-dimensional subspace of $H^1(\Omega)$.

In the terminology of Ciarlet, $\left(K, P_K, \sum_K\right)$, where

(i) $\sum_K$ is a set of m linear functionals $\psi_i, l \leq i \leq m$ defined over P_K such that given any real scalars $\alpha_i, 1 \leq i \leq m$, there exists a unique function $w \in P_K$ satisfying $\psi_i(w) = \alpha_i, 1 \leq i \leq m$;

(ii) K is a compact subset of R^n with a non-empty interior and a Lipschitz continuous boundary; and

(iii) P_K is a finite-dimensional space of real-valued functions defined over the set K with dimension m;

is called a **finite element**.

Remark 3.14. (i) It is clear that m linear functionals ψ_i are linearly independent and, in particular, there exist m functions $w_i \in P(K), 1 \leq i \leq m$, that satisfy

$$\psi_j(w_i) = \delta_{ij}, \qquad 1 \leq i \leq m,$$

and we have

$$w = \sum_{i=1}^{m} \psi_i(w) w_i \quad \text{for all } w \in P_K.$$

(ii) $\sum_K$ is often called a set of m degrees of freedom, that is, of m linear functionals, $\phi_i, 1 \leq i \leq m$ and ϕ_i are called the **degrees of freedom** of the finite element. The functions $w_i, 1 \leq i \leq m$, are called the **basis functions** of the finite element. The basis functions are also called the **shape functions** in the literature of engineering and technology.

(iii) Very often simply K is called the finite element, especially in the engineering literature (See remark Ciarlet p. 94[1991]). H_h is called the space of finite elements.

(iv) If K is an n-simplex, $(K, P_K, \sum_K)$ is called the simplitial **finite element**. Finite elements are called **triangular and rectangular** as K is triangular and rectangular, respectively, in R^2. In R^3, a finite element is called **tetrahedral** if K is tetrahedral.

(v) In practice, $\sum\limits_K$ in R^2 is the set of values of $w \in P_K$ at the vertices and middle points of the triangle or rectangle, as the case may be. P_K is a set of polynomial of degree less than or equal to 1.

3.2.4 Finite element method in concrete cases

The material presented here is mainly based on Brenner and Scott [1994], Reddy [1985], Chari and Silvester [1980], Jin [1993], Silvester and Ferrari [1990], Dautry-Lions [vol.4, 1990] and Zenisek [1990]. For basic knowledge for tackling real-life problems through finite element methods, we refer to these lucidly written text books. For a comprehensive study and numerous applications, we cite Ciarlet-Lions [1991] and Kardestuncer and Norrie [1987].

The steps involved in the finite element analysis of a problem are as follows:

(i) Formulation of the problem in the variational form (equality or inequality form).

(ii) Discretization or representation of the domain of the problem into collection of preselected finite elements.

(iii) Construction of the finite element mesh of preselected elements.

(iv) Numbering of nodes and elements.

(v) Generation of the geometric properties, for example, coordinates, cross-section required.

(vi) Derivation of discretized equations for all typical elements in the mesh, that is, equations of the form

$$\left[K^{(e)}\right]\left\{u^{(e)}\right\} = \left\{F^{(e)}\right\},$$

where an arbitrary element $u = \sum\limits_{i=1}^{n} \alpha_i w_i$.

(vii) The search of appropriate basis/shape/interpolation function/approximation function w_i.

(viii) Assemblage of element equations to obtain the equations of the whole problem.

(ix) Imposition of the boundary conditions.

(x) Solution of the boundary conditions.

(xi) Visualization of the solution and interpretation.

The steps involved will be explained by the finite element solution of the following boundary value problem:

$$\begin{cases} -\frac{d}{dx}\left(\alpha\frac{du}{dx}\right) = f, & 0 < x < a; \\ u(0) = 0, & \left(\alpha(x)\frac{du}{dx}\right)_{x=a} = \lambda, \end{cases} \tag{3.88}$$

where $\alpha = \alpha(x), f = f(x)$, and λ are the data of the problem.

It may be remarked that this boundary value problem models the following physical phenomena.

(i) In electrostatics, it models electrostatic potential u between two parallel plates, one located at $x = 0$ with $u = 0$ and the other located at $x = a$ with a value depending on λ where $\alpha(x)$ and λ are die-electric constants and electric flux, respectively.

(ii) Transverse deflection of a cable, $u =$ transverse deflection, $\alpha(x) =$ tension in cable, $f =$ distributed transverse load, $\lambda =$ axial force.

(iii) Axial deformation of a bar, $u =$ longitudinal displacement, $\alpha(x) = EA$ (where $E =$ modulus and $A =$ area of cross-section), $f =$ friction or contact force on surface of bar, $\lambda =$ axial force.

(iv) Heat transfer, $u =$ temperature, $\alpha(x) =$ thermal conductivity, $f =$ heat generation, $\lambda =$ heat flux.

(v) Flow through porous media, $u =$ fluid head, $\lambda(x) =$ coefficient of permeability, $f =$ fluid flux, $\lambda =$ flow (seepage).

Variational formulation. Multiply both sides of (3.88) by v and integrate over $(0, a)$, then we get

$$\int_0^a \alpha\frac{dv}{dx}\frac{du}{dx}dx + \left[v(-\alpha\frac{du}{dx})\right]_0^a = \int_0^a vf\, dx$$

by applying integration by part. Thus (3.88) can be written in the form

$$a(u, v) = F(v), \tag{3.88$'$}$$

where

$$a(u, v) = \int_0^a \alpha\frac{du}{dx}\frac{du}{dx}\ dx \tag{3.88$'$(i)}$$

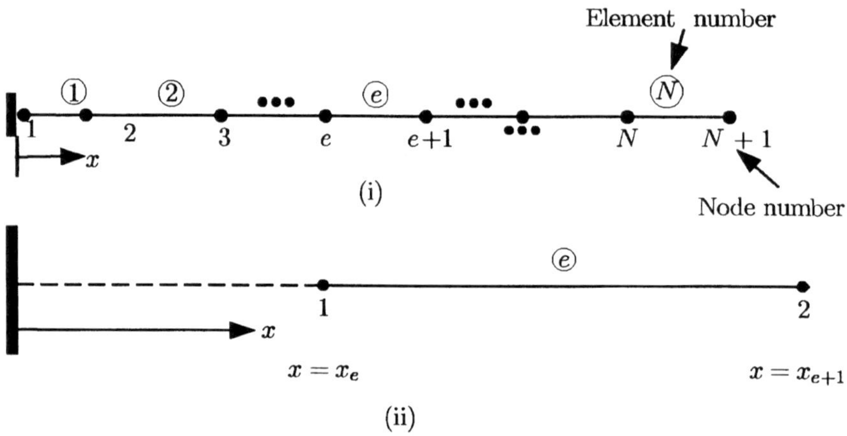

Figure 3.3(i) and 3.3(ii). Finite element representation of a line (one-dimensional domain) by line elements.

$$F(v) = \int_0^a vf\,dx + \gamma + \lambda, \tag{3.88$'$(ii)}$$

where

$$\gamma = \left(-v\alpha\frac{du}{dx}\right)_{x=0}.$$

Discretization of the domain, meshes, nodes etc. Figures 3.3(i) to 3.3(iv) give geometrical meaning of finite elements, meshes, nodes. Here nodes are points of the interval $[0,a]$ ($\Omega = (0,a), \overline{\Omega} = [0,a]$, boundary of Ω consists of points 0 and a) which is subdivided into a set of subintervals or line elements called the **finite element mesh** or **triangulation**. The mesh in Figure (3.3)(i) is a non-uniform mesh as the elements are not of equal length. The intersection of any two elements is called the inter-element boundary. The number of elements used in a problem depends mainly on the element type and accuracy desired. e^{th} elements and e^{th} nodes are shown in Figure (3.3)(ii). A typical element $\Omega^e = (x_A, x_B)$, a line element, is called e^{th} **element**. x_e will denote the coordinate of eth element (see Figure 3.3(i)). In Figure 3.3(iv), the boundary conditions of an element on the typical element are shown.

Derivation of discrete or approximate equation over e^{th} element. Let the exact solution u of (3.88) be approximated on the eth element by

$$u_e(x) = \sum_{j=1}^{m} \alpha_j^{(e)} \phi_j^{(e)}(x), \tag{3.89}$$

where α_j are the parameters to be determined and $\phi_j(x)$ are the approximation functions or the basis functions to be constructed. Substituting $u_e(x)$ for u and

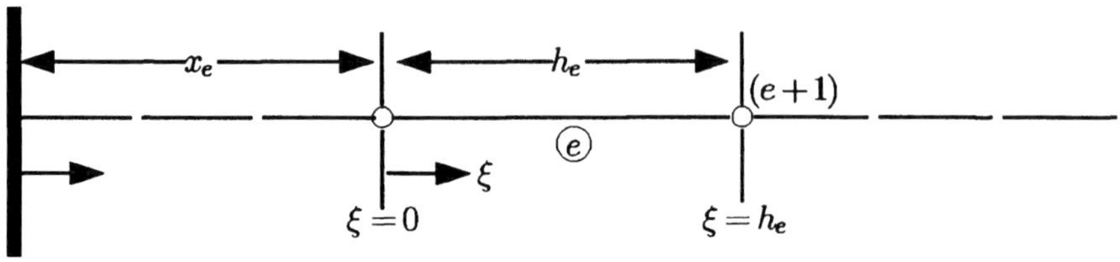

(iii)

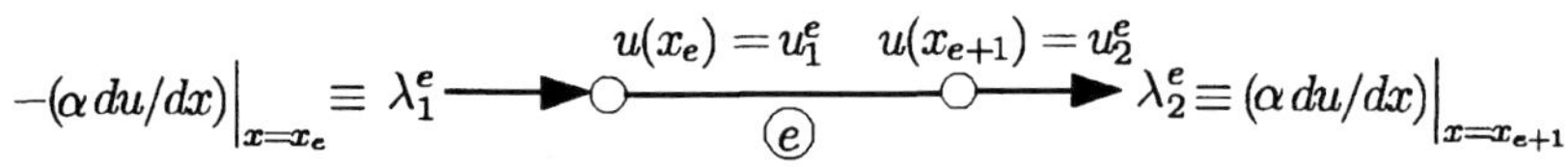

(iv)

Figure 3.3(iii) and 3.3(iv). Finite element discretization of a one-dimensional domain.

$v = \phi_j$ over the eth element, namely, (x_A, x_B) into (3.88)$'$, we get

$$\left[K^{(e)}\right]\left\{\alpha^{(e)}\right\} = \left\{F^{(e)}\right\}, \tag{3.90)(i}$$

where

$$K_{ij}^{(e)} = a(\phi_i, \phi_j) = \int_{x_A}^{x_B} a\frac{d\phi_i}{dx}\frac{d\phi_j}{dx}\,dx \tag{3.90)(ii}$$

$$F_i^{(e)} = \int_{x_A}^{x_B} \phi_i f dx + \gamma^e \phi_i(x_A) + \lambda^e \phi_i(x_B). \tag{3.90)(iii}$$

Derivation of the interpolation functions for an element. $\phi^{(e)}$ are constructed using the conditions of Rayleigh-Ritz methods test function. To satisfy these conditions, we must select ϕ_i such that equation (3.89) is differentiable at least with respect to x and satisfies the essential boundary conditions $u(x_A) = u_1^{(e)}, u(x_B) = u_2^{(e)}$. Furthermore, $\{\phi_i\}$ must be linearly independent and complete. These conditions will be met if we choose

$$u(x) = c_1 + c_2 x = c_1\phi_1 + c_2\phi_2\,. \tag{3.91}$$

The continuity is obviously satisfied, $\phi_1 = 1$ and $\phi_2 = x$ are linearly independent and the set $\{1, x\}$ is complete. To satisfy the remaining requirements on $\{\phi_i\}$, u must satisfy the essential boundary conditions of the element; that is,

$$\begin{cases} u(x_e) &= u_1^{(e)} = c_1 + c_2\, x_e; \\ u(x_{e+1}) &= u_2^{(e)} = c_1 + c_2 x_{e+1}\,, \end{cases} \tag{3.91$'$}$$

or, in matrix form

$$\begin{Bmatrix} u_1^{(e)} \\ u_2^{(e)} \end{Bmatrix} = \begin{bmatrix} 1 & x_e \\ 1 & x_{e+1} \end{bmatrix} \begin{Bmatrix} c_1 \\ c_2 \end{Bmatrix}. \tag{3.92}$$

Solving (3.92) for c_1 and c_2 in terms of $u_1^{(e)}$ and $u_2^{(e)}$, we obtain

$$c_1 = \frac{u_1^{(e)} x_{e+1} - u_2^{(e)} x_e}{x_{e+1} - x_e}, \qquad c_2 = \frac{u_2^{(e)} - u_1^{(e)}}{x_{e+1} - x_e}. \tag{3.93}$$

Putting these values of c_1 and c_2 in (3.91), we get

$$u(x) = \sum_{i=1}^{2} u_i^{(e)} \phi_i^{(e)}, \tag{3.94}$$

where

$$\phi_1^{(e)} = \frac{x_{e+1} - x}{x_{e+1} - x_e}, \phi_2^{(e)} = \frac{x - x_e}{x_{e+1} - x_e}, x_e \le x \le x_{e+1}. \tag{3.95}$$

Expression (3.94) satisfies the essential boundary conditions of the element, and $\{\phi_i\}$ are continuous, linearly independent, and complete over the element. Comparing (3.89) with (3.94), we find that $m = 2$ and $\alpha_j^{(e)} = u_j^{(e)}$.

Since the approximation functions are derived from (3.91) in such a way that $u(x)$ is equal to $u_1^{(e)}$ at node $1(x_A = x_e)$ and to $u_2^{(e)}$ at node $2(x_B = x_{e+1})$, that is interpolated, they are known as the Lagrange family of interpolation functions. The interpolation functions have the following properties besides $\phi_i^{(e)} = 0$ outside the element $\Omega^e = (0, a)$:

$$\text{(i) } \phi_i^{(e)}(x_j) = \begin{cases} 0 & \text{if } i \neq j \\ 1 & \text{if } i = j \end{cases} \quad x_1 = x_e, x_2 = x_{e+1} \tag{3.96)(i}$$

$$\text{(ii) } \sum_{i=1}^{2} \phi_i^{(e)}(x) = 1 \tag{3.96)(ii}$$

The important point is to note that one can derive discrete equation (3.90) using (3.89) and the ϕ_i will depend upon the type of element that is its geometry, number of nodes and number of primary unknowns per node.

The nature of ϕ_i will also be crucial in evaluating integrals in (3.90)(ii) and (3.90)(iii). In view of (3.96)(i), (3.90) can be written as

$$\left[K^{(e)}\right]\left\{u^{(e)}\right\} = \left\{F^{(e)}\right\}, \tag{3.97)(i}$$

where for the linear element, $x_A = x_e$ and $x_B = x_{e+1}$

$$K_{ij}^{(e)} = \int_{x_e}^{x_{e+1}} \alpha \frac{d\phi_i}{dx} \frac{d\phi_j}{dx}\, dx \tag{3.97)(ii}$$

$$F_i^{(e)} = \int_{x_e}^{x_{e+1}} \phi_i f dx + \lambda_i^{(e)}. \tag{3.97)(iii}$$

Equations (3.97) are known as the finite element model of the given boundary value problem (3.88) or equivalently (3.88)′ over an element.

A local coordinate system (a coordinate system in the element) proves more convenient in the derivation of the basis functions. If $\bar{x}$ denotes the local coordinate whose origin is at node 1 of an element, then the local coordinate is related to the global coordinate (coordinate used in the earlier formulation) by linear transformation

$$x = \bar{x} + x_e \,.$$

Putting this value of x in (3.91) and (3.92), we get

$$\begin{cases} u(\bar{x}) & = \bar{c}_1 + \bar{c}_2\bar{x} \\ u_1^{(e)} & = u(0) = \bar{c}_1 \\ u_2^{(e)} & = u(h_e) = \bar{c}_1 + \bar{c}_2 h_e \\ \begin{Bmatrix} u_1^{(e)} \\ u_2^{(e)} \end{Bmatrix} & = \begin{bmatrix} 1 & 0 \\ 1 & h_e \end{bmatrix} \begin{Bmatrix} \bar{c}_1 \\ \bar{c}_2 \end{Bmatrix} \\ h_e & = x_{e+1} - x_e \\ \bar{c}_1 & = u_1^{(e)}, \quad \bar{c}_2 = \dfrac{u_2^{(e)} - u_1^{(e)}}{h_e} , \end{cases} \tag{3.98}$$

and finally we obtain

$$\begin{cases} u & = \sum_{j=1}^{2} u_j^{(e)} \phi_j^{(e)}(\bar{x}) \\ \phi_1^{(e)}(\bar{x}) = 1 - \dfrac{\bar{x}}{h_e}, \; \phi_2^{(e)}(\bar{x}) = \dfrac{\bar{x}}{h_e}, & 0 \le \bar{x} \le h_e \,; \end{cases} \tag{3.99}$$

(3.97)(ii) and (3.97)(iii) take the following form in the local coordinate:

$$K_{ij}^{(e)} = \int_0^{h_e} \bar{a} \frac{d\phi_i}{d\bar{x}} \frac{d\phi_j}{d\bar{x}} d\bar{x} \tag{3.100)(i}$$

$$F_i^{(e)} = \int_0^{h_e} \phi_i \bar{f} d\bar{x} + \lambda_i^{(e)} , \tag{3.100)(ii}$$

where

$$\bar{a} = \text{evaluated at } (x_e + \bar{x}) = a(x_e + \bar{x}) \tag{3.100)(iii}$$

$$\bar{f} = f(x_e + \bar{x}). \tag{3.100)(iv}$$

When the linear interpolation functions are used to approximate the dependent variable of the present problem, we obtain

$$\left[K^{(e)}\right] = \frac{a_e}{h_e} \begin{bmatrix} 1 & -1 \\ -1 & 1 \end{bmatrix} \tag{3.101)(i}$$

$$\left\{F^{(e)}\right\} = \frac{f_e h_e}{2} \begin{Bmatrix} 1 \\ 1 \end{Bmatrix} + \begin{Bmatrix} \lambda_1^{(e)} \\ \lambda_2^{(e)} \end{Bmatrix} , \tag{3.101)(ii}$$

where a_e and f_e are element-wise constant.

Assembly on connectivity of elements equations: Equation (3.97) is valid for arbitrary elements. Suppose $\Omega = (0, a)$ is divided into three elements of not necessarily equal lengths. Since these elements are connected at nodes 2 and 3 and u is continuous, u_2 of element Ω^e should be the same as u_1 of element Ω^{e+1} for $e = 1, 2$. If $U_i, i = 1, 2, 3 \cdots m$ are values of u at global nodes and u_i^j are nodal values in the local coordinate, then

$$u_1^{(1)} = U_1 \quad u_2^{(1)} = U_2 = u_1^{(2)} \quad u_2^{(2)} = U_3 = u_1^{(3)} \quad u_2^{(3)} = U_4\,. \tag{3.102}$$

Relations (3.102) are called the inter-element continuity condition. The element equation (3.97) can be written for different elements in the following form:

Element 1.

$$\frac{a_1}{h_1}\begin{bmatrix} 1 & -1 & 0 & 0 \\ 1 & 1 & 0 & 0 \\ 0 & 0 & 0 & 0 \\ 0 & 0 & 0 & 0 \end{bmatrix}\begin{Bmatrix} U_1 \\ U_2 \\ U_3 \\ U_4 \end{Bmatrix} = \frac{f_1 h_1}{2}\begin{Bmatrix} 1 \\ 1 \\ 0 \\ 0 \end{Bmatrix} + \begin{Bmatrix} \lambda_1^{(1)} \\ \lambda_2^{(1)} \\ 0 \\ 0 \end{Bmatrix}. \tag{3.103(i)}$$

Element 2.

$$\frac{a_2}{h_2}\begin{bmatrix} 0 & 0 & 0 & 0 \\ 0 & 1 & -1 & 0 \\ 0 & -1 & 1 & 0 \\ 0 & 0 & 0 & 0 \end{bmatrix}\begin{Bmatrix} U_1 \\ U_2 \\ U_3 \\ U_4 \end{Bmatrix} = \frac{f_2 h_2}{2}\begin{Bmatrix} 0 \\ 1 \\ 1 \\ 0 \end{Bmatrix} + \begin{Bmatrix} 0 \\ \lambda_1^{(2)} \\ \lambda_2^{(2)} \\ 0 \end{Bmatrix}. \tag{3.103(ii)}$$

Element 3.

$$\frac{a_3}{h_3}\begin{bmatrix} 0 & 0 & 0 & 0 \\ 0 & 0 & 0 & 0 \\ 0 & 0 & 1 & -1 \\ 0 & 0 & -1 & 1 \end{bmatrix}\begin{Bmatrix} U_1 \\ U_2 \\ U_3 \\ U_4 \end{Bmatrix} = \frac{f_3 h_3}{2}\begin{Bmatrix} 0 \\ 0 \\ 1 \\ 1 \end{Bmatrix} + \begin{Bmatrix} 0 \\ 0 \\ \lambda_1^{(3)} \\ \lambda_2^{(3)} \end{Bmatrix}. \tag{3.103(iii)}$$

By superimposing, that is, adding equations (3.103)(i)–(iii), we get the global **finite element model** of (3.88); namely,

$$\begin{bmatrix} \frac{a_1}{h_1} & -\frac{a_1}{h_1} & 0 & 0 \\ -\frac{a_1}{h_1} & \frac{a_1}{h_1}+\frac{a_2}{h_2} & \frac{-a_2}{h_2} & 0 \\ 0 & -\frac{a_2}{h_2} & \frac{a_2}{h_2}+\frac{a_3}{h_3} & -\frac{a_3}{h_3} \\ 0 & 0 & -\frac{a_3}{h_3} & \frac{a_3}{h_3} \end{bmatrix}\begin{Bmatrix} U_1 \\ U_2 \\ U_3 \\ U_4 \end{Bmatrix} = \frac{1}{2}\begin{Bmatrix} f_1 h_1 \\ f_1 h_1 + f_2 h_2 \\ f_2 h_2 + f_3 h_3 \\ f_3 h_3 \end{Bmatrix} + \begin{Bmatrix} \lambda_1^{(1)} \\ \lambda_2^{(1)} + \lambda_1^{(2)} \\ \lambda_2^{(2)} + \lambda_1^{(3)} \\ \lambda_2^{(3)}\,. \end{Bmatrix} \tag{3.104}$$

Imposition of boundary conditions. Equation (3.104) is valid for any problem described by the differential equation (3.88) irrespective of the boundary conditions. After putting appropriate boundary conditions, we obtain a non-singular matrix which can be inverted to have the solution of (3.104). The detailed account of the physical meaning of the boundary conditions can be found in Reddy [1985].
Solution of the discretized equation (3.104) **and its visualization:**

MATLAB and Mathematica are two of the most popular computing environments for computation and visualization of the solution of equation (3.104). These packages have been introduced in Appendix 7.1. See also Appendix 7.5 for a solution of (3.104). Methods discussed in Chapter 4 could be useful for more accurate evaluation of integrals in (3.90), especially, in higher dimensions.

It may be remarked that the steps narrated here will remain the same for solving differential equations of higher orders or in higher dimensions, that is, in solving partial differential equations. Solutions of a fairly large number of modelling problems of science and technology through the finite element methods are discussed in Reddy [1985] and Hackbusch [1992] in a simple way. See solutions of problems for illustration of the finite element method in concrete cases.

3.3 Boundary element method

The theory of integral equations is well known through the classical work of Fredholm [1903], especially due to its relevance to the potential theory (Laplace type problems). The work of Kellog [1928], Tricomi [1985], Kupradze [1968], Mikhlin [1957,1965], etc. provides quite rich literature on the subject. However, methods of integral equations were not popular with engineers until late seventies in spite of the invention of digital computers. Researchers at Southampton University may be credited for making a systematic endeavour in eighties and nineties to solve problems of different branches of engineering by modelling them through integral equations on the boundary of a region (domain) and then by discretizing these models. This technique led to the development of the boundary element method (also called the boundary integral method in the initial stages). Thus the **boundary element method (BEM)** consists in the transformation of the partial differential equation describing the behaviour of an unknown inside and on the boundary of the domain into an integral equation relating to only boundary values, and then finding out the numerical solution. The combination of the finite element method is possible with the boundary element method where the integral equation on the boundary can be discretized through the finite element techniques treating the boundary as an element. Now-a-days, BEM is a powerful technique as it reduces the dimension of the problem by one as boundary integral equations and it permits for complex surface elements to define the external surface of the domain. One of the significant advantages of this method is that discontinuous elements can also be used and meshes of such elements can be refined in a simpler manner. This method has been applied in a variety of physical phenomena like **transient heat conduction, thermo-**

elasticity, contact problems, free boundary problems, water waves, aerodynamics, elasto-plastic material behaviour, electromagnetism, soil mechanics. There is a vast literature published on these topics in the last fifteen years which can be found in books by Brebbia [1978, 1984, 1985, 1988, 1991], the proceedings of international conferences held in different parts of the world edited by Brebbia et al. [1980, 1984, 1990, 1991], Antes and Panagiotopoulos [1992], Pozrikidis [1992], Banerjee [1994], Chen and Zhou [1992], and Hackbusch [1995]. We confine ourselves here to the basic mathematical techniques of the method and illustration of the main steps in the solution of problems by this method. This is a fast-expanding area of practical numerical methods and this can play a vital role in finding solutions of industrial problems.

3.3.1 Basic mathematical results for the boundary element method

Let H be an appropriate Hilbert space, then a bounded linear operator on H into itself is called self-adjoint, if $T = T^\star$, where $T^\star$ is the adjoint operator of T defined by the relation

$$\langle Tu, v\rangle = \langle u, T^\star v\rangle \ .$$

The self-adjointness of an operator is analogous to the symmetry of a matrix.
More frequently, H is taken to be the Sobolev space of order 1, namely, $H^1(\Omega)$. Equations (3.51) can be written as

$$Tu = -f \text{ on } \Omega \tag{3.105)(i}$$
$$Su = g \text{ on } \Gamma_1 \tag{3.105)(ii}$$
$$Lu = h \text{ on } \Gamma_2 \ , \tag{3.105)(iii}$$

where $\Gamma = \Gamma_1 + \Gamma_2 =$ boundary of the region $\Omega \subset R^n (n = 1, 2 \text{ or } 3)$ (3.105)(ii) and (3.105)(iii) are called, respectively, the **essential** and **natural boundary** conditions.
Let

$$\langle u, v\rangle = \int_\Omega uv d\,\Omega$$
$$\langle u, v\rangle_\Gamma = \int_\Gamma uv d\,\Gamma$$
$$\langle Tu, v\rangle = \int_\Omega T(u) v d\,\Omega \ .$$

Various examples of T are discussed earlier.

Applying integration by parts repeatedly, one can get the relationship of the type

$$\begin{aligned}\langle Tu, w\rangle &= \int_\Omega T(u)\ w\ d\,\Omega \\ &= \int_\Omega u\ T(w)\ d\,\Omega + \int_\Gamma [L(w)S(u) - S(w)L(u)]\, d\Gamma \ .\end{aligned} \tag{3.106}$$

(i) Let

$$T(u) = \frac{d^2u}{dx^2} - \lambda^2 u, \ \Omega = (0,1).$$

Then,

$$\Gamma = \{0,1\}, \ S(u) = \frac{du}{dx}, \ L(u) = u,$$

and the relation (3.106) takes the form

$$\int_0^1 T(u)w dx = \int_0^1 T(w)u dx + [L(w)S(u)]_0^1 - [L(u)S(w)]_0^1 .$$

(ii) Let

$$T(u) = \frac{d^2u}{dx^2}, \ \ \Omega = (0,1),$$

then (3.106) takes the form

$$\begin{aligned}\int_0^1 T(u)w dx = & \int_0^1 T(w)u dx + [L_1(w)S_1(u) - L_2(w)S_2(u)]_0^1 \\ & + [S_2(w)L_2(u) - S_1(w)L_1(u)]_0^1 ,\end{aligned}$$

where

$$\begin{aligned}L_1(u) &= u, \ L_2(u) = \frac{du}{dx} \\ S_1(u) &= \frac{d^3u}{dx^3}, \ \ S_2(u) = \frac{d^2u}{dx^2} .\end{aligned}$$

Let a particle of unit mass subject to the force of a specified field F be moved from a point x to a point y in space. Then the work done on the particle by the force, denoted by W, is given by the expression

$$W = \int_x^y F dr, \tag{3.107}$$

where F is the force field vector and dr is the differential motion of the particle on the path from x to y.

If x is a fixed point while y varies, the integral (3.107) represents a function of y alone. This scalar function

$$u(y) = \int_x^y F dr, \tag{3.108}$$

is called the **potential of the field F.**

If the field is gravitational, then the potential is Newtonian. The Newtonian potential generated by two particles of masses m_1 and m_2 located at points x (fixed) and y (variable), respectively, is of the form

$$u(x) = \int_x^y G m_1 m_2 \nabla (\frac{1}{r}) = G m_1 m_2 \frac{1}{r} + \text{constant}, \tag{3.109}$$

where G is the gravitational constant and r is the distance between x and y; that is,

$$r(x,y) = \|x-y\| = d(x,y) = \left[(x_1-y_1)^2 + (x_2-y_2)^2 + (x_3-y_3)^2\right]^{\frac{1}{2}}, \quad (3.110)$$

where $x = (x_1, x_2, x_3), y = (y_1, y_2, y_3)$.

Attraction forces of the same character as those occurring in gravitation also act between electric charges and the poles of magnets. In that case, we shall refer to sources rather than masses. Thus, we assume that a unit simple source located at a source point x in space generates at a point y the potential $\dfrac{1}{\|x-y\|}$ which is a continuous function of y_i infinitely differentiable except at the source point x.

A discrete distribution of simple sources of intensities $\varphi_1, \varphi_2 \cdots \varphi_m$ located at points $\xi_1, \xi_2, \cdots \xi_m$, respectively, generate the Newtonian potential

$$u(y) = \sum_{n=1}^{m} \varphi(\xi_n) \frac{1}{\|\xi_n - y\|}, \quad (3.111)$$

at point y.

This potential is also a continuous function of y and infinitely differentiable at all points except those coinciding with ξ_n. If we consider a continuous distribution of simple sources of volume density φ throughout the domain Ω, then the potential associated with this force field is a volume potential

$$u(x) = \int_\Omega \varphi(y) \frac{1}{\|x-y\|} d\Omega(y). \quad (3.112)$$

The integrand has a singularity when the field point x lies inside the domain Ω. It can be easily seen that $u(x)$ exists at all $x \in \Omega$ and is differentiable provided φ is bounded. If φ satisfies the Lipschitz condition of order $\alpha > 0$, that is, $|\varphi(x) - \varphi(y)| \leq M\|x-y\|^\alpha, \quad M > 0$, then $\dfrac{\partial^2 x}{\partial x_i^2}$ exists; that is,

$$\begin{aligned} \frac{\partial u(x)}{\partial x_i} &= \int_\Omega \varphi(y) \frac{\partial}{\partial x_i} \left(\frac{1}{\|x-y\|} \right) d\Omega(y) \\ &= -\int_\Gamma \varphi(y) \frac{1}{\|x-y\|} n_i(y) d\Gamma(y) + \int_\Omega \frac{\partial \varphi(y)}{\partial x_i(y)} \frac{1}{\|x-y\|} d\Omega(y). \end{aligned}$$

By differentiating this relation with respect to $x_i(x)$, the i-th coordinate of x, we get

$$\begin{aligned}
\frac{\partial^2 u(x)}{\partial x_i^2} &= -\int_\Gamma \varphi(y)\frac{\partial}{\partial x_i}\left(\frac{1}{\|x-y\|}\right) n_i(y)d\,\Gamma(y) \\
&+ \int_\Omega \frac{\partial}{\partial x_i(x)}\left[\varphi(x)-\varphi(y)\right]\frac{\partial}{\partial x_i(x)}\left(\frac{1}{\|x-y\|}\right) d\Omega(y) \\
&= -\int_\Gamma \varphi(x)\frac{\partial}{\partial x_i(x)}\left(\frac{1}{\|x-y\|}\right) n_i(y)d\Gamma(y) \\
&+ \int_\Gamma \left[\varphi(x)-\varphi(y)\right]\frac{\partial^2}{\partial x_i^2(x)}\left(\frac{1}{\|x-y\|}\right) n_i(y)d\Gamma(y) \\
&+ \int_\Omega \left[\varphi(x)-\varphi(y)\right]\frac{\partial^2}{\partial x_i^2(x)}\left(\frac{1}{\|x-y\|}\right) d\Omega(y)\,,
\end{aligned}$$

or

$$\left\{\begin{aligned}
\frac{\partial^2 u(x)}{\partial x_i^2(x)} &= \varphi(x)\int_\Gamma \frac{\partial}{\partial x_i(y)}\left(\frac{1}{\|x-y\|}\right) n_i(y)d\Gamma(y) \\
&+ \int_\Omega \left[\varphi(x)-\varphi(y)\right]\frac{\partial^2}{\partial x_i(x)}\left(\frac{1}{\|x-y\|}\right) d\Omega(y)\,.
\end{aligned}\right. \tag{3.113}$$

It is clear that for derivation, we have used the relations

$$\begin{aligned}
\frac{\partial}{\partial x_i(x)}\left(\frac{1}{\|x-y\|}\right) &= -\frac{\partial}{\partial x_i(y)}\left(\frac{1}{\|x-y\|}\right) \\
\frac{\partial}{\partial x_i(y)}\varphi(y) &= \frac{\partial}{\partial x_i(y)}\left[\varphi(x)-\varphi(y)\right],
\end{aligned}$$

[$x_i(y)$ denote the i-th coordinate of y].

Adding (3.113) for different values of i, we get

$$\left\{\begin{aligned}
\nabla^2 u(x) &= \varphi(x)\int_\Gamma \frac{\partial}{\partial n(y)}\left(\frac{1}{\|x-y\|}\right) d\Gamma(y) \\
&+ \int_\Omega \left[\varphi(x)-\varphi(y)\right]\nabla^2\left(\frac{1}{\|x-y\|}\right) d\Omega(y)\,.
\end{aligned}\right. \tag{3.114}$$

The second integral of (3.114) vanishes as the Ω can be divided into two parts, one of a small sphere of radius ϵ surrounding the point x and the other entire region denoted by $\Omega\epsilon$ and keeping in mind that $\dfrac{1}{\|x-y\|}$ satisfies Laplacian (the integral over the small sphere vanishes as $\varphi(x)$ satisfies the Lipschitz condition).
Therefore

$$\left\{\begin{aligned}
\nabla^2 u(x) &= \varphi(x)\int_\Gamma \frac{\partial}{\partial n(y)}\frac{1}{\|x-y\|}d\,\Gamma(y) \\
\nabla^2 u(x) &= \varphi(x)\int_{\Gamma_\epsilon} \frac{\partial}{\partial n(y)}\frac{\partial}{\|x-y\|}\,d\,\Gamma(y) + \varphi(x)\int_\Gamma \frac{\partial}{\partial n(x)}\frac{1}{\|x-y\|}d\Gamma(y)\,,
\end{aligned}\right. \tag{3.115}$$

where Γ_ϵ is the boundary of the sphere surrounding x and Γ is the boundary of Ω.

We have

$$\int_{\Gamma_\epsilon} \frac{\partial}{\partial n(y)} \frac{1}{\|x-y\|} \, d\,\Gamma(y) = -\frac{1}{\epsilon^2} \int_{\Gamma_\epsilon} d\,\Gamma = -4\pi.$$

Since there are no sources in the domain between Γ_ϵ, and Γ, therefore there is no flux out of the region and, consequently,

$$\int_{\Gamma_\epsilon} \frac{\partial}{\partial n}\left(\frac{1}{r}\right) d\,\Gamma_\epsilon + \int_{\Gamma} \frac{\partial}{\partial n}\left(\frac{1}{r}\right) d\,\Gamma = 0, \tag{3.116}$$

where $r = \|x-y\|$ and the normal is outward on Γ, but inward on Γ_ϵ. Therefore

$$\int_{\Gamma} \frac{\partial}{\partial n(y)} \left(\frac{1}{\|x-y\|}\right) d\,\Gamma(y) = -4\pi\,. \tag{3.117}$$

By (3.115) and (3.117), we get

$$\nabla^2 u(x) = -4\pi\varphi(x). \tag{3.118}$$

The potential associated with a continuous distribution of simple sources extending over a surface Γ and of surface density σ defined by the relation

$$u(x) = \int_{\Gamma} \sigma(y) \frac{1}{\|x-y\|} d\Gamma(y) \tag{3.119}$$

is called a **single layer potential.**

Let $\Gamma(y)$ and $\Gamma(y^{'})$ be two surfaces separated by a small distance carrying distributions of attraction of magnitude $\sigma(y)$ and $\sigma(y^{'})$, respectively, such that

$$\sigma(y) d\Gamma(y) = -\sigma(y^{'}) d\Gamma(y^{'}),$$

then the potential

$$u(x) = \int_{\Gamma} \mu(y) \frac{\partial}{\partial n(y)} \left(\frac{1}{\|x-y\|}\right) d\Gamma(y), \tag{3.120}$$

where μ is the limit of $\sigma(h)$ as $h \to 0$ and $\sigma \to \infty$, is called a **double layer potential.** The function μ is known as the surface density or the double integral. It can be verified (see, for example, Brebbia, Telles and Wrobel [1984]) that the value of double layer potential given by (3.120) is equal to

$$u^{+}(x) = -2\pi\mu(x) + \int_{\Gamma} \mu(y) \frac{\partial}{\partial n(y)} \left(\frac{1}{\|x-y\|}\right) d\Gamma(y),$$

if x approaches the surface Γ from the inside. $u(x)$ is equal to $\overline{u}(x)$, given by

$$\overline{u}(x) = 2\pi\mu(x) + \int_{\Gamma} \mu(y) \frac{\partial}{\partial n(y)} \left(\frac{1}{\|x-y\|}\right) d\,\Gamma(y),$$

in the case x approaches from outside.

Thus we get

$$u^+(u) - \overline{u}(x) = -4\pi\mu(x). \tag{3.121}$$

The concept equivalent to the Newtonian potential in two dimensions is the logarithmic potential defined as $\log \frac{1}{\|x-y\|}$, where

$$\|x-y\| = d(x,y) = \Big\{[x_1(x) - x_1(y)]^2 + [x_2(x) - x_2(y)]^2 + [x_3(x) - x_3(y)]^2\Big\}^{\frac{1}{2}},$$

and $x_i(x)$ and $x_i(y)$ are coordinates of the points x and y, respectively. The single-layer potential for two-dimensional problems is given by

$$u(x) = \int_\Gamma \sigma(y) \log \frac{1}{\|x-y\|} d\Gamma(y). \tag{3.122}$$

Then the double layer potential for two-dimensional problems is of the form

$$u(x) = \int_\Gamma \mu(y) \frac{\partial}{\partial n(y)} \left(\log \frac{1}{\|x-y\|}\right) d\Gamma(y). \tag{3.123}$$

The two-dimensional volume potential is defined as

$$u(x) = \int_\Omega \varphi(y) \log \frac{1}{\|x-y\|} d\Omega(y).$$

It satisfies Poisson's equation

$$\nabla^2 u(x) = -2\pi\varphi(x).$$

Equation (3.121) in two-dimensional problems takes the form

$$u^+(u) - \overline{u}(x) = -2\pi\mu(x). \tag{3.124}$$

3.3.2 Formulation of boundary value problems in terms of integral equations over boundary of the given domain

Indirect method. We consider Laplace's equation with the Dirichlet or Neumann boundary conditions, namely, (3.105) with $T = \nabla^2, f = 0$, where $S = I, g = \overline{u}, x \in \Gamma$, (Dirichlet boundary condition), or $L = \frac{\partial}{\partial n}$, where n is the unit outward normal to surface $\Gamma, h = \overline{q}(x), x \in \Gamma_2$ (Neumann boundary condition), $\Gamma = \Gamma_1 + \Gamma_2$, $\overline{u}$ and $\overline{q}$ are prescribed values of $u(\cdot)$.

A function u is called **harmonic** within Ω, bounded by a closed surface Γ, if (i) u is continuous on the boundary of Ω, namely, on Γ and inside Ω, (ii) u is at least twice differentiable, and (iii) u satisfies Laplace's equation in Ω.

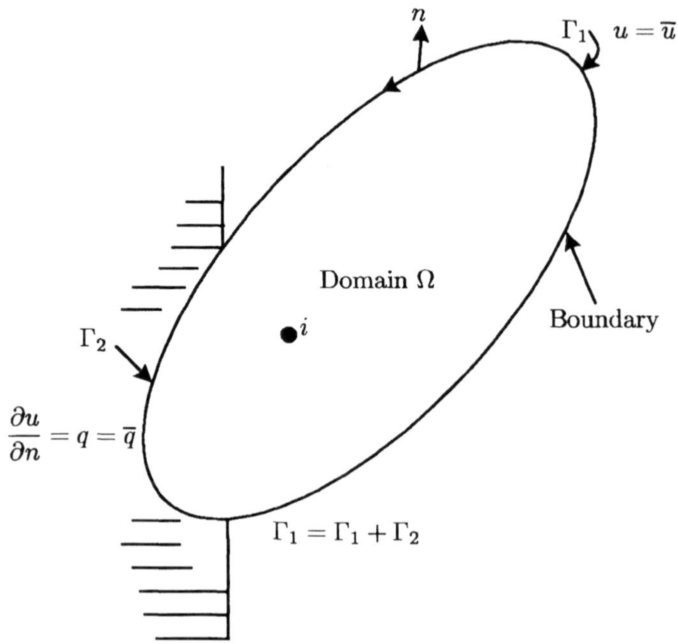

Figure 3.4. $h(x) = \overline{q}(x)$ on Γ_2 Neumann boundary condition.

It is well known that a harmonic function can be represented by a potential distribution and, conversely, every potential is a harmonic function (see, for example, Kellog [1929]). An effective method of formulating the boundary value problem of potential theory is to represent the harmonic function by a single-layer or a double-layer potential generated by continuous source distributions over Γ, provided these potentials satisfy the prescribed boundary conditions for u. This method helps us to formulate the associated integral equations which define the source densities concerned.

Thus, for formulating the Dirichlet boundary value problem in terms of an integral equation, we assume that the unknown function $u(x)$ is expressed by a single-layer potential with unknown density σ; that is,

$$u(x) = \int_{\Gamma} \sigma(x) u^{\star}(x, y) d\Gamma(y), \quad x \in \Omega, \tag{3.125}$$

where

$$u^{\star}(x, y) = \frac{1}{||x - y||}, \quad \text{is the Newtonian potential.}$$

(See Theorem 3 of Dautry and Lions [1990] in three dimensions.)

Since the kernel $u^{\star}(x, y)$ in (3.125) is continuous as x passes through the surface, taking limit in this equation as x approaches Γ, gives the integral equation on Γ

$$u(x) = \int_\Gamma \sigma(y) u^\star(x,y) d\Gamma(y), \quad x \in \Gamma. \tag{3.126}$$

In this equation, $\sigma(\cdot)$ is the only unknown. Equation (3.126) is a Fredholm equation of the first kind. While using a double-layer potential for $u(x)$, one can write the Dirichlet boundary problem in the form of a Fredholm equation of the second kind, but experts prefer a single-layer treatment as it is more convenient mathematically and more illuminating physically.

If we represent the solution of the Neumann boundary problem in three dimensions by a single-layer potential with unknown density σ; namely,

$$u(x) = \int_\Gamma \sigma(y) u^\star(x,y) d\Gamma(y), \tag{3.127}$$

we get

$$q(x) = \frac{\partial u}{\partial n(x)} = -\alpha\pi\sigma(x) + \int_\Gamma \sigma(y) \frac{\partial u^\star}{\partial n(x)}(x,y) d\Gamma(y), \tag{3.128}$$

where $\alpha = 1$ for two-dimensional problems and $\alpha = 2$ for three-dimensional problems. ($u^\star(\cdot,\cdot)$ is Newtonian potential in three dimensions and logarithmic potential in two dimensions). (3.128) is a Fredholm equation of the second kind as the unknown $\sigma(\cdot)$ appears both inside and outside the integral. (3.128) has a solution provided

$$\int_\Gamma q(x)\ d\Gamma(x) = 0. \tag{3.129}$$

The solution is unique only within an arbitrary additive constant. See results in this direction in Dautry and Lions [pp. 130–136, vol. 4, 1990].

Direct formulation of boundary value problems in the form of integral equation on boundary. We use the weighted residual method discussed in Section 3.2.1 for writing boundary value problems into integral equations on the boundary of a given domain.

Let us consider the boundary value problem

$$\begin{cases} \nabla^2 u(x) = 0 \text{ in } \Omega \\ u(x) \quad = \overline{u} \text{ on } \Gamma_1 \\ \dfrac{\partial u}{\partial n} \quad = \overline{q}(x) \text{ on } \Gamma_2, \end{cases} \tag{3.130}$$

where $\Gamma = \Gamma_1 + \Gamma_2, \Omega \subset R^3, \Gamma =$ boundary of the region Ω. The weighted residual equation (3.62) takes the following form for this boundary value problem:

$$\begin{cases} \displaystyle\int_\Omega \nabla^2 u(x) u^\star(x,y) d\Omega(x) = \int_{\Gamma_2} (q(x) - \overline{q})\, u^\star(x,y)\ d\Gamma(x) \\ \qquad\qquad\qquad\qquad\qquad\quad \displaystyle - \int_{\Gamma_1} [u(x) - u(x))\, q^\star(x,y)\ d\Gamma(x)\,, \end{cases} \tag{3.131}$$

where $u^\star(\cdot,\cdot)$ is interpreted as the weighting function and

$$q^\star(x,y) = \frac{\partial u^\star(x,y)}{\partial n(x)} \,. \tag{3.132}$$

Integrating by parts (3.131) with respect to x_i, we have

$$\begin{cases} -\int_\Omega \frac{\partial u(x)}{\partial x_i(x)} \frac{\partial u^\star(x,y)}{\partial x_i(x)} \, d\Omega(x) = -\int_{\Gamma_1} q(x)u^\star(x,y)d\Gamma(x) \\ \qquad - -\int_{\Gamma_2} \overline{q}(x)u^\star(x,y)d\Gamma(x) \\ \qquad -\int_{\Gamma_1} [u(x) - \overline{u}(x)]\, q^\star(x,y)d\Gamma(x)\,, \end{cases} \tag{3.133}$$

where $i = 1,2,3$ and Einstein's summation convention is followed for repeated indices. Using integration by parts once more, we get

$$\begin{cases} \int_\Omega \nabla^2 u^\star(x,y)u(x)d\Omega(x) = -\int_\Gamma q(u)u^\star(x,y)\; d\;\Gamma(x) \\ \qquad + \int_\Gamma u(x)q^\star(x,y)d\Gamma(x)\,, \end{cases} \tag{3.134}$$

keeping in mind that $\Gamma = \Gamma_1 + \Gamma_2$.

Recalling the following properties of the Dirac delta function $\delta(x,y)$:

$$\begin{cases} \delta(x,y) & = 0, \text{ if } x \neq y \\ & = \infty, \text{ if } x = y \\ \int_\Omega u(x)\delta(x,y)d\Omega(x) & = u(y)\,, \end{cases} \tag{3.135}$$

assuming $u^\star(x,y)$ to be the fundamental solution of Poisson's equation; namely,

$$\nabla^2 u^\star(x,y) = -2\alpha\,\pi\;\delta(x,y)\,, \tag{3.136}$$

where $\alpha = 1$ and $\alpha = 2$ are, respectively, two- and three-dimensional problems, and putting the value of $\nabla^2 u^\star(x,y)$ from (3.136) into (3.134), we have

$$2\,\alpha\,\pi\;u(y) + \int_\Gamma u(x)q^\star(x,y)d\Gamma(x) = \int_\Gamma q(x)u^\star(x,y)d\Gamma(x). \tag{3.137}$$

Considering the point y to be on the boundary and accounting for the jump of the left hand, (3.137) yields the integral equation on the boundary of the given domain Ω (boundary integral equation)

$$c(y)u(y) + \int_\Gamma u(x)\; q^\star(x,y)d\Gamma(x) = \int_\Gamma q(x)u^\star(x,y)d\Gamma(x)\,. \tag{3.138}$$

Remark 3.15. (i) If we are looking for a solution of the Neumann boundary problem, then the right hand side of (3.138) is known and for the desired solution and we have to solve a Fredholm equation of the second kind.

(ii) If we are interested in solving a Dirichlet boundary problem, the values of $u(x)$ are prescribed throughout Γ and the problem reduces to solving a Fredholm equation of first kind in unknown $q(x) = \dfrac{\partial u}{\partial n(x)}$, normal derivative.

(iii) The solution of a Cauchy (mixed) boundary problem will require solution of a mixed integral equation for the unknown boundary data.

3.3.3 Main ingredients of the boundary element method

The following steps are involved in solving a boundary value problem by the boundary element method:

(i) Conversion of the problem into the boundary integral equation (as indicated in Subsection 3.3.2).

(ii) Discretization of the boundary Γ into a series of elements over which the potential and its normal derivatives are supposed to vary according to interpolation functions. These elements could be straight lines, circular areas, parabolas, etc.

(iii) By the collocation method, the discretized equation is applied to a number of particular nodes within each element where values of the potential and its normal derivatives are associated.

(iv) Evaluation of the integrals over each element by normally using a numerical quadrature scheme.

(v) To derive a system of linear algebraic equations imposing the prescribed boundary conditions and to find its solution by direct or iterative methods.

(vi) Determination of values of the function u at the internal points of domain under consideration.

Remark 3.16. (i) The values of $u(x)$ at any internal point of the domain are determined by equation (3.137).

(ii) The values of derivatives of u, at any internal point y with Cartesian coordinates $x_i(y), i = 1, 2, 3$ can be determined from the equation

$$\frac{\partial u(y)}{\partial x_i(y)} = \frac{1}{2\alpha\pi}\left\{\int_\Gamma q(x)\frac{\partial u^\star(x,y)}{\partial x_i(y)}\, d\Gamma(x) - \int_\Gamma u(x)\frac{\partial q^\star(x,y)}{\partial x_i(y)}\, d\,\Gamma(x)\right\}. \quad (3.139)$$

It is clear that (3.139) is obtained by differentiating (3.137) under appropriate conditions.

(iii) A significant advantage of the boundary element method is the relaxation of the condition that the boundary surface is smooth (Lyapunov), that is, it can be used for surfaces having corners or edges.

We illustrate now the above mentioned steps through concrete examples over finite regions of homogeneous isotropic media with Neumann, Dirichlet and Cauchy boundary conditions. Non-homogeneous regions may be divided into homogeneous subregions with different physical properties and the method can be applied to a system of equations for each subregion introducing compatibility of potentials and equilibrium conditions in terms of normal derivatives. The method can also be extended for infinite regions. The integral equation given in (3.138) can be discretized into a large number of elements. For the sake of simplicity, we consider a two-dimensional case in which the boundary is divided into segments of constant and linear forms as shown in Figure 3.5.

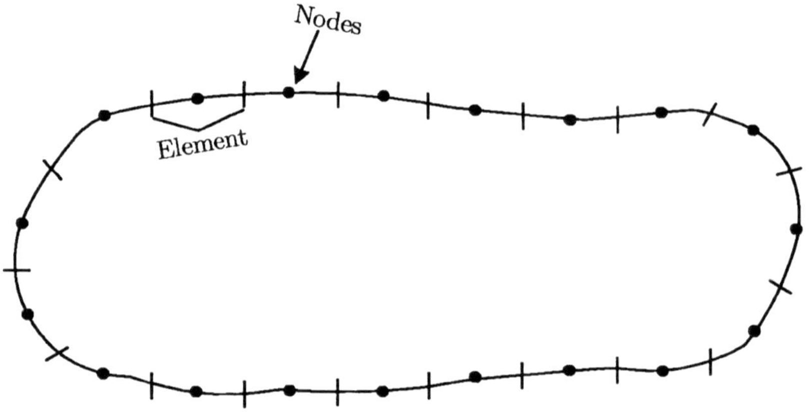

(i)

Figure 3.5(i). Constant boundary elements.

For the constant element case, the boundary is discretized into N elements. Let N_1 belong to Γ_1 and N_2 to Γ_2, where the values of u and q are taken to be constant on each element and equal to the value at the midnode of the element. We observe that in each element the value of one of the two variables u or q is known. Equation (3.138) can be written in the form

$$c_i u_i + \int_\Gamma u\, q^\star d\Gamma = \int_\Gamma q u^\star \, d\,\Gamma\,, \tag{3.140}$$

where $u^\star = \frac{1}{2\pi}\log(\frac{1}{r})$, as a two-dimensional case is considered, while $u^\star = \frac{1}{4\pi}(\frac{1}{r}), r = \|x-y\| = d(x,y)$. Here we have chosen $x_i(y) = u_i$ and $c(y) = c_i$.

(3.138) can be discretized as follows :

$$c_i u_i + \sum_{j=1}^{N} \int_{\Gamma_j} u q^\star \, d\,\Gamma = \sum_{j=1}^{N} \int_{\Gamma_j} u^\star q \, d\,\Gamma. \tag{3.141}$$

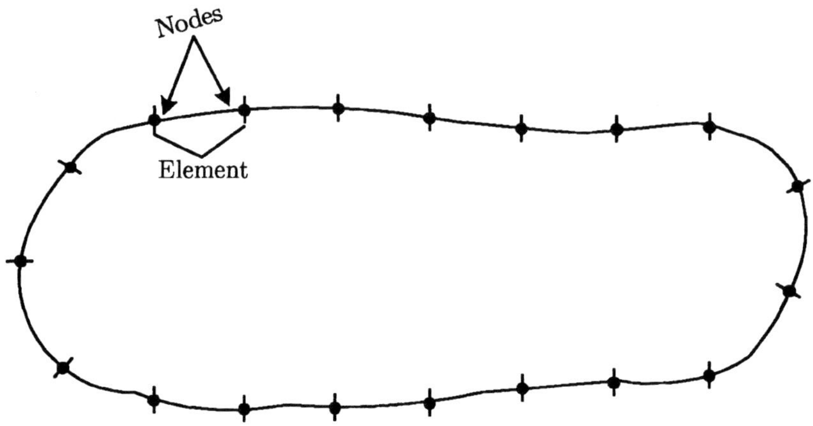

(ii)

Figure 3.5(ii). Linear boundary elements.

For a constant element, the boundary element is always smooth; hence the c_i coefficient is identically equal to $\frac{1}{2}\Gamma_j$, denote the length of element j. (3.141) represents in discrete form the relationship between node i at which the fundamental solution is applied and all the j elements, including the case $i = j$, on the boundary. The values of u and q inside the integrals in equation (3.141) are constant at one element and, consequently, we have

$$\frac{1}{2}u_i + \sum_{j=1}^{N}(\int_{\Gamma_j} q^\star \, d\,\Gamma)u_j = \sum_{j=1}^{N}\left(\int_{\Gamma_j} u^\star \, d\,\Gamma\right) q_j. \tag{3.142}$$

Let

$$\overline{H}_{ij} = \int_{\Gamma_j} q^\star \, d\,\Gamma, \text{ and } G_{ij} = \int_{\Gamma_j} u^\star \, d\,\Gamma. \tag{3.143}$$

(A symbol involving i and j is used to indicate that the integrals $\int_{\Gamma_j} q^\star \, d\,\Gamma$ relate the i-th node with the node j).

Equation (3.141) takes the form

$$\frac{1}{2}u_i + \sum_{j=1}^{N}\overline{H}_{ij}u_j = \sum_{j=1}^{N} G_{ij}\, q_j. \tag{3.144}$$

Here, the integrals in (3.141) are simple and can be evaluated analytically but, in general, numerical techniques will be employed.

Let us define

$$H_{ij} = \begin{cases} \overline{H}_{ij} & \text{if } i \neq j \\ \overline{H}_{ij} + \frac{1}{2} & \text{if } i = j\ , \end{cases} \tag{3.145}$$

then (3.144) can be written as

$$\sum_{j=1}^{N} H_{ij} u_j = \sum_{j=1}^{N} G_{ij}\ q_j,$$

which can be written in the form of matrix equation as

$$AX = F, \tag{3.146}$$

where X is the vector of unknown u's and q's (Either u's or q's will be known and X has N components), and A is a matrix of order N. The most effective and economical method for evaluating the matrix equation (3.146) can be seen in Hackbusch [1994] and Datta [1995]. A short note on such methods is also given in Appendix E. Thus all the values of potentials and fluxes on the boundary are known and one could calculate the values of potentials and fluxes at any interior point using

$$u_i = \int_\Gamma q u^\star\ d\,\Gamma - \int_\Gamma q\star\ u\ d\,\Gamma\ . \tag{3.147}$$

Equation (3.147) represents the integral relationship between an internal point, the boundary values of u and q and its discretized form is

$$u_i = \sum_{j=1}^{N} q_j G_{ij} - \sum_{j=1}^{N} u_i \overline{H}_{ij}\ . \tag{3.148}$$

The values of internal fluxes can be determined by differentiating (3.147) (See equation (3.139)) which gives us

$$\frac{\partial u}{\partial x_l} = -\int_\Gamma \int_\Gamma q\ u\ \frac{\partial q^\star}{\partial x_l}\ d\,\Gamma\ , \tag{3.149}$$

where x_l are the coordinates, $l = 1, 2$ are in two dimensions and $l = 1, 2, 3$ are in three dimensions.

Remark 3.17. (i) $\overline{H}_{ij}$ and G_{ij} can be evaluated by simple Gauss quadrature values for all elements (except one node under consideration) as follows:

$$\overline{H}_{ij} = \int_{\Gamma_j} q^\star\ d\,\Gamma = \frac{l_j}{2} \sum_{k=1}^{n} q_k^\star w_k,$$

and

$$G_{ij} = \int_{\Gamma_j} u^\star\ d\,\Gamma = \frac{l_j}{2} \sum_{k=1}^{n} q_k^\star w_k,$$

where l_j is the element length and w_k is the weight associated with the numerical integration point k.

(ii) For the integrals corresponding to singular elements, one may use the integral formulae discussed in Brebbia et al. [1984].

(iii) For the constant elements, as the present case,

$$\overline{H}_{ii} = \int_\Gamma q^\star \, d\,\Gamma = \int \frac{\partial u^\star}{\partial r}\frac{\partial r}{\partial n}\, d\,\Gamma = 0,$$

due to orthogonality between the normal and surface of the element

$$\begin{aligned} G_{ii} &= \int_{\Gamma_i} u^\star \, d\,\Gamma = \frac{1}{2\pi}\int_{\Gamma_i} \log(\frac{1}{r})\, d\,\Gamma \\ &= \frac{1}{\pi}\,|r_1\,| \left[\log\left(\frac{1}{|r_1|}\right)+1\right], \end{aligned}$$

where $r = \eta \mid r_1 \mid$, η homogeneous coordinate over a segment.

(iv) Mixed boundary conditions like

$$eu + fq = d \text{ on } \Gamma, \tag{3.150}$$

often known as Robin type conditions, can be considered where d, e, f are functions of the positions. If $f = 0$, it reduces to the Dirichlet condition and for $e = 0$, we get the Neumann condition. For $e \neq 0$, and $f \neq 0$, it can include impedance boundary conditions of electromagnetic problems and the convection boundary conditions of heat conduction problems. By applying this boundary condition at all boundary nodes, we may get the matrix equation

$$AX = F. \tag{3.151}$$

(v) If we use the indirect formulation for boundary value problems, equations (3.128) and (3.130) can be discretized as above and we get a matrix equation

$$A\sigma = F,$$

where the unknowns in the σ vector are the source intensities

$$\begin{aligned}
u_i &= \int_\Gamma \sigma u^\star \, d\,\Gamma \\
q_i &= -\frac{1}{2}\sigma_i + \int_\Gamma \sigma q^\star \, d\,\Gamma \\
u_i &= \sum_{j=1}^{n} \sigma_j \, G_{ij} \\
q_i &= -\frac{1}{2}\sigma_i + \sum_{j=1}^{N} \sigma_j \overline{H}_{ij} \\
\overline{H}_{ii} &= \overline{H}_{ii} - \frac{1}{2} \\
u_i &= \sum_{j=1}^{N} \sigma_i \, G_{ij}, \quad q_i = \sum_{j=1}^{N} \sigma_j H_{ij} \\
u_i &= \overline{u}_i \text{ at } N_1 \text{ points at } \Gamma_1 \\
q_i &= \overline{q}_i \text{ at } N_2 \text{ points at } \Gamma_2 \, .
\end{aligned}$$

Discretization by linear elements in two dimensions. We consider here a linear variation for u and q in which case nodes are located at the intersection between straight elements as shown in Figure 3.5(ii). Equation (3.140) can be written as

$$c_i u_i + \int_\Gamma u \, q^\star \, d\,\Gamma = \int_\Gamma q u^\star \, d\,\Gamma \, ,$$

or in discretized form

$$c_i u_i + \sum_{j=1}^{N} \int_{\Gamma_j} u q^\star \, d\,\Gamma = \sum_{j=1}^{N} \int_{\Gamma_j} q \, u^\star \, d\,\Gamma. \tag{3.152}$$

The values of u and q at any point of the element can be defined in terms of their nodal values and two linear interpolation functions ψ_1 and ψ_2, functions of homogeneous coordinate η, such that

$$\begin{cases} u(\eta) = \psi_1 u_1 + \psi_2 u_2 = [\,\psi_1, \psi_2\,] \begin{Bmatrix} u_1 \\ u_2 \end{Bmatrix} = \psi^T u^n \\ q(\eta) = \psi_1 q_1 + \psi_2 q_2 = [\,\psi_1, \psi_2\,] \begin{Bmatrix} q_1 \\ q_2 \end{Bmatrix} = \psi^T q^n \, . \end{cases} \tag{3.153}$$

The dimensionless coordinate η is equal to x_{l_2} and the ψ_1, ψ_2 functions are given by

$$\psi_1 = \frac{1}{2}(1-\eta), \psi_2 = \frac{1}{2}(1+\eta). \tag{3.154}$$

We have

$$\int_{\Gamma_j} u \, q^\star \, d\,\Gamma = \int_{\Gamma_j} [\psi_1, \psi_2] q^\star \, d\,\Gamma \begin{Bmatrix} u_1 \\ u_2 \end{Bmatrix} = [\,h_{ij}^1, h_{ij}^2\,] \begin{Bmatrix} u_1 \\ u_2 \end{Bmatrix}, \tag{3.155}$$

where

$$h^1_{ij} = \int_{\Gamma_j} \psi_1 q^\star \, d\,\Gamma,$$
$$h^2_{ij} = \int_{\Gamma_j} \psi_2 q^\star \, d\,\Gamma.$$

Similarly,

$$\int_{\Gamma_j} q\, u^\star \, d\,\Gamma = [\, g^1_{ij}, g^2_{ij} \,] \begin{Bmatrix} q_1 \\ q_2 \end{Bmatrix}, \tag{3.156}$$

with

$$g^1_{ij} = \int_{\Gamma_1} \psi_1 \, u^\star \, d\,\Gamma$$
$$g^2_{ij} = \int_{\Gamma} \psi_2 \, u^\star \, d\,\Gamma\,.$$

Putting these values, (3.152) can be written in terms of a matrix equation $AX = F$. For details, we refer to Brebbia et al. [1984].

3.3.4 Coupling of boundary element and finite element methods

Combining the methods of finite element and boundary element is of great interest in many practical problems. A reasonable approach could be to convert a problem on a given domain into a boundary integral equation and then to solve it by the finite element method. We present here some general results in this direction without proof (For details, we refer to Dautry-Lions [Vol. 4, 1990]).

$$\begin{aligned} \Omega \subset R^3, H^2(\Omega) &= \{f \in L_2(\Omega)/D^\alpha f \in L_2(\Omega), \alpha = (\alpha_1,\alpha_2,\alpha_3), |\alpha| \\ &= \alpha_1 + \alpha_2 + \alpha_3 \leq 2\} \\ \langle f, q\rangle &= \sum_{|\alpha|\leq 2} \langle D^\alpha f, D^\alpha g\rangle L_2(\Omega), \\ \|f\|_{H^2(\Omega)} &= \sum_{|\alpha|\leq 2} \|D^\alpha f\|_{L_2}(\Omega). \end{aligned}$$

Consider the Dirichlet problem

$$\begin{cases} \nabla u(x) = 0, x \in \Omega \\ u_{/\Gamma} \quad = 0, \Gamma \text{ is boundary of } \Omega. \end{cases} \tag{3.157}$$

This is called the interior Dirichlet problem, while the following problem is called the exterior Dirichlet problem:

$$\begin{cases} \nabla u(x) = 0, x \in R^3 \setminus \Omega = \Omega' \\ u_{/\Gamma} \quad = 0\,. \end{cases} \tag{3.158}$$

Theorem 3.5. *Let $u_0(x)$ be an element in $H^{\frac{1}{2}}(\Gamma)$ and let*

$$\frac{1}{4\pi}\int_\Gamma \frac{q(y)}{||x-y||}d\gamma(x) = u_0(x), x \in \Gamma, \tag{3.159}$$

where $d\gamma$ is the element of surface $\Gamma_0 = \Omega$, be a boundary integral with unknown $q(\cdot)$. (3.159) has a unique solution in the space $H^{-\frac{1}{2}}$. This problem has the variational formulation

$$a(q,q') = F(q') \text{ for all } q' \in H^{-\frac{1}{2}(\Gamma)}, \tag{3.160}$$

where

$$a(q,q') = \frac{1}{4\pi}\int_\Gamma\int_\Gamma \frac{q(y)q'(y)}{||x-y||}d\,\gamma(x)\,d\,\gamma(y) \tag{3.161}$$

$$F(q') = \int_\Gamma u_0(y)q'(y)\,d\,\gamma(y). \tag{3.162}$$

$a(\cdot,\cdot)$ is coercive on $H_2^{-\frac{1}{2}}$, that is, $\exists\ \alpha > 0$ such that

$$a(q,q) \geq \alpha||q||_{H^{-\frac{1}{2}}(\Gamma)}.$$

The solutions of the interior and exterior Dirichlet boundary value problems (3.157) and (3.158) exist and are given by

$$u(x) = \frac{1}{4\pi}\int_\Gamma \frac{q(u)}{||x-y||}d\,\gamma(y), \text{ for all } x \in R^3, \tag{3.163}$$

where $q(\cdot)$ is the solution of (3.159).

Note: Analogous results can be proved for the Neumann boundary problems, and the Helmholtz equation with the Dirichlet and Neumann boundary conditions (See Dautry-Lions, Vol.4, Chapter-XI).

Finite element approximation of the boundary integral equation given in (3.159).

Approximate (discrete) problem of (3.160). Find a subspace V_h of $H^{\frac{1}{2}}(\Gamma)$ and $q_h \in V_h$ such that

$$a(q_h,q_h') = f(q_h') \text{ for all } q_h' \in v_h. \tag{3.164}$$

Remark 3.18. (i) In the simplest case, V_h will consist of functions that are constant on each element of the mesh T_h. Here we can construct V_h by a finite element method using a polynomial of degree k. We can use finite elements of C^0 class whenever $k > 1$. Here the space V_h will be a space of continuous functions whose restriction to each triangle is a polynomial of degree k.

(ii) If V_h is the space of functions constant on the elements of the mesh, the number of unknowns are equal to the number of these elements. The coefficients of the matrix of the linear system to be solved are

$$a_{ij} = \frac{1}{4\pi} \int_{T_i} \int_{T_j} \frac{1}{\|x-y\|} d\gamma(x) d\gamma(y), T_i, T_j \in T_h \,. \tag{3.165}$$

The matrix is symmetric and positive definite by virtue of the coercivity of the bilinear form $a(q, q')$. If $h =$ maximum diameter of the elements in the mesh T_h, then the error between the solution q_h of the discretized equation (3.164) and q, the solution of (3.160), is given by the following theorem.

Theorem 3.6. *Let q be the solution of (3.160) and q_h be the solution of (3.164), then*

$$\|q - q_h\|_{H^{-\frac{1}{2}}(\Gamma)} \le C h^{k+\frac{3}{2}} \|q\|_{H^{k+1}}(\Gamma) \,, \tag{3.166}$$

($k = 1$ if elements are from $H^1(\Omega)$ or $k = 2$ if elements belong to $H^2(\Omega)$).

Let u_h be given by

$$u_h = \frac{1}{4\pi} \int_\Gamma \frac{q_h(y)}{\|x-y\|} d\gamma(y),$$

and u by (3.163), then

$$\|u - u_h\|_{H_1(R^3)} \le \|q - q_h\|_{H^{-\frac{1}{2}}(\Gamma)} \,.$$

Results analogous to Theorem 3.6 are possible for many other classes of boundary value problems like the Helmholtz equation with Dirichlet and Neumann boundary conditions and problems of linear elasticity and Stokes system (See Dautry-Lions [Vol. 4, 1990]).

3.4 Problems

Problem 3.1 (i) What do you understand by wave propagation in uniform guides?

(ii) Write down a mathematical equation which models wave propagation in uniform guides.

(iii) Write down the variational (weak) formulation of the mathematical model.

Problem 3.2 (i) Write down the Helmholtz equation in the three-dimensional space with the Dirichlet and the Neumann boundary conditions.

(ii) What kind of physical phenomena are represented by the Helmholtz equation with well known boundary conditions?

Problem 3.3. Discuss the numerical solution of the Helmholtz equation with Dirichlet and Neumann boundary conditions by

(i) Finite element method.

(ii) Boundary element method.

Problem 3.4. Discuss the solution of the Helmholtz equation with hysteresis.

Problem 3.5. Write down Maxwell's equations into integral form.

Problem 3.6. Discuss the solution of Maxwell's equations independent of time applying wavelet techniques.

Problem 3.7. Write down the mathematical equation modelling the radio antennas. Find its finite element solution.

Problem 3.8. What is the physical meaning of the Sommerfeld's radiation condition? Write down its mathematical formulation.

Problem 3.9. Explain the meaning of the following terms:

(i) The electromagnetic energy and Maxwell stress tensor.

(ii) Spectral problems of wave guides.

(iii) Formulae providing transmission conditions at the interface between different continuous media.

Problem 3.10. Discuss the solution of Maxwell's equation by Mathematica.

Problem 3.11. Discuss the error between the exact and the finite element solution of the following differential equation

$$\begin{aligned} -\frac{d^2u}{dx^2} &= 2, \quad 0 < x < 1 \\ u(0) &= u(1) = 0\,. \end{aligned}$$

Problem 3.12. Consider the one-dimensional heat transfer in an insulated rod of cross-section area A, Length L, conductivity k, convection coefficient β, and surrounding media (ambient) temperature T_∞. Write down the differential equation modelling this phenomenon and solve it by the finite element method.

Problem 3.13. Write down the discrete form of the differential equation

$$\frac{d^2}{dx^2}\left(p(x)\, b\frac{d^2u}{dx^2}\right) + f(x) = 0, \quad 0 < x < L,$$

under the appropriate boundary condition.

Problem 3.14. Solve the Poisson equation

$$-\nabla^2 u = 1$$

in a square of side 1 with the boundary conditions

$$\begin{aligned}\frac{\partial u}{\partial x}(0,y) &= \frac{\partial u}{\partial y}(x,0) = 0\\ u(1,y) &= u(x,1) = 0\end{aligned}$$

by the finite element method using the linear elements.

Problem 3.15. Solve the integral equations of electrostatics in one dimension by the finite element method.

Problem 3.16. Discuss the finite element solution of the nonlinear equation

$$\nabla \times (v\nabla \times A) = J\,,$$

where $v = \frac{1}{\mu}$, μ is the material permeability, A is a vector potential, whose curl gives the magnetic flux density B ($\nabla \times A = B$).

Problem 3.17. Solve Poisson's equation $\nabla^2 u = b$ in a square with sides of length 2 and homogeneous boundary conditions $u = 0$, at $x = \pm 1$, and $y = \pm 1$, by the boundary element method.

Problem 3.18. Discuss the solution of the transient scalar wave equation by the boundary elment method.

Problem 3.19. Write down an essay on the finite element method in less than 1000 words introducing the topic without writing mathematical formula.

Problem 3.20. What do you understand by interpolation (approximation) function? Explain this concept with the help of examples.

Problem 3.21. Show that the Helmholtz operator $A = \nabla^2 + k^2$ is self-adjoint.

Problem 3.22. Consider Poisson's equation

$$-\nabla^2 u = 2(x+y) - u$$

in the square whose vertices are

$$\begin{aligned}u(0,y) &= y^2, u(x,0) = x^2\\ \frac{\partial u}{\partial x}(1,y) &= 2 - 2y - y^2, \quad \frac{\partial u}{\partial y}(x,1) = 2 - x - x^2.\end{aligned}$$

Solve the boundary value problems using trial functions of the form:

$$\begin{aligned} u_1(x,y) &= x^2 + y^2 + axy \\ u_2(x,y) &= x^2 + y^2 + bxy(x+y). \end{aligned}$$

Problem 3.23. State the main steps involved in solving the Poisson equation by applying triangular finite elements.

Problem 3.24. What do you understand by Eddy current problems? Describe the mathematical modelling of Eddy current and discuss a numerical solution of the model.

Problem 3.25. Discuss the finite element solution of the following boundary value problem:

$$\begin{aligned} -\frac{d^2u}{dx^2} + u(x) &= f(x), \quad x \in (0,1) \\ u(0) &= u(1) = 0. \end{aligned}$$

Problem 3.26. Find the Ritz-Galerkin solution of the following boundary value problems:

(a) $\left(\frac{d^2u}{dx^2} + \frac{d^2u}{dy^2}\right) = -1$ in the square $(0,1) \times (0,1)$

$u(x) =$ 0, on the boundary of the square

(b) $\left(\frac{d^2u}{dx^2} + \frac{d^2u}{dy^2}\right) = -\pi^2 \cos \pi x$ in the square $(0,1) \times (0,1)$

$\frac{\partial u}{\partial n} =$ 0, on the boundary of the square.

Hints for solution:

Problem 3.1. (i) The behaviour of wave guides is derived from the property of a hollow tube with a highly reflecting inside surface that channelizes electromagnetic wave energy from one end to the other. A simple example is that of shining light through a tube with a mirror interior surface. Because of the possibility of launching two alternative, independent plane waves distinct by virtue of possessing orthogonal polarization, one may get two independent sets of waves or modes in the guide, known as transverse magnetic and transverse electric waves. This phenomenon is known as the wave propagation in uniform guides.

(ii) Let us consider a cylindrical waveguide of arbitrary cross-section and with its axis aligned along the z-direction. In order that a coherent wave propagate at frequency w, we must have

$$E = E(x,y) \exp i(wt - \beta\, z),$$

$H = H(x,y)\exp i(wt - \beta\ z)$, where β is a propagation constant corresponding to wavelength $2\pi/\beta$. Putting these values in the following form of Maxwell's equations:

$$\nabla \times E = -i\omega\mu H \tag{i}$$
$$\nabla \times H = i\omega\epsilon E. \tag{ii}$$

(We get (i) and (ii) by keeping in mind that $\frac{\partial}{\partial t}$ in (3.17)(i) and (3.17)(ii) can be replaced by $i\omega$ by change of variable), we get,

$$\frac{\partial E_z}{\partial y} + i\beta E_y = -i\ w\mu H_x \tag{iii}$$
$$-i\ \beta\ E_x - \frac{\partial E_z}{\partial x} = -i\ w\mu H_x \tag{iv}$$
$$\frac{\partial E_y}{\partial x} - \frac{\partial E_x}{\partial y} = -i\ w\ \mu\ H_z \tag{v}$$
$$\frac{\partial H_z}{\partial y} + i\ \beta\ H_y = i\ w\ \epsilon\ E_x \tag{vi}$$
$$-i\beta H_x - \frac{\partial H_z}{\partial x} = i\ w\ \epsilon\ E_y \tag{vii}$$
$$\frac{\partial H_y}{\partial x} - \frac{\partial H_x}{\partial y} = i\ w\ \epsilon\ E_z\,. \tag{viii}$$

By solving (iii) and (vii) for H_x and E_y and (iv) and (vi) for H_y and E_x, we get

$$H_x = i\left(w\ \epsilon\ \frac{\partial E_2}{\partial y} - \beta\ \frac{\partial H_z}{\partial x}\right)/k^2 - \beta^2 \tag{ix}$$
$$H_y = -j\left(w\ \epsilon\ \frac{\partial E_z}{\partial x} + \beta\ \frac{\partial H_z}{\partial y}\right)/(k^2 - \beta^2) \tag{x}$$
$$E_x = -i\left(\beta\ \frac{\partial E_z}{\partial x} + w\ \mu\ \frac{\partial H_z}{\partial y}\right)/(k^2 - p^2) \tag{xi}$$
$$E_y = i\left(-\beta\ \frac{\partial E_z}{\partial y} + w\ \mu\ \frac{\partial H_z}{\partial y}\right)/(k^2 - \beta^2),\ \text{where } k^2 = w^2\ \mu\ \epsilon. \tag{xii}$$

If μ and ϵ are constants, then, by choosing $k^2 = w^2\ \mu\,\epsilon$, it can be shown that

$$\nabla^2 E + k^2 E = 0 \tag{xiii}$$
$$\nabla^2 H + k^2 H = 0. \tag{xiv}$$

See also Remark 3.10. Observing the $\exp(-i\beta z)$ dependence of the fields, the z-component equations of the above is given as

$$\nabla_T^2 E_z + \left(k^2 - \beta^2\right) E_z = 0 \tag{xv}$$
$$\nabla_T^2 H_z + \left(k^2 - \beta^2\right) = 0, \tag{xvi}$$

where $\nabla_T^2 = \dfrac{\partial^2}{\partial\, x^2} + \dfrac{\partial^2}{\partial\, y^2}$ is the transverse Laplacian operator.

E_z and the transverse E-field components determined by it from equations (ix) to (xii) may exist separately from H_z and its corresponding transverse components. Thus, the two separate modes, transverse electric (TE) where $E_z = 0, H \neq 0$ and transverse magnetic (TM) when $H_z = 0, E_z \neq 0$ are identified. Equations (xvi) and (xvii) are crucial as all other field quantities will follow from these two. It may be observed that these two equations are special cases of the well-known Helmholtz equation,

$$\begin{cases} F(E_z) = \frac{1}{2}\int_\Omega \left[(\nabla E_z)^2 - (k^2-\beta^2)\, E_z^2\right] dx\; dy \\ F(H_z) = \dfrac{1}{2}\displaystyle\int_\Omega (\nabla H_z)^2 - (k^2-\beta^2)\, H_z^2 dx\; dy. \end{cases} \tag{xvii}$$

Problem 3.2.

$$\begin{cases} \Omega \subset R^3, \nabla u(x) + k^2 u(x) = 0, \;\; x \in \Omega \\ u(x) = 0 \quad \text{on } \partial\,\Omega = \Gamma \text{ (Dirichlet Boundary)} \\ \dfrac{\partial u(x)}{\partial n} = 0 \quad \text{on } \partial\,\Omega = \Gamma\,, \end{cases} \tag{i}$$

where k is a positive real constant.

Homogeneous Helmholtz equation with Dirichlet and Neumann boundary conditions in three dimensions

$$\nabla\cdot(p\nabla u) + k^2 u = g, \tag{ii}$$

where $u = u(x, y, z)$ is a function of position whose determination constitutes the desired solution; p denotes the function $p(x, y, z)$ representing the material properties of the medium, k^2 is a constant independent with position which may or may not be known, and $g(x, y, z)$ is a given function.

Equation (ii) is called the non-homogeneous or inhomogeneous Helmholtz equation. It may be assigned Dirichlet, Neumann or both boundary conditions.

(iii) It is a model of propagation of wave in uniform guides as seen above. It models the radiation from electromagnetic sources. It has a close connection with the wave equation

$$\frac{\partial^2\omega}{\partial t^2} - \nabla^2\omega + g = 0 \quad \text{on } \Omega\times R, \Omega\subset R^3\,. \tag{iii}$$

For $g(x,t) = e^{ikt} f(x)$ on $\Omega \times R, u(x)$ is a solution of (ii), for $p = 1$, if and only if $\omega(x,t) = e^{ikt}u(x)$ is a solution of (iii).

In the one-dimensional Helmholtz equation,

$$\frac{d^2 u}{d\,x^2} + k^2\; u = 0 \tag{iv}$$,

is the model of the vibrating string whose general solution is

$$u(x) = \lambda\, e^{ikx} + \mu\, e^{-ikx}\,. \tag{v}$$

Originally, Helmholtz used this model to study acoustic resonance frequencies of halls, loud speakers, etc. Sound waves in air are represented by longitudinal disturbances in pressure, denoted by $u(\cdot)$, and perturbations of a prevailing static pressure u_0. Such disturbances are governed by the wave equation

$$\nabla^2 u = \frac{1}{c}\frac{\partial^2 u}{\partial\, t^2}\,, \tag{vi}$$

where $c = \gamma u_0/\varphi_0$ with φ_0 the static density of the air and γ its ratio of specific heats at constant pressure and volume. The disturbance velocity of the air v is related to the incremental pressure u through a linearized equation of motion

$$\nabla u = -\varphi\frac{\partial u}{\partial t}\,. \tag{vii}$$

(vi) and (vii) may be cast into complex phase form to the Helmholtz equation

$$\begin{aligned} &\nabla^2 u + k^2 u &&= 0 \\ &\nabla u &&= -i\omega\varphi_0 v \\ &\text{where } k^2 &&= \omega^2/c^2\,. \end{aligned}$$

The study of electrical heating by electromagnetic induction will also require the study of Helmholtz equation. Standing waves on a bounded shallow body of water are modelled by the Helmholtz equation with the Neumann boundary condition

Problem 3.3.

Finite element method. We consider here Helmholtz equation with the Dirichlet boundary condition. The Neumann boundary problem can be solved on similar lines keeping in mind that the integral on the boundary will be zero.

Let $\Omega \subset R^3$ be divided into m elements of r nodes each. We may express the behaviour of the unknown function $u(x)$ within each element as

$$u^{(j)} = \sum_{i=1}^{r} a_i\, u_i = [a]\,\{u\}^{(j)}\,, \tag{i}$$

where u_i is the value of $u(\cdot)$ at i-thth node. Here, we are considering nodal values of u as degrees of freedom; however, values of derivatives could also be used as degrees of freedom without changing the procedure to be followed. The quantity u_i may be thought of as a general nodal parameter. The functional

$$I(u) = \frac{1}{2}\int_\Omega \left[\left(\frac{\partial u}{\partial x}\right)^2 + \left(\frac{\partial u}{\partial y}\right)^2 + \left(\frac{\partial u}{\partial z}\right)^2 - \lambda\, u^2\right], \tag{ii}$$

for the Helmholtz equation

$$\frac{\partial^2 u}{\partial x^2} + \frac{\partial^2 u}{\partial y^2} + \frac{\partial^2 u}{\partial z^2} + \lambda u = 0, \qquad \lambda > 0 \text{ known or unknown},$$
$$u = 0 \text{ on } \Gamma,$$

will attain minima or maxima at the solution of the Helmholtz equation under consideration; that is,

$$\frac{\partial I}{\partial u_i} = 0, i = 1, 2, \cdots r, \tag{iii}$$

will be satisfied.

Since the functional $I(\cdot)$ contains only the first order derivative, a_i must be chosen to preserve at least continuity of u at element interfaces. If the approximating functions guarantee continuity, we must focus attention on one element because the integral $I(u)$ can be represented as the sum of integrals over all elements; that is,

$$I(u) = \sum_{j=1}^{m} I(u^{(j)}). \tag{iv}$$

The discretized form of the functional for one element is obtained by substituting equation (i) into (ii). Then (iii) takes the form

$$\frac{\partial I(u^{(j)})}{\partial u_i} = 0, i = 1, 2, \ldots, r. \tag{v}$$

By (i), we get

$$\frac{\partial^{(j)}}{\partial x} = \sum_{i=1}^{j} \frac{\partial a_i}{\partial x} u_i = \left[\frac{\partial a}{\partial x}\right] \{u\}^{(j)}$$
$$\frac{\partial}{\partial u_i}\left(\frac{\partial u^{(j)}}{\partial x}\right) = \frac{\partial a_i}{\partial x}, \quad \frac{\partial u^{(j)}}{\partial u_i} = a_i.$$

Thus

$$\left\{ \begin{aligned} \frac{\partial I(u^{(j)})}{\partial u_j^i} = 0 = \int_{\Omega^{(j)}} \Bigg[\left[\frac{\partial a}{\partial x}\right] \{u\} \frac{\partial a_i}{\partial x} + \frac{\partial a}{\partial y} \{u\} \frac{\partial a_i}{\partial y} \\ + \left[\frac{\partial a}{\partial z}\right] \{u\} \frac{\partial a_i}{\partial z} - \lambda a_i \Bigg] d\,\Omega^j (v'), \end{aligned} \right. \tag{v'}$$

Combining all equations like equation (v), we get

$$\left\{\frac{\partial I}{\partial u^{(j)}}\right\}^{(j)} = \left\{\begin{array}{c} \dfrac{\partial I(u_1^{(j)})}{\partial u_1} \\ \vdots \\ \vdots \\ \vdots \\ \dfrac{\partial I(u_r^{(j)})}{\partial u_r} \end{array}\right\} = [K]^{(j)} \{u\}_{\lambda}^{(j)} \pm \lambda \{S\}^{(j)} = 0, \qquad \text{(vi)}$$

where the coefficients of the matrices $[K]^{(j)}$, and $\{S_i\}^{(j)}$ are given by the equations

$$\left\{\begin{array}{ll} k_{ij} = \displaystyle\int_{\Omega^{(j)}} \left(\frac{\partial a_i}{\partial x}\frac{\partial a_j}{\partial x} + \frac{\partial a_i}{\partial y}\frac{\partial a_j}{\partial y} + \frac{\partial a_i}{\partial z}\frac{\partial a_j}{\partial z}\right) d\,\Omega^{(j)} \\ s_{ij} = \displaystyle\int_{\Omega^{(j)}} a_i\, a_j\, d\,\Omega^{(j)}, i = 1,2,3\cdots r, j = 1,2,3\cdots r\,. \end{array}\right. \qquad \text{(vii)}$$

Thus we have a matrix equation.

For a solution domain of m elements, the systems of matrix equations are of the form

$$\left\{\begin{array}{ll} [K] & = \lambda\,[S]\,\{u\} \\ \text{where} & [K] = \displaystyle\sum_{j=1}^{m} [K]^{j}\,, \text{ and } [S] = \sum_{j=1}^{m} [S]^{(j)}\,, \end{array}\right. \qquad \text{(viii)}$$

$\{u\}$ is the column vector of nodal values of u. Equations in (viii) are a set of, say n, linear homogeneous algebraic equations in nodal values of u. The problem which we have solved here is known as the eigenvalue or characteristic value problem; the values of λ are termed eigenvalues or characteristic values. For each different value of λ_i, there is a different column vector $\{u\}$, that satisfies (viii). The vector $\{u\}_i$ that corresponds to a particular value of λ_i is called an eigenvector or a characteristic vector. We know that $\{u\} \neq \{0\}$, that is, (viii) has a non-trivial solution if and only if the characteristic determinant is zero; that is,

$$|\,[K] - [H]\,| = 0\,. \qquad \text{(ix)}$$

This equation is used to find the eigenvalues. We get the following n-th order polynomial by expanding

$$c_n\lambda^n + c_{n-1}\lambda^{n-1} + \cdots\cdots c_1\lambda + c_0 = 0\,. \qquad \text{(x)}$$

By the fundamental theorem of algebra, this polynomial has n roots λ_i. Substituting these values of λ_i in (viii), we solve n sets of equations for the n eigenvectors $\{\phi\}_i$. $[K]$ and $[H]$ are symmetric and definite positive and so eigenvalues are distinct and real and u_i's are linearly independent. The usual procedure for solution is to assign an arbitrary value to one component of the vector $\{u\}$ and then solve the remaining $(n-1)$ equations for the other components. The consequence of this fact is that the natural eigenvectors $\{\phi\}_i$ of the wave motion are known only in relation to one another, not in absolute terms. Iterative numerical methods are also known to find the solution of (viii) (See Ciarlet [1989]).

(ii) See Dautry-Lions [1990, Vol. 4 pp. 143–145], Brebbia [1984, pp. 121–123 and 416–417] and Balder and Zerger [1996].

Problem 3.5.

$$\oint_C E \cdot dl = -\int_S \frac{\partial B}{\partial t} \cdot dS$$
$$\oint_C H \cdot dl = \int_S \left(J + \frac{\partial D}{\partial t}\right) \cdot d\,S$$
$$\oint_{S^1} D \cdot d\,S = \int_\Omega \varphi d\,\Omega$$
$$\oint_S B \cdot dS = 0\,,$$

where the integrals are taken over an open surface S or its boundary contour C or over a closed surface S^1.

Problem 3.8. The Sommerfeld condition expresses the fact that the energy flux across every sphere of radius R very large, in the sense of $|x|$ increasing, is positive for the reflected electromagnetic wave satisfying the condition; namely,

$$R\left(\frac{\partial u}{\partial n} + iku\right),$$

is bounded when $R = |x| \to \infty$.

Problem 3.9. Electromagnetic energy for the field $\{E, B\}$ is defined as

$$F(t) = \frac{1}{2}\int_{R^3} \left(|E(x,t)|^2 + |B(x,t)|^2\right) dx, x \in R^3, t \in R\,.$$

Problem 3.15. See Silvester and R.L. Ferrari [1983, Chapter 4].

Problem 3.16. The equation is a mathematical model of the vector potential A in ferromagnetic materials. Ferromagnetic materials exhibit strong magnetic effects and are the most important magnetic substances. The permeability of these materials is not a constant but a function of both the applied field and the previous magnetic history of the specimen. Iron, nickel, and cobalt are ferromagnetic materials.

Two-dimension case.

$$\frac{\partial}{\partial x}\left(v\,\frac{\partial A}{\partial x}\right)+\frac{\partial}{\partial\, y}\left(v\,\frac{\partial A}{\partial\, y}\right)=-J\,.$$

Corresponding energy functional is

$$F(U)=\int_{\Omega}W(U)d\,\Omega-\int_{\Omega}JU\;d\,\Omega\,,$$

where

$$W(U)=\int H\cdot d\,B$$
$$B=\nabla\times U\ \text{ and }\ H=\mu B$$
$$U=\sum_i \alpha_i(x,y)U_i\,.$$

By the finite element approximation, we get the matrix equation

$$SU=J\,,$$

where U is the vector of nodal potential values, that is, values of U at nodes,

$$J=\int_{\Omega^{(e)}}J\alpha_k\;d\,\Omega^{(e)}$$
$$S=(S_{ij}),\quad S_{ij}=\int_{\Omega^{(e)}}v\mu\left(\frac{\partial\,\alpha_i}{\partial\,x}\frac{\partial\,\alpha_j}{\partial\,x}+\frac{\partial\,\alpha_i}{\partial\,y}\frac{\partial\,\alpha_j}{\partial\,y}\right)\,d\,\Omega\,.$$

The matrix equation can be solved by the iterative method. For further details, one may see Silvester and Ferrari [1983].

Problem 3.20. In the finite-element method, the functions used to represent behaviour of a field variable within an element are called interpolation, shape and approximation functions. In mathematical terminology these are the basis elements of the approximating subspace H_h of the Hilbert space over which the continuous problem is defined. Many types of functions could serve as interpolation functions, but mainly polynomials have been widely used in practical applications. However, well known orthonormal systems and recently invented wavelets could also be used as interpolation functions. Let us consider the two-dimensional field variable $\phi(x,y)$. The nodal values of ϕ can uniquely and continuously define $\phi(x,y)$ throughout the domain of interest in the x-y plane. Suppose the domain of $\phi(x,y)$ is divided into triangles. Then the plane passing through the three nodal values of ϕ associated with element $\Omega^{(e)}$ is described by the equation

$$\begin{cases}\phi^{(e)}(x,y)=\alpha_1^{(e)}+\alpha_2^{(e)}\,x+\alpha_3^{(e)}y\,;\\ \qquad\ \ \phi_i^{(e)}=\alpha_1^{(e)}+\alpha_2^{(e)}\,x_i+\alpha_3^{(e)}y_i\,;\\ \qquad\ \ \phi_j^{(e)}=\alpha_1^{(e)}+\alpha_2^{(e)}\,x_j+\alpha_3^{(e)}y_j\,;\\ \qquad\ \ \phi_k^{(e)}=\alpha_i^{(e)}+\alpha_2^{(e)}\,x_k+\alpha_3^{(e)}y_k\,.\end{cases}\tag{i}$$

Solving these equations yields

$$\begin{aligned}
\alpha_1^{(e)} &= \frac{\phi_i(x_j\, y_k - y_j x_k) + \phi_j(x_k\, y_i - x_i\, y_k) + \phi_k(x_i\, y_j - x_j\, y_i)}{2\,\Delta} \\
\alpha_2^{(e)} &= \frac{\phi_i(y_i - y_k) + \phi_j(y_k - y_i) + \phi_k(y_i - y_j)}{2\,\Delta} \\
\alpha_3^{(e)} &= \frac{\phi_i(x_k - x_j) + (x_i - x_k) + \phi_k(x_j - x_i)}{2\,\Delta} \\
2\,\Delta &= \begin{vmatrix} 1\, x_i & y_i \\ 1\, x_j & y_j \\ 1\, x_k & y_k \end{vmatrix}.
\end{aligned}$$

Putting these values in (i) and rearranging the terms, we have

$$\begin{aligned}
\phi^{(e)}(x,y) = {} & \frac{a_i + b_i x \ + \ c_i y}{2\,\Delta}\ \phi_i \ + \ \frac{a_i \ + \ b_j x \ + \ c_j y}{2\,\Delta}\ \phi_j \\
& + \frac{a_k + b_k x + c_k y}{2\,\Delta}\ \phi_k\,,
\end{aligned}$$

where

$$\begin{aligned}
a_i &= x_j\, y_k - x_k\, y_j, b_i = y_j - y_k, c_i = x_k - x_j \\
N_n^{(e)} &= \frac{a_n + b_{nx} + c_n y}{2\,\Delta}, n = i, j, k.
\end{aligned}$$

The functions $N^{(e)}$ are interpolation functions.

Problem 3.18. See Brebbia et al. (1984 [352–354]).

Problem 3.24. An electric current induced within the body of the conductor, when the conductor either moves through a non-uniform magnetic field or is in a region where there is a change in magnetic flux, is known as the eddy current. Since iron is a conductor, there will be an eddy current induced in it, when the flux is changed. Because of ohmic resistance, the eddy current causes a loss of energy. Hysteresis and eddy current loss form the two components of iron loss. For a long time, these currents were considered harmful but new technologies have recently arisen which make use of eddy currents for production of heat or forces like

(i) the linear induction motor, in which eddy currents are produced in a reaction rail to create a propulsive force;

(ii) in electrodynamic systems, where eddy currents provide the lift force, although create also a drag force; and

(iii) electrical heating by electro-induction, where one deals with the generation of a given eddy current distribution.

Approximate mathematical models can be derived from Maxwell's equations and it will be a Helmholtz equation linear or non-linear. For details, see Chapter 6 and 10 in Chari and Silvester [1983]. See also Balder and Zerger [1996].

Problem 3.25. We present here a finite element solution of the following boundary value problem [Datta, 1995].

$$-\frac{d^2u}{dx^2} + u(x) = f(x), \quad for\ x \in (0,1), \tag{i}$$

$u(x) = 0$, for $x \in \{0,1\}$.

Let $\mathrm{H} = \mathrm{H}_0^1(0,1)$. If we now multiply the equation $-u'' + u = f(x)$ by an arbitrary function $v \in \mathrm{H}$ (v is called a test function) and integrate the left-hand side by parts, we get

$$\int_0^1 (-u''(x) + u(x))v(x)dx = \int_0^1 f(x)v(x)dx,$$

that is,

$$\int_0^1 (u'v' + uv)dx = \int_0^1 f(x)v(x)dx. \tag{ii}$$

We write equation (ii) as:

$$a(u,v) = (F,v) \text{ for every } v \in H,$$

where

$$a(u,v) = \int_0^1 (u'v' + uv)dx \text{ and } (F,v) = \int_0^1 f(x)v(x)dx.$$

(Notice that the form $a(\cdot,\cdot)$ is symmetric (that is, $a(v,u) = a(u,v)$) and bilinear.)

It can be shown that u is a solution of equations (ii) if and only if u is a solution of (i)

The discrete problem. We now discretise the problem in (ii). We start by constructing a finite-dimensional subspace H_n of the space H.

Here, we consider only the simple case where H_n consists of continuous piecewise linear functions. For this purpose, let $0 = x_0 < x_1 < x_2 < \cdots < x_n < x_{n+1} = 1$ be a partition of the interval [0,1] into subintervals $I_j = [x_{j-1}, x_j]$ of length $h_j = [x_j - x_{j-1}]$, $j = 1, 2, \ldots, n+1$.

With this partition, we associate the set H_n with the set of all functions $v(x)$ that are continuous on the interval [0,1], linear in each subinterval $I_j, j = 1, 2, \ldots, n+1$, and satisfy the boundary conditions $v(0) = v(1) = 0$.

We now introduce the basis functions $\{\phi_1, \phi_2, \ldots, \phi_n\}$, of H_n. We define $\phi_j(x)$ by

1. $\phi_j(x_i) = \begin{cases} 1 \text{ if } i = j \\ 0 \text{ if } i \neq j \end{cases}$,

2. $\phi_j(x)$ is a continuous piecewise linear function.

$\phi_j(x)$ can be computed explicitly to yield:

$$\phi_j(x) = \begin{cases} \frac{x - x_{j-1}}{h_j}, \text{ when } x_{j-1} \leq x \leq x_i \\ \frac{x_{j+1} - x}{h_{j+1}}, \text{ when } x_j \leq x \leq x_{j+1}\,. \end{cases}$$

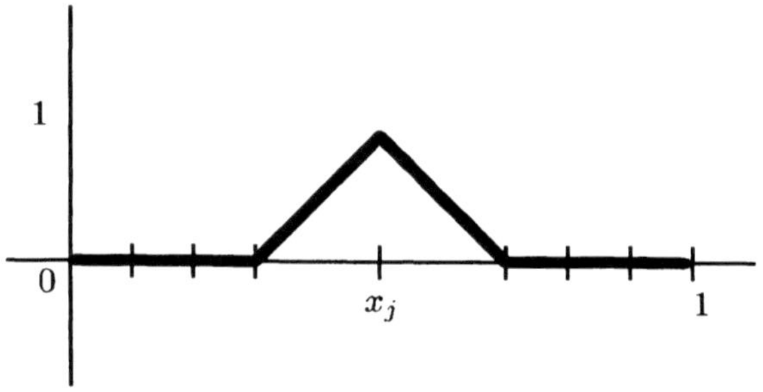

Because $\phi_1, \ldots, \phi_n$, are the basis functions, any function $v \in H_n$ can be written uniquely as

$$v(x) = \sum_{i=1}^{n} v_i \phi_i(x), \quad \textit{where} \quad v_i = v(x_i).$$

We easily see that $H_n \subset H$. The discrete analogous of (ii) then reads: Find $u_n \in H_n$ such that

$$a(u_n, v_n) = (F, v_n) \text{ for evrey } v_n \in H_n. \tag{iii}$$

Now, if we let $u_n = \sum_{i=1}^{n} c_i \phi_i(x)$ and notice that equation (iii) is true, particularly for every function $\phi_j(x)$, j = 1,…,n, we get n equations; namely,

$$a(\sum c_i \phi_i, \phi_j) = (F, \phi_j) \quad \text{for every } j = 1, 2, \ldots, n.$$

Using the linearity of $a(\cdot, \phi_j)$ leads to n linear equations in n unknowns

$$\sum_{i=1}^{n} c_i a(\phi_i, \phi_j) = (F, \phi_j) \quad \text{for every } j = 1{,}2{,}\ldots, n,$$

which can be written in a matrix form as

$$Ac = (F_n)\,, \tag{iv}$$

where $(F_n)_i = (F, \phi_j)$ and $A = (a_{ij})$ is a symmetric matrix given by $a_{ij} = a_{ji} = a(\phi_i, \phi_j)$ and $c = (c_1, \ldots, c_n)^T$.

The entries of the matrix A can be computed explicitly. We first notice that $a_{ij} = a_{ji} = a(\phi_i, \phi_j) = 0$ if $|i-j| \geq 2$. (This is due to the local support of the function $\phi_i(x)$). A direct computation now leads to

$$\begin{aligned}
a_{j,j} = a(\phi_j, \phi_j) &= \int_{x_{j-1}}^{x_j} \left[\frac{1}{h_j^2} + \frac{(x - x_{j-1})^2}{h_j^2}\right] dx \\
&+ \int_{x_j}^{x_{j+1}} \left[\frac{1}{h_{j+1}^2} + \frac{(x_{j+1} - x)}{h_{j+1}^2}\right]^2 dx \\
&= \left[\frac{1}{h_j} + \frac{1}{h_{j+1}}\right] + \frac{1}{3}[h_j + h_{j+1}] \\
a_{j,j-1} &= \int_{x_{j-1}}^{x_j} \left[-\frac{1}{h_j^2} + \frac{(x_j - x)}{h_j}\frac{(x - x_{j-1})}{h_j}\right] dx = -\frac{1}{h_j} + \frac{h_j}{6}.
\end{aligned}$$

Hence, system (iv) can be written as

$$\begin{pmatrix} a_1 & b_1 & & 0 \\ b_1 & \ddots & \ddots & \\ & \ddots & \ddots & b_{n-1} \\ 0 & & b_{n-1} & a_n \end{pmatrix} \begin{pmatrix} c_1 \\ c_2 \\ \vdots \\ c_n \end{pmatrix} = \begin{pmatrix} (F_n)_1 \\ (F_n)_2 \\ \vdots \\ (F_n)_n \end{pmatrix}, \tag{v}$$

where, $a_j = \frac{1}{h_j} + \frac{1}{h_{j+1}} + \frac{1}{3}[h_j + h_{j+1}]$ and $b_j = -\frac{1}{h_j} + \frac{h_j}{6}$. In the special case of uniform grid $h_j = \mathrm{h} = \frac{1}{(n+1)}$, the matrix A then takes the form

$$\mathbf{A} = \frac{1}{h} \begin{pmatrix} 2 & -1 & \ddots & & 0 \\ -1 & 2 & \ddots & \ddots & \\ & \ddots & \ddots & \ddots & -1 \\ 0 & & \ddots & -1 & 2 \end{pmatrix} + \frac{h}{6} \begin{pmatrix} 4 & 1 & 0 & 0 \\ 1 & \ddots & \ddots & \\ & \ddots & \ddots & 1 \\ 0 & & 1 & 4 \end{pmatrix}. \tag{vi}$$

Methods for solving (v) are given in Appendix 7.5.

Chapter 4

Monte Carlo Methods

The written history of the Monte Carlo methods began in 1949 after the paper of Metropolis and Ulm [1949], but at that time the method had already been used for several years in secret defense projects of the United States of America for simulating the behaviour of nuclear reactors. It was definitely known to J. Von Neumann, N. Metropolis, S.M. Ulm, H. Kahn and their coworkers at the Los Alamos Scientific Laboratory much before its publication. The name "Monte Carlo" comes from the random or chance character of the method and the famous casino in Monaco (South of France). It may be observed that finding random numbers is a crucial step of the Monte Carlo methods. The numbers which are generated at the roulette table in Monte Carlo are truly random, but the numbers which are actually utilized in Monte Carlo methods are generated on the computer using deterministic formulae which are called pseudo-random numbers or quasi-random numbers. A fundamental difficulty of the Monte Carlo method stems from the requirement that the nodes be independent random samples. The problem arises for generating independent random samples concretely. The users of Monte Carlo methods avoid this problem by using pseudo-random numbers in place of truly random samples. The first paper with the term quasi-Monte Carlo came to light by the Tech-report in 1951 by Richtmyer. This method can be described in a simple way as the deterministic version of a Monte Carlo method in the sense that the random samples in the Monte Carlo method are replaced by well-selected deterministic points. An excellent exposition of the quasi-Monte Carlo method can be found in Niederreiter [1992,1978] and Halton [1970]. Application-oriented treatment can be found in other references especially Koonin et al. [1990], Press et al. [1992] and Bratley, Fox and Niederreiter [1994]. For an updated account of the Monte Carlo and quasi-Monte Carlo methods, we refer to Caflish [1998]. The particle methods which are used to simulate rarefied gas flows now-a-days have a lot in common with the original work of Metropolis and Ulm, see, for example, Neunzert and Struckmeier [1997]. The general idea of the particle methods is the approximation of densities by discrete measure. An elegant description of the particle method along with applications is presented in Neunzert and Struckmeier [1995]. The Industrial Mathematics group of the University of Kaiserslautern led by Neunzert has applied the particle methods in areas like:

(i) Space flight computation: Simulation of pressure probes in space flight experiments, chemical effects in rarefied gas flows during the reentry phase of space vehicles.

(ii) Traffic flow simulation models used for the implementation in a traffic guidance systems on highways.

(iii) Glass: Investigations on the influence of radiation on the glass melting process.

(iv) Free surface flow as in air-paper industries and granular flow.

Details of these applications can be found in the report of the Industrial Mathematics Group of the Kaiserslautern University website
URL: http://www.mathematik.uni-kl.de/ harvvest/bzw.
http://www.mathematik.uni-kl.de/ wwwtecm/

Deficiencies of the Monte Carlo method

(i) There is only a probabilistic bound on the integration error.

(ii) The regularity of the integrand is not reflected.

(iii) Generating random samples is difficult.

The main goal of this chapter is to give a brief introduction of Monte Carlo, quasi-Monte Carlo, and Particle methods along with their typical applications. Section 4.1 is devoted to the Monte Carlo method while Section 4.2 deals with the quasi-Monte Carlo method. The particle methods are introduced in Section 4.3. A current study of the particle methods is mentioned in Section 4.4. Problems are mentioned in Section 4.5. Applications of the quasi-Monte Carlo method to the pricing of derivatives are discussed by Ninomiya and Tezuka [1996], Papageorgiou and Traub [1996], Paskov and Traub [1995], Tezuka [1998] and Wilmott [Chapter 49, 1998]. An introduction to the Black-Scholes financial world is presented in Appendix 7.6.

4.1 Monte Carlo method

4.1.1 Motivation

First, let us recall classical methods of numerical integration.

(i) The rectangular rule. Let

$$f : R \to R \text{ and } I = \int_a^b f(x)dx .$$

Divide the interval $[a, b]$ into n equal subintervals by the points $a = x_0 \le x_1 \le x_2 \le \cdots\cdots \le x_n = b$ such that if the length is '$b - a$', then each subinterval is of length $h = \frac{b-a}{n}$ and $x_i = a + ih, i = 0, 1, 2 \cdots, n$ and $x_{i+1} - x_i = h, i = 0, 1, 2 \cdots, n$. Let $y_i = f(x_i), i = 0, 1, 2 \cdots, n$.

To find I is equivalent to finding the area enclosed by the graph of the function between 'a' and 'b' which is approximately equal to $\sum_{i=1}^{n} hf(x_i)$ denoted by I_R. We have the error between I and I_R denoted by E_f.

$$I - I_R = \int_a^b f(x)dx - \frac{(b-a)}{n}\sum_{i=0}^{n-1} f(x_i), \tag{4.1}$$

or,

$$E_f = I - \frac{b-a}{n}\sum_{i=0}^{n-1} f(x_i)\,. \tag{4.2}$$

It can be easily checked that the error E_f is proportional to $\frac{h}{2}(b-a)f'(\eta)$ for some $\eta \in (a, b)$. In view of this E_f is exact; that is, $E_f = 0$ if f is a constant function. Further, the error is inversely proportional to n and so, for a given function f, the error will decrease as the number of subintervals will increase, that is, n will increase.

(ii) The trapezoidal rule.

$$y_i = f(x_i)$$
$$x_{i+1} - x_i = h$$
x_i's are as in the previous case.

In the trapezoidal rule, we choose

$$\left\{\begin{aligned} I_T &= \text{Approximate integral} = h\left[\frac{f(x_0) + f(x_1)}{2}\right] \\ &\quad + h\left[\frac{f(x_1) + f(x_2)}{2}\right] + \cdots\cdots + h\left[\frac{f(x_{n-1}) + f(x_n)}{2}\right] \\ &= \frac{1}{2}h\left[f(x_0) + 2f(x_1) + 2f(x_2) + \cdots\cdots + 2f(x_{n-1}) + f(x_n)\right] \\ &= \frac{b-a}{2}\sum_{i=0}^{h} t_i f(t_i)\,, \end{aligned}\right. \tag{4.3}$$

$t_i = 1$ for $i = 0$ and n and $t_i = 2$ for $i \ne 0$ and n

$$I = I_T + E_f\,. \tag{4.4}$$

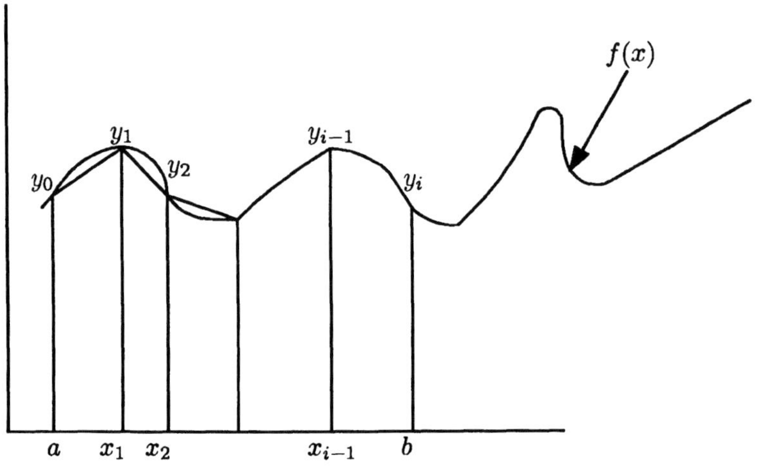

Figure 4.1. The trapezoidal rule.

It can be checked that the error in the trapezoidal rule, E_f is proportional to $-\frac{h^2}{12}(b-a)f''(\eta), \eta \in (a,b)$. The rule is exact for polynomials of degree less than or equal to 1.

(iii) Quadrature rules. A common feature of the above two rules is that the intervals (a,b) are subdivided in such a way that the x_k's, $k = 0, 1, 2, \cdots, n$ are equally spaced. Such a choice is not always possible, for example, the integrand could be singular; that is, $f(x)$ could behave as $(x-b)^\alpha$ near $x = b$, with $\alpha > -1$. A more refined approach is to try and find the values of x_k for which the error is minimized. Methods based on this approach are called the quadrature rules. Let

$$I = \int_a^b f(x)dx = \int_a^b w(x)g(x)dx, \tag{4.5}$$

where $w(x)$ is a non-negative integrable function, called weight function and $g(x) = \frac{f(x)}{w(x)}$.

We approximate the right hand side integral of (4.5) by $\sum_{i=1}^{n} w_i \; g(x_i)$ where the weight w_i and the abscissae can be found from standard literature. It can be seen that x_i's are zeros of the $(n+1)$-th degree polynomial which is orthogonal to all polynomials of a lower degree with respect to the weight $w(x)$ over $[a,b]$. The error involved, that is, $I - \sum_{i=0}^{n} w_i \; g(x_i)$, is proportional to the $(2n+2)$-th derivative of $g(x)$. The method is exact, that is, the error is zero for all polynomials of degree

$\leq (2n+1)$. For details see, for example, Stroud [1974].

Remark 4.1. (i) Until late forties, the above methods were known for finding approximate values of integrals. The quadrature rules provided fairly good approximations for integrals in lower dimensions, that is, for functions of one or two variables, but in higher dimensions, the error was high and this motivated the invention of a new technique for evaluating integrals numerically.

(ii) Suppose, for evaluating an integral with a quadrature rule requires n points, that is, to obtain sufficient accuracy we need to consider a quadrature approximation based on some n-th degree orthogonal polynomial. Then the analogous m-dimensional integral would require n points in each dimension, that is, we would need n^m points in total. In view of this, the quadrature rules are not appropriate for evaluating integrals numerically as it will involve large points.

(iii) The order of approximation for the m-dimensional quadrature rules is

$$O(n^{-2/m}).$$

4.1.2 Monte Carlo method in m-dimensional Euclidean space

Let B be a subset of R^m satisfying $0 < \mu(B) < \infty$, where $\mu(\cdot)$ is a Lebesgue measure on R^m, K is a probability space with the probability measure $d\lambda = du/\mu(B)$, then for $f \in L_1(\mu)$, we have

$$\int_B f(u)\, du = \mu(B) \int_B f \, d\,\lambda = \mu(B)E(f), \tag{4.6}$$

where $f : R^m \to R$, and $E(f)$ is the expected value of the random variable f. The problem of numerical integration is nothing but finding approximate value of expected value of a random variable. This concept can be developed in a more general setting, namely, an arbitrary probability space $(A, \mathcal{A}, \lambda)$. The Monte Carlo estimate for the expected value $E(f)$ is obtained by taking N independent λ-distributed random samples $a_1, a_2, \cdots, a_n \in A$ and letting

$$E(f) \simeq \frac{1}{n} \sum_{k=1}^{n} f(a_k)\,. \tag{4.7}$$

The strong law of large numbers guarantees that this procedure converges almost certainly in the sense that

$$\frac{1}{n}\sum_{k=1}^{n} f(a_k) = E(f) \qquad \lambda^\infty - \text{a.e.} \tag{4.8}$$

where λ^∞ is the product measure of denumerable many copies of λ. The variance of f is denoted by $\sigma^2(f) = \displaystyle\int_A (f - E(f))^2 d\,\lambda$.

It can be proved that [Niederreiter, 1992] for each $f \in L_2(\lambda)$,

$$\int_A \int_A \cdots \int_A (\frac{1}{n}\sum_{k=1}^{n} f(a_k) - E(f))^2 d\lambda(a_1) d\, \lambda(a_2) \cdots d\, \lambda(a_n) = \frac{\sigma^2(f)}{n}. \tag{4.9}$$

(4.7) and (4.9) state that $\left| E(f) - \frac{1}{n}\sum_{k=1}^{n} f(a_n) \right|$ is on an average $\sigma(f)n^{-\frac{1}{2}}$, where $\sigma(f) = (\sigma^2(f))^{\frac{1}{2}}$ is the standard deviation of f. The probabilistic information about the error is obtained from the Central Limit Theorem, which states that if $0 < \sigma(f) < \infty$, then

$$\lim_{n\to\infty} \text{Prob}\left(\frac{c_1\sigma(f)}{\sqrt{n}} \leq \frac{1}{n}\sum_{i=1}^{n} f(a_i) - E(f) \leq c_2 \frac{\sigma(f)}{\sqrt{n}} \right)$$
$$= \frac{1}{\sqrt{2\pi}} \int_{c_1}^{c_2} e^{-t^2/2}\, dt,$$

for any constants $c_1 < c_2$, where $\text{Prob}(\cdot)$ is the λ^∞-measure of the set of all sequences $a_1, a_2 \cdots, a_n$ of elements of A that have the property indicated between the parentheses.

By (4.6) and (4.8), we obtain the **Monte Carlo estimate**

$$\int_B f(x)dx \simeq \frac{\mu(B)}{n}\sum_{k=1}^{n} f(x_k), \tag{4.10}$$

where $x_1, x_2 \cdots, x_n$ are n independent λ-distributed random samples from B and

$$\left| \int_B f(x)dx - \frac{\mu(B)}{n}\sum_{k=1}^{n} f(x_k) \right| \tag{4.11}$$

is on the average $\mu(B)\sigma(f)n^{-\frac{1}{2}}$.

Remark 4.3. (i) In view of (4.11) or, on the basis of the Central Limit Theorem, the Monte Carlo method for numerical integration yields a probabilistic error bound of the form $0(n^{-\frac{1}{2}})$ in terms of the number n of nodes. The most important fact is that this order of magnitude does not depend on the dimension (number of variables of the function). In view of this, it is preferable over quadrature rules in higher dimensions, especially in dimension ≥ 5.

(ii) If the domain of integration B is so complicated that we cannot calculate $\mu(B)$, then by change of variables, it suffices to consider the situation where B is contained in the unit cube $\overline{I}^m$. Then we can write

$$\int_B f(x)dx = \int_{\overline{I}^m} f(x)\chi_B(x)dx, \tag{4.12}$$

where χ_B is the characteristic function of B.

If the integral in (4.12) is estimated according to (4.10), then the Monte Carlo estimate is

$$\int_B f(x)dx \simeq \frac{1}{n}\sum_{k=1}^{n} f(x_k), \quad x_k \in B, \tag{4.13}$$

where $x_1, x_2 \cdots, x_n$ are n independent random samples from the uniform distribution on $\overline{I}^m$. The error estimation will be of the order $n^{-\frac{1}{2}}$ as in the above case.

As discussed above, the Monte Carlo method comprises the following basic elements:

(i) The statistical modelling of the given numerical problem in terms of the random variable which is nothing but writing the integral in terms of the expected value of a random variable.

(ii) The analysis of random variables of the model with regard to the statistical properties, such as law of distribution, statistical dependence or independence.

(iii) The generation of random samples reflecting the above mentioned statistical properties.

As mentioned earlier, the Monte Carlo Methods have been used in fairly large number of real-world problems in such fields as semi-conductor devices, transport theory, structural mechanics, reliability theory, system theory. However, in view of certain deficiencies of the method mentioned earlier, the 'quasi-Monte Carlo method' in which the points $x_1, x_2 \cdots, x_n$ are deterministic has been investigated systematically in the last two decades. For advantages of the new method, we refer to Morokoff and Caflisch [1995, 1993] and Sarkar and Prasad [1987]. It has been shown, for example, that there is a significant improvement in both magnitude of error and convergence rate over Monte Carlo simulation for certain low-dimensional problems. In 1988, an improved Monte Carlo scheme has been developed by the group headed by Neunzert which reduces fluctuations and is often referred to as the Kaiserslautern Monte Carlo (KMC) scheme. A fully deterministic method for solving the Boltzmann equation is proposed in Lêcot [1991]. Some recent advances in this area can be found in Neunzert et al. [1995, 1996] and some of these results are mentioned in Sections (4.3) and (4.4).

4.2 Quasi-Monte Carlo methods

4.2.1 Basic results

The basic idea of a quasi-Monte Carlo method is to replace random samples in a Monte Carlo method by properly chosen deterministic points. The deterministic sequence of points $\{x_n\}$ should be chosen judiciously so as to guarantee small errors between $\frac{1}{n}\sum_{k=1}^{n} f(x_k)$ and $\int_B f(u)du$. The usage of this method has been highlighted in Scientific American, Science and SIAM News in the last couple of years. This method is also an excellent example of merging disciplines. Hardy and Ramanujan would have never imagined that the techniques of Number Theory would be real tools to solve real-world problems via the quasi-Monte Carlo method. In all important cases, the selection criterion for deterministic points leads to the concepts of **uniformly distributed sequence** and **discrepancy**. We introduce here a quasi-Monte Carlo method and discuss the role of the discrepancy in this method indicating deterministic bounds for the integration error in terms of discrepancy.

For the sake of convenience and better understanding, we consider $\overline{I}^m = [0,1]^m =$ m-products of $[0,1]$, the closed m-dimensional unit cube as the domain of integration. We consider the following **quasi-Monte Carlo approximation** for the integral $\int_{\overline{I}^m} f(x)dx$; that is,

$$\int_{\overline{I}^m} f(x)dx \simeq \frac{1}{n}\sum_{k=1}^{n} f(x_k), \tag{4.14}$$

where $x_1, x_2, \cdots, x_n \in \overline{I}^m$.

We are interested in those points of $\overline{I}^m$ for which the quasi-Monte Carlo approximation is convergent; that is,

$$\int_{\overline{I}^m} f(x)dx = \lim_{n\to\infty} \frac{1}{n}\sum_{k=1}^{n} f(x_k)\ . \tag{4.15}$$

A sequence of points $x_1, x_2, \cdots, x_n$ in $\overline{I}^m$ is called **uniformly distributed** if

$$\lim_{n\to\infty} \frac{1}{n}\sum_{k=1}^{n} \chi_B(x_k) = \mu(B), \tag{4.16}$$

for all subintervals B of $\overline{I}^m$, where χ_B is the characteristic function of B and μ denotes the m-dimensional Lebesgue measure.

It may be observed that (4.15) holds even for Riemann integrable functions f on $\overline{I}^m$ in case the sequence $\{x_k\}$ is uniformly distributed. Broadly speaking, the concept of uniform distribution means that the points $x_1, x_2, \cdots, x_k$ are "evenly distributed" over $\overline{I}^m$. For an arbitrary subset K of $\overline{I}^m$, we define

$$H(K,P) = \sum_{k=1}^{n} \chi_B(x_k), \tag{4.17}$$

where P is the set of points of $x_1, x_2, \cdots, x_n \in \overline{I}^m$ and χ_B is the characteristic function of B.

$H(K,P)$ is the counting function that indicates the number $k, 1 \leq k \leq n$, for which $x_k \in K$. If K is a non-empty family of Lebesgue measurable subsets of $\overline{I}^m$, then **the discrepancy of the point set** P is defined as

$$D_n(K,P) = \sup_{B\in K} \left| \frac{H(B,P)}{n} - \mu(B) \right|, \tag{4.18}$$

where μ is the Lebesgue measure of B.

It can be easily seen that $0 \leq D_n(B,P) \leq 1$. The star discrepancy $D_n^\star(P) = D_n^\star(x_1, x_2, \cdots, x_n)$ of the points set P of points $x_1, x_2, \cdots, x_n$ is defined as $D_n^\star(P) = D_n(K,P)$, , where K is the family of all subintervals of $\overline{I}^m$ of the form $\prod_{i=1}^{m} [0, u_i)$, $\overline{I}^m$ = product of m semi-closed intervals

$$[0,1) = \overbrace{[0,1) \times [0,1) \times \cdots \times [0,1)}^{m-\text{times}} .$$

The extreme discrepancy $D_n(P) = D_n(x_1, x_2, \cdots, x_n)$ of P is defined as $D_n(P) = D_n(K,P)$, where K is the family of all subintervals of $\overline{I}^m$ of the form $\Pi_{i=1}^m [u_i, v_i)$.

The following results have been proved see, for example, Niederreiter [1992].

Lemma 4.1.

(i) *For any P consisting of points in $\overline{I}^m$, we have*

$$D_n^\star(P) \leq D_n(P) \leq 2^m D_n^\star(P) . \tag{4.19}$$

(ii) *If $0 \leq x_1 \leq x_2 \leq \cdots\cdots x_n \leq 1$, then*

$$D_n^\star(x_1, x_2, \cdots, x_n) = \frac{1}{2\,n} + \max_{1 \leq k \leq n} \left| x_k - \frac{2k-1}{2n} \right| . \tag{4.20}$$

(iii) *If $0 \leq x_1 \leq x_2 \leq \cdots\cdots \leq x_n \leq 1$, then*

$$D_n(x_1, x_2 \cdots, x_n) = \frac{1}{n} + \max_{1 \leq k \leq n} \left(\frac{k}{n} - x_k \right) - \min_{1 \leq k \leq n} \left(\frac{k}{n} - x_k \right) . \tag{4.21}$$

(iv) *For a sequence S of points of $\overline{I}^m$, if $D_N(S)$ and $D_N^\star(S)$ denote the discrepancy and star discrepancy of the first n terms of S, respectively, then the following statements are equivalent:*

(a) *S is uniformly distributed in $\overline{I}^m$*

(b) $\lim_{n \to \infty} D_n(S) = 0$

(c) $\lim_{n \to \infty} D_n^\star(S) = 0$.

Remark 4.4. It is clear from the above lemma that the discrepancy and the star discrepancy are quantifications of the concept of distributed sequences in $\overline{I}^m$. For a domain of integration more general than $\overline{I}^m$, the interested reader may find details in Niederreiter [1992 and 1978] and references therein.

We prove here two theorems on the error bounds for the quasi-Monte Carlo approximation in one dimension and state the extension of these results in higher dimensions without proofs which are nicely presented in Niederreiter [1992].

Theorem 4.1. *Let f be a function of bounded variation with the total variation $V(f)$ on $[0,1]$; then, for any $x_1, x_2, \cdots, x_n \in [0,1]$, we have*

$$\left| \frac{1}{n} \sum_{k=1}^{n} f(x_k) - \int_0^1 f(x)dx \right| \leq V(f) D_n^{\star}(x_1, x_2, \cdots, x_n) . \tag{4.22}$$

Theorem 4.2. *If f is continuous on $[0,1]$ and $w(f,t) = \sup_{u,v \in [0,1]} |f(u) - f(v)|, t \geq 0, |u - v| \leq t$ denote its modulus of continuity then, for any $x_1, x_2, \cdots, x_n \in [0,1]$,*

$$\left| \frac{1}{n} \sum_{k=1}^{n} f(x_k) - \int_0^1 f(x)dx \right| \leq \omega\left(f, D_n^{\star}(x_1, x_2, \cdots, x_n)\right) . \tag{4.23}$$

Proof of Theorem 4.1. Let us suppose that $x_1 \leq x_2 \leq x_3 \leq \cdots \leq x_n, x_0 = 0$ and $x_{n+1} = 1$. By applying the formula for summation by parts (Abel's rule) and integration by parts, we obtain

$$\left\{ \begin{aligned} \frac{1}{n}\sum_{k=1}^{n} f(x_k) - \int_0^1 f(x)dx &= -\sum_{k=0}^{n} \frac{k}{n} \left(f(x_{k+1}) - f(x_k)\right) + \int_0^1 x df(x) \\ &= \sum_{k=0}^{n} \int_{x_k}^{x_{k+1}} (x - \frac{k}{n}) df(x) . \end{aligned} \right. \tag{4.24}$$

By relation (4.20), for fixed k with $0 \leq k \leq n$, we have

$$|x - \frac{k}{n}| \leq D_n^{\star}(x_1, x_2, \cdots, x_n) \text{ for } x_k \leq x \leq x_{k+1} . \tag{4.25}$$

By (4.24), (4.25) and definition of the total variation, we get the desired result.

Proof of Theorem 4.2. Let $x_1 \leq x_2 \leq \cdots \leq x_n$. By the Mean Value Theorem for integrals, we have

$$\int_0^1 f(x)dx = \sum_{k=1}^{h} \int_{(k-1)/n}^{k/n} f(x)dx = \frac{1}{n}\sum_{k=1}^{n} f(t_k),$$

with $(k-1)/n < t_k < k/n$. This implies that

$$\frac{1}{n}\sum_{k=1}^{n} f(x_k) - \int_0^1 f(x)dx = \frac{1}{n} \sum_{k=1}^{n} (f(x_k) - f(t_k)) . \tag{4.26}$$

By (4.20), $|x_k - t_k| \le D_n^\star(x_1, x_2, \cdots, x_n)$ for $1 \le k \le n$ and the result follows from (4.26).

The variation of f on $\overline{I}^m$ in the sense of Vitali is defined by

$$V^{(m)}(f) = \sup_P \sum_{J \in P} |\Delta(f, J)|\,, \tag{4.27}$$

where $\Delta(f, J)$ is an alternating sum of the values of f at the vertices of a subinterval J of $\overline{I}^m$ (function values at adjacent vertices have opposite signs) and the supremum is taken over all partitions P of $\overline{I}^m$ into subintervals. More conveniently, it is defined as

$$V^s(f) = \int_0^1 \cdots \int_0^1 \left| \frac{\partial^m f}{\partial u_1 \partial u_2 \cdots \partial u_m} \right| du_1, \cdots, du_m\,, \tag{4.28}$$

and holds whenever the indicated partial derivative is continuous on $\overline{I}^m$.

For $1 \le k \le s$ and $1 \le i_1 < i_2 < \cdots < i_k \le s$, let $V^{(k)}(f, i_1, i_2, \cdots, i_k)$ be the variation in the sense of Vitali of the restriction of f to the k-dimensional face

$$\left\{(u_1, u_2, \cdots, u_m) \in \overline{I}^m / u_j = 1 \text{ for } j \ne i_1, \cdots, i_k\right\}\,.$$

Then

$$V(f) = \sum_{k=1}^{s} \sum_{1 \le i_1 < i_2 < \cdots < i_k \le s} V^{(k)}(f, i_1, i_2, \cdots, i_k) \tag{4.29}$$

is called the variation of f on $\overline{I}^m$ **in the sense of Hardy and Krause**, and f is of bounded variation in this sense if $V(f)$ is finite.

Theorem 4.3 [Koksma-Hlawka Inequality]. *Let $\overline{I}^m = ([0,1])^m$ = m-Cartesian product of $[0,1]$. If f has a bounded variation $V(f)$ on $\overline{I}^m$ in the sense of Hardy and Krause then, for any $x_1, x_2, \cdots, x_n \in \overline{I}^m$, we have*

$$\left| \frac{1}{n} \sum_{k=1}^{n} f(x_k) - \int_{\overline{I}^m} f(x)dx \right| \le V(f) D_n^\star(x_1, x_2 \cdots, x_n)\,. \tag{4.30}$$

Theorem 4.4. *For any $x_1, x_2, \cdots, x_n \in \overline{I}^m$ and any $\epsilon > 0$, there exists a function $f \in C^\infty(\overline{I}^m)$ with $V(f) = 1$ and*

$$\left| \frac{1}{n} \sum_{k=1}^{n} f(x_k) - \int_{\overline{I}^m} f(x)dx \right| > D_n^\star(x_1, x_2, \cdots, x_n) - \epsilon\,. \tag{4.31}$$

For a continuous function f on $\overline{I}^m$, we define its modulus of continuity by

$$\omega(f; t) = \sup_{\|u - v\| \le t, u, v \in \overline{I}^m} |f(u) - f(v)| \text{ for } t \ge 0,$$

where $\| u \| = \max_{1 \le i \le m} |u_i|$ *for* $u = (u_1, u_2, \cdots u_m) \in R^m$.

Theorem 4.5. *If f is continuous on* $\overline{I}^m$*, then for any* $x_1, x_2, \ldots, x_n \in \overline{I}^m$*, we have*

$$\left| \frac{1}{n} \sum_{k=1}^{n} f(x_k) - \int_{\overline{I}^m} f(u)du \right| \le 4\omega(f, D_n^\star(x_1, x_2, \cdots, x_n)^{\frac{1}{m}}) . \tag{4.32}$$

Remark 4.5. The error estimates given in the theorems mentioned above indicate that the point sets with small star discrepancy guarantee small errors in the quasi-Monte Carlo integration over $\overline{I}^m$.

4.2.2 Properties of discrepancy and star discrepancy

Since the upper and lower bounds of the error of an integral by the quasi-Monte Carlo sum are given in terms of discrepancy and star discrepancy, it is essential to know their properties in order to study the applications of the quasi-Monte Carlo method for numerical integration.

A point set P consisting of n elements of $\overline{I}^n$ is called a **low-discrepancy** point set if $D_n^\star(P)$ or $D_n(P)$ is small. A sequence S of elements of I^m for which $D_n^\star(S)$ or $D_n(S)$ are small for all $n \ge 1$ is called a **low-discrepancy sequence.** Low-discrepancy point sets/sequences are called **quasi-random points/quasi-random sequences.**

From these discussions, it is quite clear that the accuracy of the quasi-Monte Carlo method will depend on the selection of quasi-random sequences, that is, low-discrepancy sequences. Four classes of such sequences, namely, van der Corput sequence, generalized van der Corput sequence or Faure sequence, Halton sequence (Higher-dimensional analogue of van der Corput sequence) and Hammersly sequence are introduced here.

For an integer $b \ge 2$, we put $Z_b = \{0, 1, \ldots, b-1\}$, that is, Z_b is the least residual system mod b. Every integer $n \ge 0$ has a unique digit expansion

$$n = \sum_{j=0}^{\infty} a_j(n) b^j \tag{4.33}$$

in base b, where $a_j(n) \in Z_p$ for all $j \ge 0$ and $a_j(n) = 0$ for all sufficiently large j, that is, the sum in (4.33) is actually finite.

For an integer $b \ge 2$, the **radical inverse function** ϕ_b **in base** b is defined by

$$\phi_b(n) = \sum_{j=0}^{\infty} a_j(n) b^{-j-1} \quad \text{for all integers } n \ge 0, \tag{4.34}$$

where n is given by its digit expansion (4.33) in base b.

$\phi_b(n)$ is obtained from n by a symmetric reflection of the expansion (4.33) in the decimal point. It is clear that $\phi_n(b) \in I = [0, 1)$ for all $n \geq 0$.

In the construction of many low-discrepancy sequences, it is convenient to index the terms by $n = 0, 1, 2 \ldots$. In one-dimensional low-discrepancy sequences S, we denote the terms by $x_0, x_1, \ldots$ and write

$$D_n(S) = D_n(x_0, x_1, \cdots, x_{n-1}),$$

for the discrepancy of the first n terms of S. The star discrepancy can be written similarly.

van der Corput sequence in base b. For an integer $b \geq 2$, the **van der Corput sequence** in base b is the sequence $x_0, x_1, \ldots$ with $x_n = \phi_b(n)$ for all $n \geq 0$. A sequence $x_0, x_1, x_2 \ldots$ is called the **generalized van der Corput sequence** in base b, if x_n's are given by

$$x_n = \sum_{j=0}^{\infty} \sigma(a_j(n)) b^{-j-1} \quad \text{for all } n \geq 0, \tag{4.35}$$

where n is given by (4.33).

van der Corput had studied the case $b = 2$, but the general case was studied by Faure and therefore the general case is often called the **Faure sequence.**

For a given dimension $m \geq 1$, let $b_1, b_2, \ldots, b_m$ be integers ≥ 2. A sequence $x_0, x_1, x_2, \cdots$ in $\overline{I}^m$, where

$$x_k = (\phi_{b_1}(k), \phi_{b_2}(k), \cdots, \phi_{b_m}(k)) \text{ for all } k \geq 0, \tag{4.36}$$

and ϕ_b are given by (4.34), is called the **Halton sequence** in the bases $b_1, b_2, \cdots, b_m$.

It is clear that the van der Corput sequence is a special case $m = 1$, of the Halton sequence. A Halton sequence is called the **Hammersley sequence** if an arbitrary term, say, x_k is given by

$$x_k = \left(\frac{k-1}{n}, \phi_{b_1}(k), \phi_{b_2}(k), \ldots, \phi_{b_{m-1}}(k) \right) \in I^m, \tag{4.37}$$

for $k = 0, 1, \ldots, n-1$.

Estimates of star discrepancy and discrepancy.

(i) For the van der Corput sequence S_b

$$D_n^\star(S_b) = O(n^{-1} \log\, n)\,, \tag{4.38}$$

for $n \geq 2$ with a constant depending only on b. For $b = 2$ we have $nD_n^\star(S_2) = nD_n(S_2) \leq \dfrac{\log n}{\log 8} + 1$ for all $n \geq 1$ and

$$\lim_{n\to\infty} \left(n\, D_n(S_2) - \frac{\log n}{\log 8} \right) = \frac{4}{9} + \frac{\log 3}{\log 8}\,. \tag{4.39}$$

(ii) Let S be the Halton sequence in the pairwise relatively prime bases $b_1, b_2, \ldots, b_m$, then

$$D_n^\star(S) < \frac{m}{n} + \frac{1}{n}\Pi_{k=1}^m \left(\frac{b_k - 1}{2\log b_k} \log\ n + \frac{b_i+1}{2} \right) \text{ for all } n \geq 1\,. \tag{4.40}$$

(iii) For any dimension $m \geq 1$, there exists an infinite sequence of points in I^m such that

$$D_n = O(n^{-1}\,(\log n)^m)\,. \tag{4.41}$$

Furthermore, for every $n \geq 2$, there exists a finite sequence of n points in I^m such that

$$D_n = O\left(n^{-1}(\log n)^{m-1}\right)\,. \tag{4.42}$$

(iv) For the Hammersley sequence containing n terms, we have

$$D_n^\star(S_n) < \frac{m}{n} + \frac{1}{n}\Pi_{i=1}^{m-1} \left(\frac{b_{i-1}}{2\log b_i} \log n + \frac{b_{i+1}}{2} \right).$$

Remark 4.6. Relation (4.41) is very important in the sense that it guarantees that for any dimension, there exist quasi-Monte Carlo techniques that perform substantially better than the Monte Carlo method.

Example 4.1. The Halton sequence in one dimension is generated by choosing a prime number p and expanding the sequence of integers $0, 1, 2, \cdots, n$ into base p notation. The n-th term of the sequence is given by

$$z_n = \frac{a_0}{p} + \frac{a_1}{p^2} + \frac{a_2}{p^3} + \ldots + \frac{a_m}{p^{m+1}},$$

where the a_i's are integers taken from the base p expansion of $n-1$, $[n_{-1}]_p = a_m a_{m-1} \ldots a_2\ a_1\ a_0$, with $0 \leq a_i < p$. For example, if $p = 3$, the first 12 terms of the sequence ($n = 1, 2, \ldots, 12$) are

$$\left\{0, \frac{1}{3}, \frac{2}{3}, \frac{1}{9}, \frac{4}{9}, \frac{7}{9}, \frac{2}{9}, \frac{5}{9}, \frac{8}{9}, \frac{1}{27}, \frac{10}{27}, \frac{19}{27}\right\}.$$

Example 4.2. The approximation of a surface integral by a low-discrepancy sequence: Let us consider a ball k_R with radius R where the corresponding sphere is denoted by σ_R. Then, we can write

$$\begin{aligned}\int_{\sigma_R} \phi d\omega &= \int_0^{2\pi} \left(\int_0^{\pi} \phi\left(R\sin\theta\cos\phi, R\sin\theta\sin\phi, R\cos\theta\right) R^2 \sin\theta d\theta \right) d\phi \\ &= 2\pi^2 R^2 \int_0^1 \int_0^1 \phi\left(R\sin\pi\theta'\cos 2\pi\phi', R\sin\pi\theta'\sin 2\pi\phi', R\cos\pi\theta'\right) \\ &\quad \times \sin\pi\theta' d\theta' d\phi'\,.\end{aligned}$$

One has to consider a uniform distribution with respect to the variables (θ', ϕ') in $[0,1]^2$, where the integrand is the following:

$$2\pi^2 R^2 \phi(\cdots) \sin \pi\theta' \ .$$

One can also say that it is given a distribution with the corresponding density

$$\rho(\theta', \phi') = \sin \pi\theta' \ ,$$

and integrate $2\pi R^2 \phi(\cdots)$ with respect to this density.

We choose a low-discrepancy sequence in

$$[0,1]^2 : \{(\theta_1', \phi_1'), \cdots, (\theta_N', \phi_N')\},$$

and put

$$\underline{P}_i = \left(R \sin \pi\theta_i' \cos 2\pi\phi_i', R \sin \pi\theta_i' \sin 2\pi\phi_i', R \cos \pi\theta_i'\right) ,$$

such that

$$\frac{2\pi^2 R^2}{N} \sum_{i=1}^{N} \sin \pi\theta_i' \phi(\underline{P}_i)$$

is an approximation for $\int_{\sigma_R} \phi d\omega$.

Now, one has to consider the surface from more spheres and must take into account the weight $2\pi R^2 \sin \pi\theta_i'$.

One selects N points on the sphere σ_1 and another N points on the σ_2 and chooses $2N$ points such that they lie on the union of the surfaces of spheres. For this, one should realize not only the angles but also the radii. See Figure 4.2.

The algorithm is the following: Consider M spheres $\sigma_1, \cdots, \sigma_M$ with radii $(r_1, \cdots, r_M)$. One determines N points on every sphere, where the radius and $\sin \pi\theta_i$ always take, as usual, the weight

$$\left\{(r_i, P_1^{(i)}), \cdots, (r_i, P_N^{(i)})\right\}, i = 1, 2, \cdots, M \ .$$

From NM points, we choose those that lie on $\partial \cup_{\sigma=1}^{M} K_\sigma$.

Remark. One must check if, for a given point $P_j^{(i)}$, the following is true:

$$\|P_j^{(i)} - \underline{x}_k\| \geq r_k,$$

for all $k = 1, \cdots, M$. If the above inequality is true, the point belongs to $\partial \cup_{\sigma=1}^{M} K_\sigma$; otherwise, not.

So, one gets a point set which consists of maximal NM points $(r_i, P_l^{(i)})$, $i \in \{1, \cdots, M\}$, $l \in \{1, \cdots, N\}$. We sum

$$\frac{2\pi^2 r_i^2}{N} \sin \pi\theta_l^{(i)'} \phi\left(P_l^{(i)}\right) \ ,$$

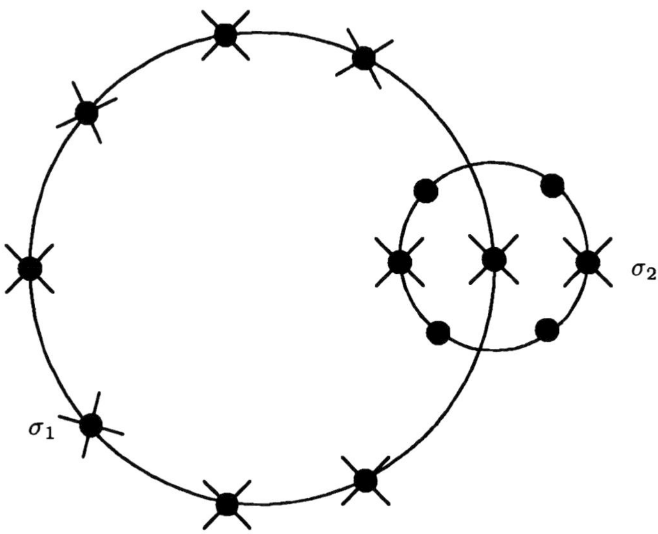

Figure 4.2

over these sets such that we get evidently an approximation for $\int_{\partial \cup K_\sigma} \phi d\omega$.

The evaluation of integrals by the Monte Carlo and quasi-Monte Carlo methods depends on the generation of random numbers (often called pseudo-random numbers) and quasi-random numbers or quasi-random or pseudo-sequences, respectively. We shall briefly mention here how they are generated.

Now-a-days, most computers contain routines that generate random numbers uniformly distributed between 0 and 1. The method of generation most commonly used in the computer routines for the generation of uniformly distributed random numbers is the multiplicative congruential method where the i-th element r_i of the sequence is given by the previous element r_{i-1} by a relation such as

$$\begin{cases} r_i = pr_{i-1}(\text{mod } q), (r_i \text{ is the remainder} \\ \qquad \text{when } pr_{i-1} \text{ is divided by q}), \end{cases} \tag{4.43}$$

where p and q are appropriate constants. The first element of the sequence must be given by the user. The numbers r_i of the sequence in equation (4.43) are obtained with a precise mathematical algorithm and therefore they are not at all random; in reality, given the first element called seed, all other terms are easily predictable. However, for a good choice of the constants p and q, the sequences of r_i behave randomly in the sense that they pass a large number of statistical tests of randomness. Numbers r of such a type are called **pseudo-random numbers** and sequences of such numbers are called **pseudo-random sequences**. These numbers have the advantage over truly random numbers of being generated in a fast way and of being reproducible when desired for program debugging.

Relation (4.43) can be replaced by a more general formula

$$r_i = pr_{i-1} + l(\text{mod q})\,. \tag{4.44}$$

Relations (4.43) and (4.44) are called congruential methods of generating pseudo-random numbers; in particular (4.43) is called **multiplicative congruential method**. For more details, one may see Hammersley and Handscomb [1967] and Press, Teukolsky [1992]. van der Corput, Faure, Halton and Hammersley sequences discussed in the preceding sections are examples of quasi-random sequence. The current research papers of Morokoff and Calflish (see, for example, paper of 1995) and references therein may provide important clues for generating quasi-random sequences.

4.3 The particle methods

4.3.1 Introduction

The particle methods are numerical methods of approximation (discretization) of partial differential equations of the type

$$\frac{\partial u}{\partial t} + \sum_{i=1}^{n} \frac{\partial}{\partial x_i}(a_i u) + a_0 u = g, \quad x \in R^n, \ t > 0, \tag{4.45}$$

where $u(x,0) = u_0(x)$ is known as the hyperbolic equation (in conservation form).

There are important partial differential equations which belong to this class, some of which are:

1. The Boltzmann equation for dilute gases

   $$f_t + v\nabla_x f + F\nabla_v f = \frac{1}{\epsilon}Q(f)\,.$$

   Here, $f = f(t,x,v)$ denotes the density of a dilute gas

   $$f : R_+ \times \Omega \times R^3 \longrightarrow R_+$$

   where $t \geq 0$ is the time variable, $x \in \Omega \subset R^3$ the space coordinate, and $v \in R^3$ the velocity, $F = F(t,x)$ denotes an external force (gravity) or is related to a self-consistent potential Φ such that $F = \nabla\Phi$.

2. Euler equations

   $$\frac{\partial \rho}{\partial t} + \operatorname{div}(\rho u) = 0$$
   $$\frac{\partial \rho u_i}{\partial t} + \operatorname{div}(\rho u_i u) = \frac{\partial P}{\partial x_i}, \ i = 1,2,3,$$

   where $P = P(\rho)$ is the pressure, which is given as a function of ρ.

In analogy to the methods like finite element, finite difference, the particle methods are also known as the point-set methods. These methods can be studied from:

(a) measure-theoretic approach,

(b) number-theoretic approach,

(c) functional analytic techniques, and

(d) statistical methods.

The Monte Carlo and quasi-Monte Carlo are, respectively, essentially statistical and number-theoretic approaches of the particle methods discussed in the previous sections. In Subsection 4.3.2, we introduce the measure-theoretic approach while in Subsection 4.3.3, the functional analytic method is presented. Our presentation here is mainly based on Neunzert and Struckmeier [1995], and Raviart [1983].

4.3.2 Particle approximations – measure-theoretic approach

A particle is characterized by its position x, velocity v and mass (or charge) α. In order to simplify the notation, we put $P = (x, v)$. A particle ensemble (or finite point set) is given by

$$\omega_N = \{(\alpha_1, P_1), \cdots, (\alpha_N, P_N)\},$$

or, in another notation, by

$$\delta_{\omega_N} = \sum_{i=1}^{N} \alpha_i \delta_{P_i}.$$

Here, δ denotes the Dirac measure and N is the number of particles.

We consider sequences of particle ensembles

$$\omega_N^N = \{(\alpha_1^N, P_1^N), \cdots, (\alpha_N^N, P_N^N)\},$$

or

$$\delta_{\omega_N^N} = \sum_{i=1}^{N} \alpha_i^N \delta_{P_i}^N.$$

Often, P_i^N are taken from a sequence of $P_1, P_2 \cdots$, that is, more and more particles are brought into the game. Then

$$\{P_1^N, \cdots, P_N^N\} = \{P_1, P_2, \cdots, P_N\}.$$

Now, for a given density $f \in L_+(R^3)$, we say that $\delta_{\omega_N^N}$ converges to f if

$$\lim_{N\to\infty} \sum_{i=1}^{N} \alpha_i^N \phi(P_i^N) = \int f\phi dv dx,$$

for all $\phi \in C^b(R^3 \times R^3)$. This means that the discrete measure $\delta_{\omega_N^N}$ weak* converges to $fdvdx$, where C^b is a set of all continuous and bounded functions. One may interpret this as an integration rule where we integrate the function ϕ with respect to the measure $fdvdx$. We would like to measure the distance between ω_N^N and f. This might be done by any distance in measure spaces such as the Prohorov metric or bounded Lipschitz distance, but also since the limit $fdvdx$ is absolutely continuous with respect to the Lebesgue measure, with the help of the discrepancy. A pertinent result in this direction is by Neunzert and Wick (see, for example, Babovsky and Illner [1989])

$$\delta_{\omega_N^N} \longrightarrow f \text{ iff } D(\omega_N^N, f) \longrightarrow 0\,.$$

It may be recalled that the bounded Lipschitz distance $d_L(\mu,\nu)$ between two measures μ and ν is defined as

$$d_L(\mu,\nu) = \sup_{\phi \in Lip_1} \left| \int \phi D(\mu,\nu) \right| ,$$

where Lip_1 denotes the set of Lipschitz continuous functions with Lipschitz constant 1, whereas the discrepancy between two measures μ and ν is defined by

$$D(\mu,\nu) = \sup_A |\mu(A) - \nu(A)|\,.$$

Here, A denotes a rectangle parallel to an axis of the underlying space.

Given the measure μ with density f and an appropriate distance between measures, an important question is how to construct δ_{ω_N} such that the distance between μ and δ_{ω_N} is as small as possible if N, the number of particles, is fixed. For the construction of appropriate δ_{ω_N}, one may see Neunzert and Struckmeier [1995].

4.3.3 Functional analytic approach

We present here the particle method of approximation of problem (4.45). Let us assume that $g = 0$, then the method consists in approximating the initial condition u_0 by a linear combination of Dirac measures

$$u_h^0 = \sum_{j \in J} \alpha_j \delta(x - x_j), \tag{4.46}$$

for some set $(x_j, \alpha_j)_{j \in J}$ of points $x_j \in R^n$ and weights $\alpha_j \in R_+$. Replacing u_0 by u_h^0, we look for the solution of the problem

$$\begin{cases} \dfrac{\partial u_h}{\partial t} + \displaystyle\sum_{i=1}^{n} \frac{\partial}{\partial x_i}(a_i u_h) + a_0 u_h = 0\,; \\ \qquad\qquad\qquad\qquad u_h(\cdot, 0) = u_h^0\,. \end{cases} \tag{4.47}$$

By a well-known result of Raviart [1983], (4.47) has a unique solution

$$u_h(x,t) = \sum_{j \in J} \alpha_j(t) \delta(x - X_j(t)),$$

where

$$\begin{aligned} X_j(t) &= X(t; x_j, 0) \\ \alpha_j(t) &= \alpha_j \exp(-\int_0^t a_0(X_j(s), s)ds) . \end{aligned}$$

Thus, we have obtained a particle approximation u_h of the solution u of problem (4.45) in the case $f = 0$. Raviart has also studied the approximation of function by Dirac measures. He has considered the following problem: Given a function $u_0 \in C^0(R^n)$, how to choose

$$u_h^0 = \sum_{j \in J} \alpha_j \delta(x - x_j) ,$$

so that it represents a good approximation of u_0. This problem has to be understood as an approximation in the sense of measure on R^n. Thus, for a given ϕ in $C_0^0(R^n)$, we have to compare

$$\langle u_0, \phi \rangle = \int_{R^n} u_0 \phi dx \quad \text{and} \quad \langle u_h^0, \phi \rangle = \sum_{j \in J} \alpha_j \phi(x_j) .$$

We interpret this now as the classical problem of numerical quadrature. Given a parameter $h > 0$, we cover R^n with a uniform mesh with size h. For all $j = (j_1, j_2, \cdots, j_n) \in Z^n$, we denote by B_j the cell

$$B_j = \{x \in R^n \mid (j_i - \frac{1}{2})h \le x_i \le (j_i + \frac{1}{2})h, \ 1 \le i \le n\} ,$$

and by $x_j = (j_i h)_{1 \le i \le n}$ the center of B_j. Then we set

$$u_h^0 = \sum_{j \in Z^n} \alpha_j \delta(x - x_j) , \tag{4.48}$$

where α_j is some approximation of $\int_{B_j} u_0 dx$. Since $u \in C^0(R^n)$, we can choose

$$\alpha_j = h^n u_0 x_j . \tag{4.49}$$

Raviart [1983] has estimated the error between $\int_{R^n} u_0 \phi dx$ and $\sum_{j \in Z^n} \alpha_j \phi(x_j)$.

4.4 A current study of the particle method

4.4.1 Introduction

In kinetic theory, a gas is assumed to be in an equilibrium state when it does not exchange mass and energy with other bodies, and its state does not change with time. The trend towards a Maxwellian distribution in the space homogeneous case is expressed by Boltzmann's H-theorem. This indicates that this particular distribution is a good candidate to describe a gas in a (statistical) equilibrium state. The parameters, density ρ, mean velocity u and the temperature T of a Maxwellian distribution may depend on the time and space variables. In this case, a Maxwellian distribution is called a **local Maxwellian distribution** or a **local equilibrium state**.

Finding a criterion of local equilibrium is one of the fundamental tasks for coupling of the Boltzmann and hydrodynamics equations. A typical application of coupling of these equations arises in the re-entry of the space shuttle in the atmosphere. In this region, the Boltzmann equation also gives the correct description of the flow but, in the numerical solution, the computational effort is very high. In fact, the mean free path between the molecules is very small and, in any numerical code of the Boltzmann equation like the finite point set method (FPM), the mesh size should be smaller than the mean free path. On the other hand, in this region, the Boltzmann equation can be replaced by the Euler equations if the particle distribution function is a Maxwellian distribution or close to it. But the Euler equations are not valid everywhere, especially in the shock region or on the solid boundary. Therefore, in those regions where Euler equations are not valid, one must solve the Boltzmann equation.

In order to determine the domain of validity of these equations, we need a criterion to determine whether a particle system is near enough to a Maxwellian distribution or not. This means that one needs a distance between a Maxwellian distribution and a particle system.

The most widely used method for simulation of the evolution of a rarefied gas is the particle method. As indicated earlier, the Euler equations can also be solved by the particle method, see Tiwari and Rjasanow [1997].

As mentioned in the previous subsection, the particle methods are applied to evolution equations for densities $f(t,x,v)$, $(x,v) \in \Omega$, of particle number, mass, charge, momentum in the phase space Ω, which is normally the position-velocity space. In other words, particle methods are applied to appropriate conservation laws for quantities given by measure μ, describing the particle number, defined by the relation

$$\mu(A) = \int_A f dx dv,$$

where A denotes a measurable set in Ω. Typically, conservation equations are evolution equations for measures, which are a posteriori transformed into (partial) differential equations for the corresponding density function f.

Since the discrete measure is a sum of Dirac measures, it is not easy to use, for example, the relative entropy which also measures the distance between the particle system and the density function. Another difficulty is that we need a distance in velocity space which is three-dimensional and the number of data is about 50. Therefore, using the statistical tools is not simple. It was proposed by Neunzert that

H^s-norm as a distance between a particle system and the density function should be used. Tiwari and Rjasanow [1997] have effectively implemented his ideas. It was proved in a Ph.D. thesis written under the supervision of Prof. Neunzert (see Schreiner [1994]) that the weak* convergence of δ_{ω_N} to μ is equivalent to

$$\|\delta_{\omega_N} - \mu\|_{H^s} \longrightarrow 0$$

as long as $s < -d/2$, where d is the dimension of the velocity space which is equal to three in our case. Therefore, we choose $s = -2$ for our computation. We present below a summary of the results of Tiwari and Rjasanow [1997].

4.4.2 Derivation of distance

Let $s \in R$, then $H^s(R^d)$, $d = 1, 2, 3$ is the subspace of $S'(R^d)$, the space of tempered distributions f with property

$$(1 + |\xi|^2)^{s/2}\, \hat{f} \in L^2(R^d), \quad \xi \in R^d, \tag{4.50}$$

where $\hat{f}(\xi)$ is the Fourier transform of f defined by

$$\hat{f}(\xi) = \int_{R^d} e^{\,-i<v,\xi>}\, f(v)dv. \tag{4.51}$$

$H^s(R^d)$ is a Hilbert space with scalar product

$$\langle f, g\rangle_{H^s} = \int_{R^d} (1 + |\xi|^2)^s \hat{f}(\xi)\overline{\hat{g}(\xi)}d\xi\,,$$

and the corresponding norm is

$$\|f\|_{H^s} = \left(\int_{R^d} (1 + |\xi|^2)^s |\hat{f}(\xi)|^2 d\xi\right)^{1/2}. \tag{4.52}$$

The Dirac measure $\delta_a = \delta(x - a)$, at $a \in R^d$ belongs to $H^s(R^d)$, if $s < -d/2$.

For $s = m \in N$, the space $H^s(R^d)$ coincides with the Sobolev space $W^{m,2}(R^d)$, where

$$\begin{cases} W^{m,p}(R^d) := \{u \in L^p(R^d) : D^\alpha u \in L^p(R^d), \\ \qquad \text{for } \alpha \leq m, 1 \leq p < \infty, m \geq 0\}. \end{cases}$$

We are interested in calculating the distance (Sobolev norm of the difference) between the local Maxwellian distribution $f_M[\rho, U, T](v)$, which we denote by f_M, and a given particle distribution, which can be interpreted by the discrete measure δ_{ω_N}.

The local Maxwellian distribution function is defined by

$$f_M\,[\rho, U, T]\,(v) = \frac{\rho}{(2\pi RT)^{3/2}}\, e^{-\frac{|v-U|^2}{2RT}}\,, \tag{4.53}$$

where ρ is the density, U the mean velocity, T the temperature and R the gas constant and ρ, U, T depend on time t and position x.

We normalize the Maxwellian distribution by defining the new velocities by

$$w = \frac{v - U}{\sqrt{RT}}\,, \tag{4.54}$$

such that the new temperature is equal to 1 and the new mean velocity is equal to 0. Then the Maxwellian distribution (4.53) is given by

$$f_M(w) = \frac{\rho}{(2\pi)^{3/2}}\, e^{-\frac{|w|^2}{2}}. \tag{4.55}$$

We normalize f_M and δ_{ω_N} to 1, that is, we divide f_M and δ_{ω_N} by the density ρ. We have to compute the H^s-norm for $s < -3/2$ for the normalized Maxwellian distribution

$$f_M(w) = \frac{1}{(2\pi)^{3/2}} e^{-\frac{|w|^2}{2}}, \tag{4.56}$$

and the discrete measure

$$\delta_{\omega_N} = \frac{1}{N}\sum_{j=1}^{N} \delta(w - w_j). \tag{4.57}$$

The Fourier transform of the Maxwellian distribution (4.53) is again a Maxwellian distribution and is given by

$$\hat{f}_M(\xi) = e^{-\frac{|\xi|^2}{2}}\,, \tag{4.58}$$

and the Fourier transform of the discrete measure (4.57) is given by

$$\hat{\delta}_{\omega_N}(\xi) = \frac{1}{N}\sum_{j=1}^{N} e^{-i\langle\ xi, v_j\rangle 1}. \tag{4.59}$$

Since δ_{ω_N} belongs to $H^s(R^d)$ if $s < -\frac{d}{2}$, it is also true that f_M belongs to $H^s(R^d)$ for all $s < -\frac{d}{2}$. As we have already mentioned that we are looking for a distance in the three-dimensional velocity space, a simple choice for s is -2.

We define the distance between f_M and δ_{ω_N} by

$$||f_M - \delta_{\omega_N}||_{H^{-2}} = \left(\int_{R^3} \frac{|\hat{f}_M(\xi) - \hat{\delta}_{\omega_N}(\xi)|^2}{(1+|\xi|^2)^2} d\xi\right)^{1/2}. \tag{4.60}$$

In order to compute (4.60), we consider

$$||f_M - \delta_{\omega_N}||^2_{H^{-2}} = ||f_M||^2_{H^{-2}} - 2\langle f_M, \delta_{\omega_N}\rangle_{H^{-2}} + ||\delta_{\omega_N}||^2_{H^{-2}}\,. \tag{4.61}$$

Now, we compute each term on the right hand side of (4.61). First, we consider

$$||f_M||^2_{H^{-2}} = \int_{R^3} \frac{|\hat{f_M}(\xi)|^2}{(1+|\xi|^2)^2} d\xi = \int_{R^3} \frac{e^{-|\xi|^2}}{(1+|\xi|^2)^2} d\xi\,.$$

Transforming ξ into spherical coordinates

$$\begin{cases} \xi_1 = r\sin\theta\cos\phi,\ \xi_2 = r\sin\theta\sin\phi, \xi_3 = r\cos\theta \\ \quad \times(0 \le r < \infty, 0 \le \theta \le \pi,\ 0 \le \phi \le 2\pi), \end{cases}$$

and integrating with respect to θ and ϕ, we get

$$\|f_M\|^2_{H^{-2}} = 4\pi \int_0^\infty \frac{r^2 e^{-r^2}}{(1+r^2)^2} dr.$$

Integrating by parts, we get

$$\begin{aligned} \|f_M\|^2_{H^{-2}} &= 2\pi \left[3\int_0^\infty \frac{e^{-r^2}}{1+r^2} dr - 2\int_0^\infty e^{-r^2} dr \right] \\ &= 2\pi \left[\frac{3\pi}{2}\, e \operatorname{Erfc}(1) - \sqrt{\pi} \right], \end{aligned}$$

where Erfc(x) is the complementary error function and is related to the error function by the expression

$$\operatorname{Erfc}(x) = 1 - \operatorname{Erf}(x) = 1 - \frac{2}{\sqrt{\pi}} \int_0^x e^{-t^2} dt.$$

Therefore,

$$\|f_M\|^2_{H^{-2}} = 3\pi^2 e \operatorname{Erfc}(1) - 2\pi\sqrt{\pi}. \tag{4.62}$$

After appropriate calculations, we arrive at

$$\begin{aligned} \langle f_M, \delta_{\omega_N} \rangle 1_{H^{-2}} &= \sqrt{e}\frac{\pi^2}{2N} \sum_{j=1}^{N} \left[e^{-|v_j|}(1 - \frac{1}{|v_j|})\operatorname{Erfc}(\frac{1-|v_j|}{\sqrt{2}}) \right. \\ &\quad \left. + e^{|v_j|}(1 + \frac{1}{|v_j|})\operatorname{Erfc}(\frac{1+|v_j|}{\sqrt{2}}) \right] \end{aligned} \tag{4.63}$$

$$\|\delta_{\omega_N}\|^2_{H^{-2}} = \frac{\pi^2}{N^2} \sum_{j=1}^{N} \sum_{l=1}^{N} e^{-|v_j - v_l|}, \tag{4.64}$$

and

$$\|f_M - \delta_{\omega_N}\|_{H^{-2}} = \left(\|f_M\|^2_{H^{-2}} - 2\langle f_M, \delta_{\omega_N} \rangle_{H^{-2}} + \|\delta_{\omega_N}\|^2_{H^{-2}} \right)^{1/2}, \tag{4.65}$$

where $\|f_M\|^2_{H^{-2}}$, $\langle f_M, \delta_{\omega_N} \rangle_{H^{-2}}$, and $\|\delta_{\omega_N}\|^2_{H^{-2}}$ are given by (4.62), (4.63) and (4.64), respectively.

4.4.3 Computational results

As a first test, we measure the distance between the Maxwellian distribution and the particle system generated by it. Table 1 shows the H^{-2}-norm as a distance between the Maxwellian distribution and Maxwellian distributed particles for different numbers of particles.

Table 1. Distance between Maxwellian and its distributed data

Number of particles	H^{-2}-Norm
10	0.7552
50	0.2724
100	0.1794
500	0.0619
1000	0.0397

Table 1 shows that the norm decreases if the number of particles increases. The norm is approximately of order $2/\sqrt{N}$. If we compute the distance between the Maxwellian distribution and non-Maxwellian distributed particles, the distance should be larger than the distance between the Maxwellian distribution and its distributed data. Therefore, we have generated the particles according to the sum of two Maxwellians

$$f(v) = \frac{\rho}{2(2\pi RT)^{3/2}} \left(e^{-\frac{|v-U|^2}{2RT}} + e^{-\frac{|v+U|^2}{2RT}} \right),$$

and then computed the distance between $f(v)$ and the Maxwellian distribution according to such a particle system with the help of H^{-2}-norm. As in the previous case, we have computed for different numbers of particles.

Table 2 shows that the distance between the particle system, generated according to the sum of two Maxwellian distributions and the Maxwellian distribution with parameters given by the particle system, is bigger compared with that in Table 1. In this case also, the norm decreases if the particles number increases but at a very slow rate and is bounded from below.

Table 2. Distance between sum of two Maxwellians and the corresponding Maxwellian

Number of particles	H^{-2}-Norm
10	1.7867
50	1.7516
100	1.7494
500	1.7484
1000	1.7393

4.4.4 Spatially homogeneous Boltzmann equation

One of the properties of the spatially homogeneous Boltzmann equation is the Boltzmann's H-theorem which states that the tendency of a gas is to approach a equilibrium distribution (see Cercignani [1989]). The Maxwellian distribution can be considered as an equilibrium distribution. We consider first the spatially homogeneous Boltzmann equation with non-equilibrium distribution at time $t = 0$.

The spatially homogeneous Boltzmann equation for $f(t, v)$ is given by

$$\frac{\partial f}{\partial t} = J(f, f), \tag{4.66}$$

where

$$\begin{aligned} J(f,f) = &\int_{R^3} \int_{\eta \in S^2_+} |v-u|\sigma(|v-u|,\eta)\{f(t,v')f(t,w') \\ &- f(t,v)f(t,w)\}d\omega(\eta)dw \\ v' &= v - \langle v-w, \eta\rangle\eta \\ w' &= w - \langle w-v, \eta\rangle\eta\,. \end{aligned}$$

We solve equation (4.66) with the following initial condition:

$$f(0,v) = \frac{\rho}{2(2\pi RT)^{3/2}}\left(e^{-\frac{|v-U|^2}{2RT}} + e^{-\frac{|v+U|^2}{2RT}}\right). \tag{4.67}$$

As time t tends to infinity, the solution of (4.66)–(4.67) tends to an equilibrium distribution with the density, mean velocity, and the temperature obtained from the initial function $f(0, v)$. Therefore, $\|f_M - \delta_{\omega_N}\|_{H^{-2}}$, defined in (4.60), decreases to a constant as time t tends to infinity. We verify this behaviour by using the particle code, developed in AG Technomathematik, University of Kaiserslautern. At every time step, we compute the H^{-2}-norm. Figure 4.3 shows the decrease in H^{-2}-norm as time increases. The simulation is done with 100 particles. Note that the asymptotic value for the distance is approximately the same as that between the Maxwellian distribution and the corresponding particles system which can be seen from Table 1.

4.5 Problems

Problem 4.1. Compute $I = \int_0^1 x dx$ by the Monte Carlo method.

Problem 4.2. Show that the sequence $x_n = (n\theta), n = 1, 2, 3\ldots,$ where θ is an irrational number, is uniformly distributed in $[0, 1]$.

Problem 4.3. Let α be the root of polynomial of degree n that has integer coefficients and is irreducible over the rationals. Then show that there exists a constant

$c > 0$ such that, for every pair of integers,

$$p, q \text{ with } q > 0, \quad |\alpha - p/q| \geq \frac{c}{q^n}\,.$$

Problem 4.4. Let $f(u)$ be periodic in $[0,1]$ and be of class $C^3[0,1]$ so that we have $f(0) = f(1), f^{'}(0) = f^{'}(1), f^{''}(0) = f^{''}(1)$. Let θ be a quadratic irrational number, then prove that

$$|\frac{1}{N}\sum_{n=1}^{N} f\left((n\theta)\right) - \int_0^1 f(x)d\,x| \leq \frac{c}{N}\,.$$

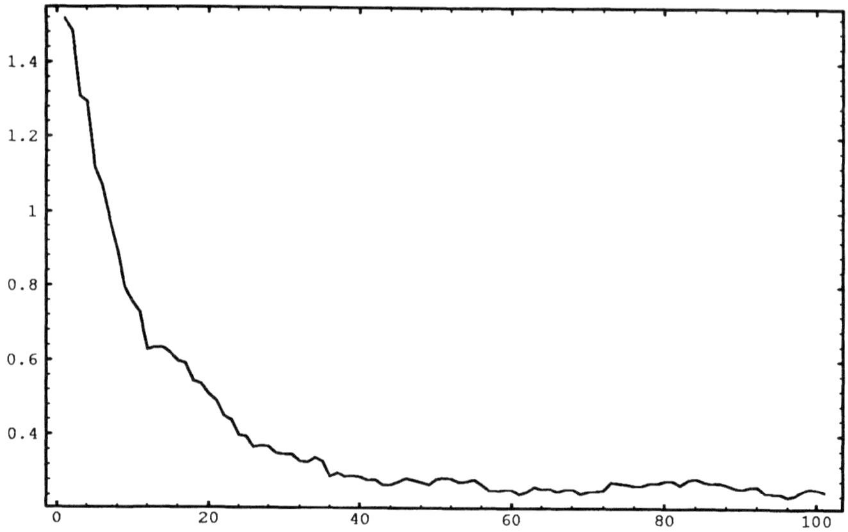

Figure 4.3 H^{-2}. Norm for 100 particles.

Problem 4.5. Compute $\int_0^1 \int \cdots \int_0^1 e^{x_1 x_2 x_3 x_4}\; dx_1 dx_2 dx_3 dx_4$ using a uniformly distributed sequence of points.

Problem 4.6. (i) Show that $D_n^{\star}(x_1, x_2 \cdots x_n) \geq \frac{1}{2\,n}$, in general, and

$$D_n^{\star}(x_1, x_2, \cdots, x_n) = \frac{1}{2n} \text{ if } x_k = \frac{2\,k-1}{2\,n} \text{ for } 1 \leq k \leq n\,,$$

in particular.

(ii) $D_n(x_1, x_2, \cdots, x_n) \geq \frac{1}{2\,n}$ if $x_k = \frac{2\,k-1}{2\,n}$ for $1 \leq k \leq n$.
(iii) $D_n(S) \geq Cn^{-1}\log\, n, D_n^{\star}(S) \geq (0.06)n^{-1}\log\, n$ for infinitely many n.
(iv) $D_n^{\star}(S) = O(n^{-1}\log\, n)$ for all $n > 2$.

Problem 4.7. Show that the sequence $w_n = (x_1^n, x_2^n, \cdots, x_n^n)$ is uniformly distributed if and only if

$$\lim_{n\to\infty} \sum_{i=0} n\phi(x_n^n) = \int_0^1 \phi(x)dx,$$

for every real-valued continuous function ϕ on [0, 1].

Problem 4.8. Verify the results (4.63) and (4.64).

Chapter 5

Image Processing and Fourier – Wavelet Methods

The image processing and signal analysis whose ingredients are modelling, transforms, smoothing and sharpening, restoration, encoding (image data compression, image transmission, feature extraction), decoding, segmentation, representation, archiving and description, have played a vital role in understanding the intricacies of nature, for providing comforts, luxuries and pleasure to all of us, and even probing mission of a space-craft. Image processing is used in telecommunications (telephone and television), photocopying machine, video camera, fax machine, transmission and analysis of satellite images, medical imaging (echography, tomography and nuclear magnetic resonance), warfare, artificial vision, climatology, meteorology and film industry for imagery scenes. In short, it is one of the most important disciplines for industrial development and unveils the secrets of nature. Different kinds of techniques and tools like Fourier series, Fourier transform, Walsh-Fourier transform, Haar-Fourier transform, Hotelling transform, Hadamard transform, entropy encoding and, more recently, wavelets, wavelet packets, and fractal methodology have been used to tackle the problems of this field. It is difficult to say authoritatively which method is superior to the other in a particular situation. However, a combination of the wavelets and fractal methods promises for a great future. This chapter comprises five sections, namely, **image model and methods of image processing, classical Fourier analysis, wavelet techniques in image processing, Fractal image processing** and **problems**. Several subsections of Section 5.1 are devoted to image modelling, image enhancement, image smoothing, image restoration, degradation and image analysis. In Section 5.2, basic results concerning Fourier series, Fourier coefficients and Fourier transform, fast Fourier algorithm, and sampling theorems – the backbone of digital image processing via Fourier techniques – are discussed. The celebrated theory of wavelet analysis, developed in the last decade, along with its applications to the digital image processing is presented in Section 5.3.

The main objective of this chapter is to familiarize the readers with the mathematical models and mathematical methods for studying modern techniques of image processing. Our treatment is mainly based on the work of Weaver [1989], Gonzalez and Woods [1993], Maaß and Strark [1994], Wickerhauser [1994], Barnsley and Hurd [1993], Lu [1997], and Fisher [1995].

5.1 Image model and methods of image processing

5.1.1 Image model

Signal or image or data is considered as a function. With sound or radio waves, the function expresses intensity or amplitude in terms of time. For still images, the free variable is spatial. For video, both time and space are varied. When the domain and range of such a function are defined as intervals of real numbers, and the function can attain all of the values in its range, the signal or data is said to be **analogue.** When the domain and range consist of finite sets of possible values, represented by integers, the data is said to be digital.

Assume that an image of an object is represented by a function of two variables $f(x,y)(f : R\times R \longrightarrow R)$, where the value or amplitude of f at $(x,y) \in R\times R = R^2$ denotes the intensity (brightness) of the image at that point. Since light is a form of energy, $f(x,y)$ must be non-zero and finite; that is,

$$0 < f(x,y) < \infty.$$

We may write

$$f(x,y) = i(x,y)r(x,y), \tag{5.1}$$

where $i(x,y)$ is called the **illumination component** and $r(x,y)$ is called the **reflectance component**. We may assume that

$$0 < i(x,y) < \infty,\ 0 < r(x,y) < 1\ .$$

The nature of $i(x,y)$ is determined by the light source, while $r(x,y)$ is determined by the characteristics of the objects in a scene. The intensity of a monochrome image f at coordinates (x,y) will be called the **grey level** (l) of the image at that point. If $l=0$, then the image is considered as black and if $l=L$, then it is considered as white.

Very often, we take

$$L = i_{\max}\cdot r_{\max}\ (\text{Product of the maximum value of}\, i\, \text{and}\, r)\ .$$

Suppose that a continuous image $f(x,y)$ is approximated by equally spaced samples arranged in the form of an $N\times N$ arrays. Then it is called a digital image. In other words, a digital image is an image $f(x,y)$ that has been discretized both in the spatial coordinate and in brightness. We may consider a digital image as a matrix whose row and column indices identify a point in the image and the corresponding matrix element value identifies the grey level at that point. The elements of such a digital array are called image elements, picture elements, **pixels** or **pels**, with the last two names being commonly used abbreviations of 'picture elements'. There are several advantages of selecting sequence arrays with sizes and number of grey levels that are integer powers of 2. A typical size comparable in quality to a monochrome TV image is a 512×512 array with 128 grey levels. Elements of digitized $f(x, y)$ or digital images are given in equation (5.2).

$$f(x,y) = \begin{bmatrix} f(0,0) & f(0,1) & \cdots & f(0,N-1) \\ f(1,0) & f(1,1) & \cdots & f(1,N-1) \\ \cdots & \cdots & \cdots & \cdots\cdots\cdots \\ \cdots & \cdots & \cdots & \cdots\cdots\cdots \\ f(N-1,0) & f(N-1,1) & \cdots & f(N-1,N-1) \end{bmatrix} . \tag{5.2}$$

Each element of this matrix is called a 'pixel'. Usually, $N = 2^n$, and $G = 2^m$, where G denotes the number of grey levels. The number b of bits required to store a digitized image is given by

$$b = N \times N \times m. \tag{5.3}$$

The quality of an image will vary along with variation of N and m. At some point in the acquisition of a digital image, a spatial resolution (pixel width and length) and a number of values (pixel depth) are imposed on the image. The resolution, that is, the degree of discernible detail of an image is strongly dependent on both N and m. The more these parameters are increased, the closer the digitized array will approximate the original image. However, relation (5.3) clearly indicates that storage and consequently processing requirements increase rapidly as a function of N and m. The effect of variation of N and m on the quality of the image has been studied but no conclusive result is available.

5.1.2 Image enhancement

Enhancement methods are related to improvements of a given image for a particular purpose. These methods are problem-oriented, that is, a method may be quite useful for enhancing X-ray images but it may not be the best approach for enhancing pictures of Mars transmitted by a space probe.

We discuss here two broad categories of image enhancement methods: (1) frequency domain methods and (2) spatial domain methods. The frequency domain method is related to modifying the Fourier transform of an image. The spatial domain method is based on direct manipulation of the pixel in an image.

Frequency domain methods. The frequency domain methods are based on the convolution theorem discussed in the next section. Let $g(x, y)$ be an image formed by the convolution of an image $f(x, y)$ and a position invariant operator $h(x, y)$; that is,

$$g(x,y) = h(x,y) * f(x,y) . \tag{5.4}$$

Then by the Convolution Theorem, we have

$$G(u,v) = H(u,v)\,F(u,v), \tag{5.5}$$

where G, H and F are the Fourier transforms of g, h and f, respectively. $H(u,v)$ is known as the transform function of the process. The goal is to find $H(u,v)$ for a given image $f(x,y)$ so that the desired image given by

$$g(x,y) = \mathcal{F}^{-1}[H(u,v)\,F(u,v)] \tag{5.6}$$

exhibits some highlighted feature of $f(x,y)$. For example, [It is clear that $F(u,v)$ can be calculated] edges in $f(x,y)$ can be accentuated by using a function $H(u,v)$ that emphasizes the high-frequency components of $F(u,v)$.

Spatial domain methods. The term **spatial domain** refers to the aggregate of pixels composing an image, and spatial domain methods are procedures that operate directly on these pixels. Image processing functions are the spatial functions in the spatial domain and may be expressed by

$$g(x,y) = T[f(x,y)], \tag{5.7}$$

where $f(x,y)$ is an input image, $g(x,y)$ is the processed image and T is an operator operating on f, defined over a neighbourhood of (x,y), or T may be considered as defined on a set of input images, such as performing the pixel-by-pixel sum of K images for noise reduction.

Let us consider a neighbourhood about (x,y) as shown in Figure 5.1, a square centred at (x,y).

Rectangles and circles may be used to describe neighbourhoods but rectangles are more appropriate from implementation point of view. The centre of the subimage is moved from pixel to pixel starting, say, at the top corner, and applying the operator at each location (x,y) to yield the value of g at that location.

If the neighbourhood is a square of the unit length, then T is of its simplest form and g depends only on the value of f at (x,y) and T becomes a grey-level transformation action or mapping of the form

$$s = T(r), \tag{5.8}$$

where r and s denote the grey level of $f(x,y)$ and $g(x,y)$ at any point. If $T(r)$ has the form as shown in Figure 5.2, the effect of this transformation is to produce an image of higher contrast than the original by darkening the levels below a value m and brightening the levels above m in the original pixel spectrum. In this technique, known as **contrast stretching** (different foundation), the levels of r below m are compressed by the transform into a narrow range of s towards the dark end of the spectrum. The opposite effect takes place for values of r above m.

In the limiting case shown in Figure 5.3, $T(r)$ produces a two-level (binary) image.

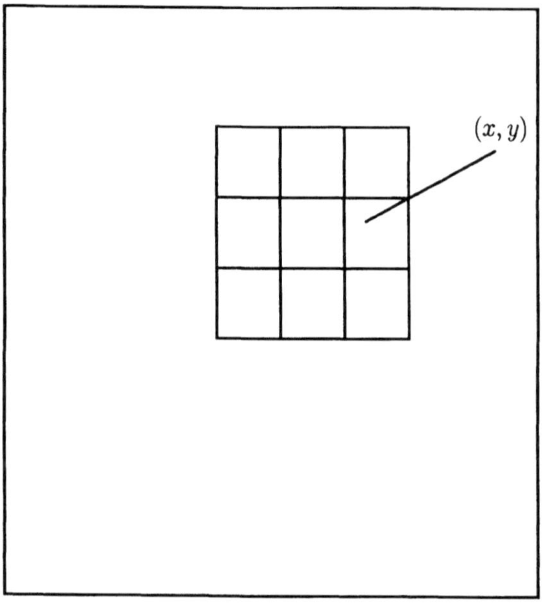

Figure 5.1. A 3×3 neighbourhood about a point (x, y) in an image.

The general approach is to let the values of f in a predefined neighbourhood of (x, y) determine the value of g at those coordinates.

The notion of masks (also referred to as templates, windows, or filters) which are small two-dimensional arrays, for example, 3×3 are quite useful for this method (for more details, one may see Gonzalez and Woods [1993]).

5.1.3 Image smoothing

Smoothing operators are employed mainly for diminishing or illuminating false effects that may be present in an image as a result of poor sampling performance or transmission channel. The present methods for smoothing are:

(1) low-pass filtering,

(2) neighbourhood averaging,

(3) median filtering,

(4) Butterworth filter, and

(5) averaging multiple images.

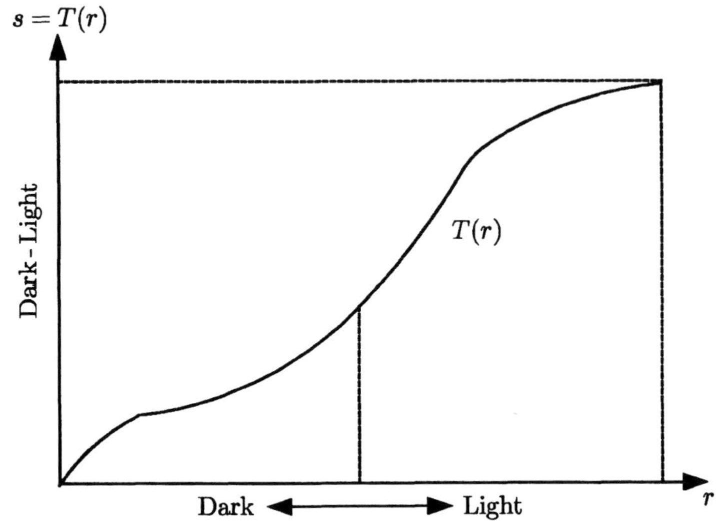

Figure 5.2. Dark $\longleftrightarrow$ Light.

We present here the low-pass filtering technique for image smoothing and we refer to the bibliography for other methods.

Low-pass filtering. Edges and other transitions such as noise in the grey levels of an image contribute heavily to the high frequency content of its Fourier transform. It therefore follows that blurring can be achieved via the frequency domain by attenuating a specified range of high-frequency components in the transform of a given image. By (5.5), we have

$$G(u,v) = H(u,v)F(u,v),$$

where $F(\cdot,\cdot)$ is the Fourier transform of the image $f(\cdot,\cdot)$ which we want to smooth. The problem is to select a function $H(\cdot,\cdot)$ that yields $G(\cdot,\cdot)$ by attenuating the high-frequency components of $F(\cdot,\cdot)$. The inverse Fourier transform of $G(\cdot,\cdot)$ will then yield the desired smoothed image $g(\cdot,\cdot)$. Since high-frequency components are 'filtered out' and information in the low-frequency range is passed without attenuation, this process is known as **low-pass filtering**. $H(\cdot,\cdot)$ in this reference is referred as the **filter transfer function**. An **ideal low-pass filter** (ILPF) in two-dimensional space is defined by the relation

$$H(u,v) = \begin{cases} 1 \; if \; D(u,v) = D_0 ; \\ 0 \; if \; D(u,v) > D_0 , \end{cases} \tag{5.9}$$

where D_0 is a specified nonnegative quantity and $D(u,v)$ is the distance from point (u,v) to the origin of the frequency plane; that is,

$$D(u,v) = (u^2+v^2)^{1/2}. \tag{5.10}$$

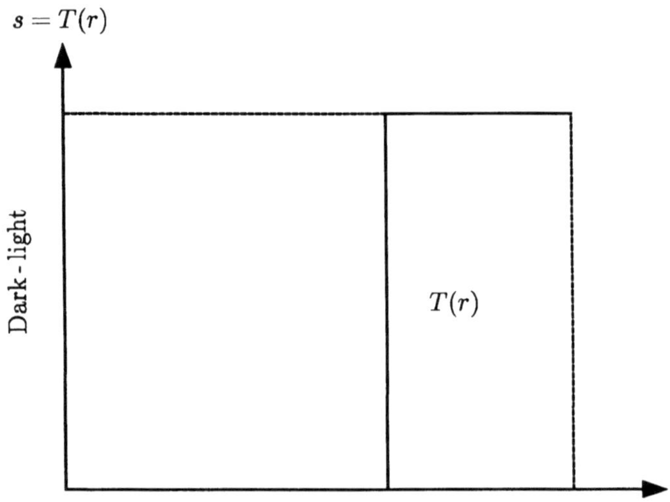

Figure 5.3. Dark ⟷ Light.

The concept of the blurring and ringing can be explained with the help of the **convolution theorem**. Since

$$G(u,v) = H(u,v)\, F(u,v) \qquad \text{[See (5.5)]},$$

it follows from the Convolution Theorem that

$$g(x,y) = h(x,y) * f(x,y),$$

where $g(\cdot,\cdot)$, $h(\cdot,\cdot)$ and $f(\cdot,\cdot)$ are the inverse Fourier transforms of $G(\cdot,\cdot)$, $H(\cdot,\cdot)$ and $F(\cdot,\cdot)$, respectively.

Enhancement based on an image model. As we have seen earlier, an image, $f(x,y)$ can be written as

$$f(x,y) = i\,(x,y)\, r(x,y), \qquad \text{(Equation (5.1))}.$$

Let $z(x,y) = log\, f(x,y)$, then

$$z(x,y) = \log i(x,y) + \log r(x,y), \tag{5.11}$$

or

$$\mathcal{F}(z(x,y)) = \mathcal{F}(\log i(x,y)) + \mathcal{F}(\log r(x,y)), \quad \text{by Theorem 5.7},$$

or

$$Z(u,v) = I(u,v) + R(u,v), \tag{5.12}$$

where $Z(u,v)$, $I(u,v)$ and $R(u,v)$ are the Fourier transform of $Z(x,y)$, $\log i(x,y)$ and $\log r(x,y)$, respectively. If we process $Z(u,v)$ by means of a filter function $H(u,v)$, it follows from equation (5.5) that

$$S(u,v) = H(u,v)\,Z(u,v) = H(u,v)\,I(u,v) + H(u,v)\,R(u,v), \tag{5.13}$$

where $S(u,v)$ is the Fourier transform of the result [compare with Equations (5.4) and (5.5)]. In the spatial domain, we have the relation

$$s(x,y) = \mathcal{F}^{-1}\{S(u,v)\} = \mathcal{F}^{-1}\{H(u,v)I(u,v)\} + \mathcal{F}^{-1}\{H(u,v)R(u,v)\}. \tag{5.14}$$

By letting

$$i'(x,y) = \mathcal{F}^{-1}\{H(u,v)I(u,v)\}, \tag{5.15}$$

and

$$r'(x,y) = \mathcal{F}^{-1}\{H(u,v)R(u,v)\}. \tag{5.16}$$

Equation (5.14) can be written in the form

$$s(x,y) = i'(x,y) + r'(x,y). \tag{5.17}$$

The derived enhanced image $g(x,y)$ is given by

$$g(x,y) = \exp\{s(x,y)\} \tag{5.18}$$

$$= e^{i'(x,y)} \cdot e^{r'(x,y)}. \tag{5.19}$$

This process is summarized in Figure 5.4.

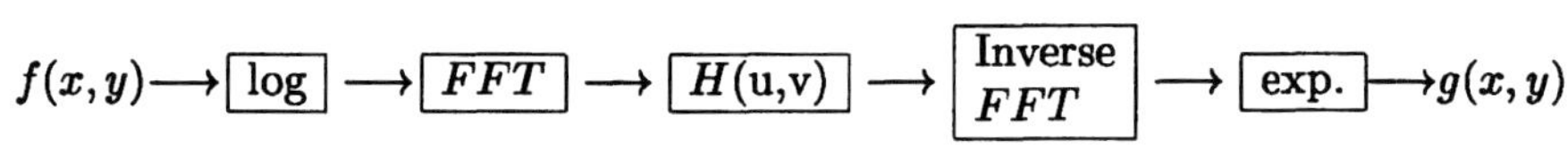

Figure 5.4.

A relatively recent and potentially powerful area of image processing is the use of pseudo-colour for image display and enhancement. Fast Fourier transforms and different kinds of filters have been successfully used in this area; see, for example, Gonzalez and Woods [1993].

5.1.4 Image restoration

As in the case of image enhancement, the final goal of the image restoration methods is to improve a given image in a certain sense. For the sake of differentiation, we consider restoration as a process of endeavours to reconstruct or recover an image that has been degraded by using a prior knowledge of the degradation phenomenon. Thus, restoration methods are nothing but modelling of the degradation and applying the inverse process to recover the original image. We discuss here models which will employ Fourier transform and convolution theory of Section 3, techniques of the minimum norm problem, and the method of Lagrange multipliers.

Image degradation model and Fourier analysis. The input-output relationship in a model of the image degradation process can be written as

$$g(x,y) = H\,f(x,y) + \eta(x,y)\,, \tag{5.20}$$

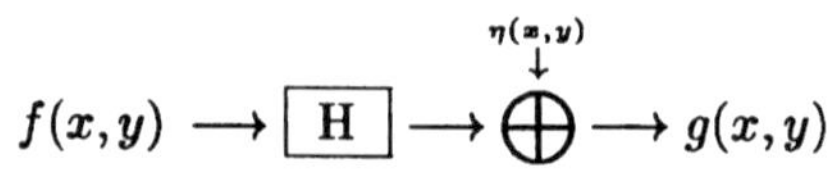

Figure 5.5. A model of the image degradation process.

H is an operator on an input image $f(x,y)$ to produce a degraded image $g(x,y)$ minus noise term represented by $\eta(x,y)$. If $\eta(x,y)=0$, that is, there is no noise, the operator maps images to $g(x,y)$. If H is linear, then the system is called a linear system. An operator having input-output relationship $g(x,y) = Hf(x,y)$ is said to be **position** (space) **invariant** if

$$Hf(x-\alpha,\, y-\beta) = g(x-\alpha,\, y-\beta), \tag{5.21}$$

for any $f(x,y)$ and any α and β.

For example, we can express $f(x,y)$ in the form

$$f(x,y) = \int_{-\infty}^{\infty}\int_{-\infty}^{\infty} f(\alpha,\beta)\,\delta(x-\alpha,\, y-\beta)\,d\alpha\,d\beta, \tag{5.22}$$

Then, if $\eta(x,y)=0$ in (5.20),

$$g(x,y) = Hf(x,y) = H\left[\int_{-\infty}^{\infty}\int_{-\infty}^{\infty} f(\alpha,\beta)\,\delta(x-\alpha,y-\beta)\,d\alpha\,d\beta\right]. \tag{5.23}$$

If H is linear and additive property is valid for integrals, then

$$g(x,y) = \int_{-\infty}^{\infty}\int_{-\infty}^{\infty} H[f(\alpha,\beta)\,\delta(x-\alpha,\, y-\beta)]\,d\alpha\,d\beta \tag{5.20}$$

$$= \int_{-\infty}^{\infty}\int_{-\infty}^{\infty} f(\alpha,\beta)\,H\delta(x-\alpha,\, y-\beta)\,d\alpha\,d\beta\,, \tag{5.21}$$

by the homogeneity property of H.

The term $h(x,\alpha,y,\beta) = H\delta(x-\alpha,\ y-\beta)$ is called the **impulse response** of H. In other words, if $\eta(x,y)=0$ in equation (5.20), we find that $h(x,\alpha,y,\beta)$ is the response of H to an impulse at coordinates (α,β). If H is position invariant, then from (5.21) we get

$$H\delta(x-\alpha,\, y-\beta) = h(x-\alpha,\, y-\beta)\,.$$

Thus, (5.25) can be written as

$$g(x,y) = \int_{-\infty}^{\infty}\int_{-\infty}^{\infty} f(\alpha,\beta)\,h(x-\alpha,y-\beta)d\alpha\,d\beta, \tag{5.26}$$

or

$$g(x,y) = f(x,y) * h(x,y).$$

In the presence of the noise factor $\eta(x,y)$, we have

$$g(x,y) = f(x,y) * h(x,y) + \eta(x,y), \tag{5.27}$$

or $G(u,v) = F(u,v)\,H(u,v) + N(u,v)$ by the Convolution Theorem, where $G(\cdot,\cdot)$, $F(\cdot,\cdot)$, $H(\cdot,\cdot)$, $N(\cdot,\cdot)$ are the Fourier transforms of the functions $g(\cdot,\cdot)$, $f(\cdot,\cdot)$, $h(\cdot,\cdot)$ and $\eta(\cdot,\cdot)$, respectively. Relation (5.27) and its equivalent form is valid for the discrete case also.

The degradation problem can be found in terms of matrices (associated with the discrete Fourier transform) in the following manner: Let f, g, and n be MN-dimensional column vectors formed by stacking the rows of the $M \times N$ functions $f_e(x,y)$, $g_e(x,y)$ and $\eta_e(x,y)$, then the equation

$$g_e(x,y) = \sum_{m=0}^{M-1}\sum_{n=0}^{N-1} f_e(m,n)h_e(x-m,\, y-n) + \eta_e(x,y) \tag{5.28}$$

can be written in a matrix form

$$g = Hf + \eta. \tag{5.29}$$

Thus, **image degradation problem** is to find out the ideal image $f(x,y)$, given $g(x,y)$ and having a knowledge of $h(x,y)$ and $\eta(x,y)$, that is, estimate f while g, H and η are given. The fast Fourier transform algorithm can be used to solve it.

Restoration through the minimum norm problem. By (5.29), we have

$$\eta = g - Hf\,.$$

In the absence of any knowledge about η, a meaningful criterion function is to find $\hat{f}$ such that $\|\eta\|^2 = \|g - H\hat{f}\|^2$ is minimum where $\|\eta\|^2 = \eta\eta'$, $\|g - H\hat{f}\|^2 = (g - H\hat{f})'(g - H\hat{f})$ or, equivalently, we want to minimize $J(\hat{f}) = \|g - H\hat{f}\|^2$. For this, the necessary condition is

$$\frac{\delta J}{\delta \hat{f}} = 0 = -2H'(g - H\hat{f}).$$

The relation holds if

$$\hat{f} = (H'H)^{-1}H'g.$$

If $M = N$ and H^{-1} exists, then

$$\hat{f} = H^{-1}g, \tag{5.30}$$

where "$'$" stands here for transpose.

Constrained restoration. Let J be a linear operator which transmits the image represented by f. The **least square restoration** problem is the process of finding $\hat{f}$ for which $||J(\hat{f})||^2$ is minimum, subject to the constraint $||g - H\hat{f}||^2 = ||\eta||^2$. This problem can be solved by using the method of Lagrange multipliers. The method relates to express the constraint in the form $\alpha(||g - H\hat{f}||^2)$ and add it to $||J(\hat{f})||^2$. In other words, find $\hat{f}$ that minimizes the functional

$$L(\hat{f}) = ||J(\hat{f})||^2 + \alpha(||g - H\hat{f}||^2 - ||\eta||^2),$$

where α is a constant called the Lagrange multiplier. Differentiating $L(\hat{f})$ and by solving the equation

$$\frac{\partial L(\hat{f})}{\partial \hat{f}} = 0\,,$$

we get

$$\hat{f} = (H'H + rJ'J)^{-1}H'g,$$

where $r = \frac{1}{\alpha}$ and J' denote the adjoint (transpose) of J. This method introduces considerable flexibility as one can find different solutions for different choices of J.

5.1.5 Image analysis

Broadly speaking, image analysis can be divided into two major areas, **image handling** and **image understanding**. Image handling refers to the manipulation of visual information in order to prepare visual data for the human observer. This manipulation can be led by simply aesthetic or artistic goals as in computer graphics, or it may aim at extracting the essential information for some skilled observer. Noise removal, sharpening of contrast and, to some extent, edge detection are nice examples of this theme. Image understanding requires the interpretation of a given image in terms of human knowledge. Among the challenging industrial applications are problems of quality control such as the automatic inspection for completeness of mounted electronic devices. More difficult tasks are scene interpretation and identification problems. Different problems need methods of different complexity; hence, they are characterized as low- or high-level procedures.

As mentioned earlier, image raw data consists of a single-valued (grey scale image) or a three-component (colour pictures) two-dimensional intensity function $f(x, y)$ on some usually rectangular domain $\Omega \subset R^2$. In a digital computer, (x, y) as well as $f(x, y)$-values are discretized and the image raw data forms a two-dimensional array, the pixel matrix. **Low-level procedures** transform these numerical data into other numerical data which might again be a pixel matrix or values $p_i(f)$ of some (non)linear functional p_i on the image data. The functional p_i's are called **attributes**.

The purpose of a low-level transformation is to create numerical data which is better suited for the classification or scene interpretation task done by a human observer or a high-level procedure. The compression techniques aim at reducing storage requirements for pictorial data and at speeding up read or write operations to or from disk. A good example of low-level procedures is **multiscale analysis** which looks at the image f through a microscope whose resolution gets coarser and coarser; thus it associates to f a sequence of smoothed version f^a, labelled by a scale parameter a which increases from some minimal value.

The capability of compressing images is essential for fast transmission of digitized images or their efficient storage. Both applications require a representation of the image matrix $f_{i,j}$, $i = 1, 2, \cdots, n$, $j = 1, 2, \cdots, n$, with fewer parameters affecting the picture quality visibly. A typical algorithm proceeds in three steps:

Step 1: Compute the coefficients $(f_T)(u,v) = (Tf)(u,v)$, where T is a linear, invertible transformation.

Step 2: A standard procedure for quantization of the coefficients consist in specifying a quantization matrix Q and computing

$$f_T^q(u,v) = \text{Integer Round}\left\{\frac{f_T(u,v)}{Q(u,v)}\right\}.$$

Step 3: The most commonly used coding methods are entropy coders which aim at storing frequency appearing values as few possible bits. The compressed image is denoted by f_c.

Step 1 defines a one-to-one mapping without resulting in any data reduction. Compression is achieved by the subsequent two steps. Both steps are chosen such that the number of bits needed to encode f_c is considerably smaller than the number of bits in the original image. The quotient of both numbers is called the compression rate.

Until recently, the JPEG standard, an effort to standardize transform compression algorithm by making choices and recommendations for Steps 1 to 3 above, was mainly in practice. The JPEG standard is based on a cosine transform in Step 1. In recent years, other standards like EPIC and fractal compression are gaining in popularity. See, Section (5.3) for details. In the remaining subsections, we briefly introduce the basic concept of image compression model, notions of fidelity and entropy.

Image compression models. As indicated in Figure 5.6(a), a compression system consists of two distinct structural blocks, namely, an **encoder** and a **decoder**. An input image $f(x,y)$ is fed into the encoder which creates a set of symbols from the input data. After transmission over the channel, the encoded representation is fed to the decoder where a restructured output image $\hat{f}(x,y)$ is generated. In general, $\hat{f}(x,y)$ may or may not be an exact replica of $f(x,y)$. Both the encoder and decoder shown in this figure consist of two relatively independent functions or sub-blocks. The source encoder is responsible for reducing or eliminating any coding inter-pixel or psycho-visual redundancies in the output image. As Figure 5.6(b) shows, each operation is designed to reduce the redundancies. Figure 5.6(c) depicts the corresponding source coder. Figures 5.6(d) and 5.6(e) represent the transform coding system.

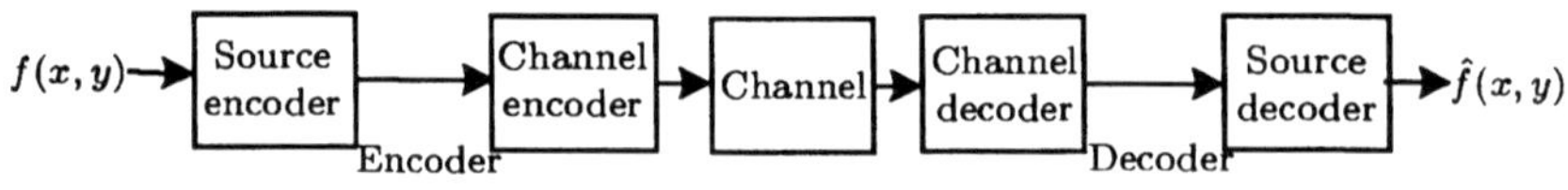

Figure 5.6(a).

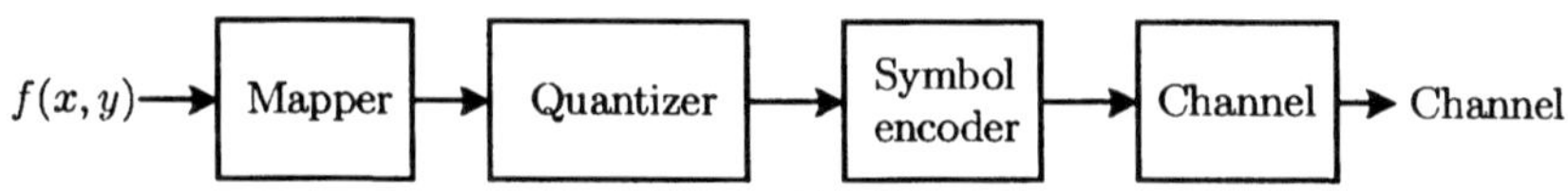

Figure 5.6(b).

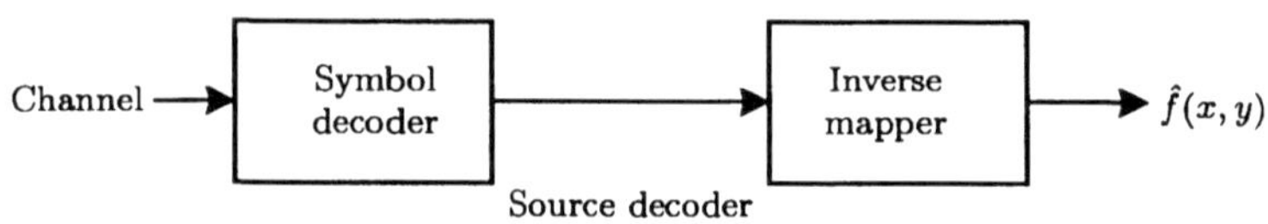

Figure 5.6(c).

A compression technique which provides error-free compressed image is called the error-free compression method. The Huffman coding is an example of the error-free compression method, details of which can be found in Gonzalez and Woods [1993]. The transform coding is an example of the lossy compression technique. Lossy compression is based on the concept of compromising the accuracy of the constructed image in exchange for increased compression. If the resulting distortion, which may or may not be visually apparent, can be tolerated, the increase in compression can be significant. In fact, many lossy encoding techniques are capable of reproducing recognizable monochrome images from data that have been compressed by more than 30:1 and images that are virtually indistinguishable from the originals at 10:1 to 20:1. Error-free encoding of monochrome images seldom results in more than a 3:1 reduction in data. However, it may be observed that in the recent past, the wavelets and fractal techniques gave much better results. See, for example, the discussion towards the end of Section 5.3.

Fidelity criteria. Removal of psycho-visually redundant data results in a loss of real or quantative visual information because information of interest may be lost, a

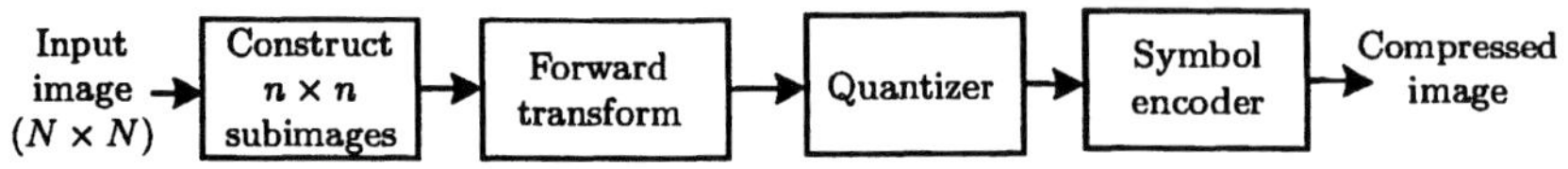

Figure 5.6(d).

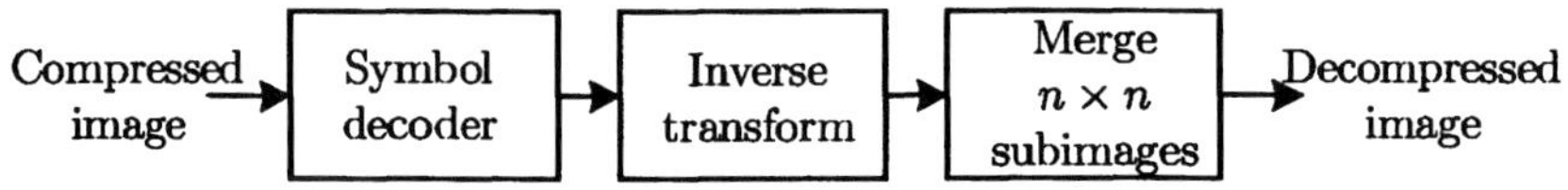

Figure 5.6(e).

repeatable or reducible means of quantifying the nature and extent of information lost is highly desirable. Two general criteria are used as the basis of such an assessment: one is called **objective fidelity** and the other is called **subjective fidelity**. When the level of information loss can be expressed as a function of original or input image and the compressed and subsequently decompressed output image, it is said to be based on objective fidelity criteria. A good example is the root mean square (RMS) error between an input and output image. This is defined as follows: Let $f(x,y)$ represent an input image and let $g(x,y)$ denote an approximation of $f(x,y)$ that results from compressing and subsequently decompressing the input. For any value of x and y, the error $e(x,y)$ between $f(x,y)$ and $g(x,y)$ can be defined as $e(x,y) = f(x,y) - g(x,y)$, so that the total error between the two images is

$$\sum_{x=0}^{M-1} \sum_{y=0}^{N-1} [f(x,y) - g(x,y)],$$

where the images are of size $M \times N$.

Following are the distortion measures between the images:

1. The **mean absolute error**

$$e_{\text{ma}} = \frac{1}{MN} \sum_{x=0}^{M-1} \sum_{y=0}^{N-1} |f(x,y) - g(x,y)| \, .$$

2. The **mean square error,** e_{ms}, between $f(x,y)$ and $g(x,y)$ is the squared error averaged over the $M \times N$ array, or,

$$e_{\text{ms}} = \frac{1}{MN} \sum_{x=0}^{M-1} \sum_{y=0}^{N-1} [f(x,y) - g(x,y)]^2 \, .$$

The root mean square error (e_{rms}) is the square root of e_{ms}; i.e.,

$$e_{\text{rms}} = \sqrt{e_{\text{ms}}} \, .$$

3. The **mean-square signal-to-noise ratio** of the output image, denoted by SNR_{ms}, is

$$SNR_{ms} = \frac{\sum_{x=0}^{M-1} \sum_{y=0}^{N-1} g^2(x,y)}{\sum_{x=0}^{M-1} \sum_{y=0}^{N-1} [f(x,y) - g(x,y)]^2} .$$

The root mean-square signal-to-noise ratio, SNR_{rms}, is simply the square root of SNR_{ms}; i.e.,

$$SNR_{rms} = \sqrt{SNR_{ms}} .$$

4. The **peak signal-to-noise ratio** $PSNR$ is defined as follows:

$$PSNR = 10 \log_{10} \left(\frac{255 \times 255}{e_{ms}} \right) .$$

where 255 is the highest pixel value in a grey scale image.

Transform coding. Let a transformation A be defined with the help of an $n \times n$ matrix as follows:

$$y = Ax, \tag{5.31}$$

where

$$x = \begin{bmatrix} x_1 \\ x_2 \\ \vdots \\ x_n \end{bmatrix} \quad y = \begin{bmatrix} y_1 \\ y_2 \\ \vdots \\ y_n \end{bmatrix}$$

$$A = \begin{bmatrix} a_{11} & a_{12} & \cdots & a_{1n} \\ a_{21} & a_{22} & \cdots & a_{2n} \\ \vdots & \vdots & \vdots & \vdots \\ a_{n1} & a_{n2} & \vdots & a_{nn} \end{bmatrix} .$$

In general, A is not invertible. In this case, the vector of pixels x is transformed into a vector of coefficients y. For some sets of vectors x and some transformations A, fewer bits are required to encode the n coefficients of y than the n pixels of x. In particular, if the elements $x_1, x_2, \cdots, x_n$ are highly correlated and the transformation matrix A is chosen such that the coefficients $y_1, y_2, \cdots, y_n$ are less correlated, then the y_i's can be individually coded more efficiently than the x_i's. A difference mapping, explained below, is obtained if we choose A as

$$A = \begin{pmatrix} 1 & 0 & 0 & 0 & 0 & 0 \\ 1 & -1 & 0 & 0 & 0 & 0 \\ 0 & 1 & -1 & 0 & 0 & 0 \\ 0 & 0 & 1 & -1 & 0 & 0 \\ 0 & 0 & 0 & 1 & -1 & 0 \\ 0 & 0 & 0 & 0 & 1 & -1 \end{pmatrix}, \tag{5.32}$$

in equation (5.31). The first element of y is $y_1 = x_1$ and all subsequent coefficients are given by $y_i = x_{i-1} - x_i$. If the grey levels of the adjacent pixels are similar, then the differences $y_i = x_{i-1} - x_i$ will, on the average, be smaller than the grey levels so that it should require fewer bits to code them. This mapping is invertible.

If A is a unitary matrix, then A^{-1} exists and $A^{-1} = A'$, whereas if A' is the transpose of A, then (5.31) can be written as

$$x = A'y. \tag{5.33}$$

It is quite clear from (5.31) that each coefficient y_k is a linear combination of all pixels; that is,

$$y_k = \sum_{i=1}^{n} a_{ki}x_i, \tag{5.34}$$

for $k = 1, 2, 3, \cdots, n$.

Similarly, by equation (5.33), each pixel x_i is a linear combination of all the pixels

$$x_i = \sum_{k=1}^{n} b_{ik}y_k, \tag{5.35}$$

for $i = 1, 2, 3, \cdots, n$.

Equations (5.34) and (5.35) are similar to the expressions defining the forward and the inverse transformation kernels, respectively, whereas a_{ki} is the forward transformation kernel and b_{ik} is the inverse transformation kernel.

For the two-dimensional case, (5.34) and (5.35) take the form

$$y_{kl} = \sum_{i=1}^{n}\sum_{j=1}^{n} x_{ij}a_{ijkl}, \tag{5.36}$$

and

$$x_{ij} = \sum_{k=1}^{n}\sum_{l=1}^{n} y_{kl}b_{ijkl}. \tag{5.37}$$

Here, a_{ijkl} and b_{ijkl} are forward and inverse transformation kernels, respectively.

In equations (5.36) and (5.37) the Fourier, Walsh and Hadamard transforms are commonly used for encoding purposes, which fit directly in this setting and they give fairly good results. For example, the Fourier kernel is given by

$$a_{ijkl} = \frac{1}{N}e^{-j2\pi(ik+jl)/N}.$$

Another interpretation of (5.37) is possible. Let us write (5.37) in the form

$$X = \sum_{k=1}^{n}\sum_{l=1}^{n} y_{kl} B_{kl}, \tag{5.38}$$

and interpret this as a series expansion of the $n \times n$ subimage X into $n^2 n \times n$ basis images

$$B_{kl} = \begin{bmatrix} b_{kl11} & b_{kl12} & \cdots & b_{kl1n} \\ b_{kl21} & b_{kl22} & \cdots & b_{kl2n} \\ \cdots & \cdots & \cdots & \cdots \\ \cdots & \cdots & \cdots & \cdots \\ b_{kln1} & b_{kln2} & \cdots & b_{klmn} \end{bmatrix}, \tag{5.39}$$

with the y_{kl} for $k, l = 1, 2, \cdots,\ n$ being the coefficients (weights) of the expansion. Hence, (5.39) gives the image X as a weighted sum of the basis B_{kl}. The coefficients of the expansion are given by (5.36) which may be written in the form

$$y = A_{kl} X, \tag{5.40}$$

where A_{kl} is formed in the same manner as B_{kl} except that the forward kernel is used.

The quantizer. , Let us consider the number of all possible values of the coefficients y_i in equation (5.31) where

$$y_i = \sum_{j=1}^{n} a_{ij} x_j.$$

If each element x_j can have any 2^m different values, then each $a_{ij}x_j$ term can also have any of 2^m different values and the sum of n such terms could have any of $(2^m)^n = 2^{mn}$ different values. Therefore, a natural binary representation would require mn bit code words to assign a unique word to each of the possible 2^{mn} values of y_i. Since only m bit words would be required to code any x_j, and our objective is to use fewer bits to code the y_i, we must round off the y_i to a fewer number of allowed levels.

A **quantizer** is a device whose output can have only a limited number of possible values. Each input is forced to one of the permissible output values. For more details, we refer to Gonzales and Woods [1993].

The coder. As shown in Figure 5.6, the input to the coder are the n elements of the vector

$$v = \begin{bmatrix} v_1 \\ v_2 \\ \vdots \\ v_n \end{bmatrix}.$$

Let us suppose that each v_i, $i = 1, 2, \cdots, n$ takes one of M values (levels) $w_1, w_2, \cdots, w_M$. For each input v_i, the coder outputs a binary word whose value depends on the value w_k of the input. The coder input-output relationship is one-to-one in that a unique code word c_k is assigned to each possible input value w_k. This process is reversible and error-free. If the coder is required to handle M possible input values, then designing the coder amounts to choosing M unique binary code words and assigning one of them to each input.

An equal-length code is a set of code words each of which has the same number of bits, along with a rule for assigning code words to quantizer output levels. One example of an equal-length code is the natural binary code. One possible assignment rule for the natural code is to order the code words according to their binary values. For example, suppose that there are eight possible coder input values (quantizer output levels) ordered w_1, $w_2 \cdots$, w_8; then the natural code is $c_1 = 000$, $c_2 = 011$, $\cdots$, $c_8 = 111$, as illustrated in Table 5.1. There are 8! possible assignments of the eight-code words to the eight inputs. The **reflected binary** or **grey code**, illustrated in Table 5.1, has the property that any two adjacent code words in the set differ in only one bit position.

Table 5.1 Some typical codes

Input	Natural Code	Grey Code
y_1	000	111
y_2	001	110
y_3	010	100
y_4	011	101
y_5	100	001
y_6	101	000
y_7	110	010
y_8	111	011

A **uniquely decodable code** is a code with the property that a sequence of code words can be decoded in only one way. The code $c_1 = 0$, $c_2 = 1$, $c_3 = 1$, $c_4 = 10$ is not unique because of bits 0011 could be decoded as $c_1c_1c_2c_2$ or as $c_1c_3c_2$. The codes presented in Table 5.1 are uniquely decodable.

An **instantaneous code** is one that can be decoded instantaneously. That is, if we look at the sequence of incoming bits one at a time, we know the value of the input when we come to the end of a code word. We do not have to look ahead at any future incoming bits in order to decode the bit stream. The codes in Table 5.1 are instantaneous.

5.1.6 Variational methods in image processing

In the recent past, certain important algorithms and functional analytic techniques have been developed for understanding the automatic segmentation of digital images and visual perception. Perception theory [see, for example, Koendrink 1985] deals with the complex interaction between regions and edges (or boundaries) in an image. **Segmentation** of a digital image means search of homogeneous regions and edges by a numerical algorithm. The homogeneous regions are assumed to correspond to meaningful parts of objects in the real-life problem, and edges to their apparent contours. It is interesting to note that in most of the segmentation algorithms, one tries to minimize segmentation energy. Segmentation energy is the measure of smoothness of the region and tells us how faithful the analyzed image to the original image and the obtained edges to the image discontinuities are. A fairly general segmentation energy model is discussed in Morel and Solimini [1995, Chapter 4]. However, we briefly mention here the **Mumford-Shah segmentation energy model**. This model defines the segmentation problem as a joint smoothing/edge detection problem, namely, "Given an image $g(u)$, one seeks simultaneously a piecewise smooth image $u(x)$ with a set K of abrupt discontinuities, the edges of g". The best segmentation of a given image is obtained by minimizing the functional

$$E(u,K) = \int_{\Omega\setminus K} (|\nabla u(x)|^2 + (u-g)^2)dx + \text{ length } (K).$$

The first term imposes that u is smooth outside the edges, the second that the piecewise smooth image $u(x)$ indeed approximates $g(x)$, and the third that the discontinuity set K has minimal length and, therefore, is as smooth as possible. The model is minimal in the sense that removing one of the above three terms would yield a trivial solution. Mumford and Shah [1985] conjectured the existence of minimal segmentations made of a finite set of C^2 curves. So far, this conjecture is not fully proved but enough useful results have been obtained in this direction [See Morel and Solimini 1995 for detailed updated work].

Multiscale analysis is a fairly recent concept in image processing. It deals with the analysis of pictorial data on different resolution levels and was systematically treated for the first time by Mallat [1989] through wavelet techniques. Multiscale analysis looks at the image f through a microscope whose resolution gets coarser and coarser. We define a picture as a **grey level** or **brightness** real function $u_0(x)$ defined at each point (pixel) of a domain Ω, which generally is a plane, a square or a rectangle. Multiscale analysis is concerned with the generation, from a single initial picture f_0, of a sequence of simplified pictures $f_\lambda(x)$, where $f_\lambda(x)$ appears to be a rougher and sketchier version of f_0 as λ increases.

In f_λ, details and features like edges are kept if their "scale" exceeds λ. Let H_λ be the map carrying f_0 to $H_\lambda(f_0) = f_\lambda$, then $H_\lambda(f_0)$ is either a picture, denoted by f_λ or, more generally, a pair (K_λ, f_λ) where K_λ is a set of **boundaries** or **edges at scale** λ.

Any **multiscale analysis** must satisfy the properties:

(i) Fidelity, namely, $f_\lambda \to f_0$ as $\lambda \to 0$.

(ii) Causality, namely, $H_\lambda(f_0)$ only depends on $H_{\lambda'}(f_0)$ if $\lambda > \lambda'$.

(iii) Euclidean invariance, namely, if A is an isometry, then $H_\lambda(f_0 \circ A) = H_\lambda(f_0) \circ A$.

(iv) Strong causality (in case of boundary detection); namely,

$$K_\lambda \subset K_{\lambda'} \text{ if } \lambda > \lambda' .$$

In view of (i) and (ii), there must exist an energy functional $E(f_\lambda)$ such that $E(f_\lambda)$ decreases as λ increases. By virtue of (iii), this functional is likely to be a Lebesgue integral of some bidimensional or mono-dimensional terms depending on f_λ and K_λ in the case of boundary detection.

Morel and Solimini [1995] have discussed numerous examples of multiscale analysis and their variational formulation. Koendernik [1985] observed for the first time the relationship between images and solution of partial differential equations. He noticed that the convolution of the signal with Gaussian at each scale was equivalent to the solution of the heat equation with the signal as the initial datum. Let f_0 be this datum, then the scale space analysis associated with f_0 consists in solving the system

$$\frac{\partial f(x,t)}{\partial t} = \Delta f(x,t)$$
$$f(x,0) = f_0(x) .$$

The solution of this equation for an initial datum with bounded quadratic norm is $f(x,t) = G_t(*)f_0$, where

$$G_\sigma(x) = \frac{1}{4\pi\sigma} \exp\left(-\frac{||x||^2}{4\sigma}\right)$$

is the Gaussian function.

The sequence $f_\lambda = f(\cdot, \lambda)$ clearly defines a multiscale analysis satisfying the properties like fidelity, causality and invariance. In Perona and Malik theory (See Perona and Malik [1990], Morel and Solimini [1995]), the heat equation has been replaced by a nonlinear equation of the type

$$\frac{\partial f}{\partial t} = \text{div } (g(|\nabla f|)\nabla f)$$
$$f(0) = f_0,$$

where g is a smooth non-increasing function with $g(0) = 1, g(s) \geq 0$, and $g(s) \to 0$ as $s \to \infty$.

Variational formulation of the Perona and Malik model and related results can be found in the references cited above. Berger et al. [1996] contains valuable research papers on image processing through the numerical treatment of partial differential equations. An updated account of the wavelet theory providing a stable mathematical foundation for understanding of multiscale analysis can be found in Mallat [1996]. Chambolle and Lions [1997] have studied the problem of image recovery via total variation minimization and have discussed the suitability of this method for recovery of images belonging to a certain class of images. Chambolle et al. [1998], De Vore [1998], March and Dozio [1997], and Chipot, March and Vitulano [1999] contain valuable material in this area.

5.2 Introduction to Fourier analysis

Fourier analysis (studies of Fourier series and transforms) is a well-established topic which has been studied extensively and applied vigorously in different disciplines, especially engineering and physics. Renowned mathematicians have taken part in the development of this subject in its history of about 170 years which was invented by the great French Scientist Jean Baptiste Joseph de Fourier (1768-1830). Sound and vision are the most important external sensory signals which help us to interpret the world around us. In some sense, they are related but in another sense, they are very different. The essential physical model for light and sound is the same, namely waves. The wave model is also significant for the discussion of heat radiation. It has been demonstrated by Leonhar **Euler**, Daniel **Bernoulli**, Joseph Louis **Lagrange** and Jean Baptiste Joseph de **Fourier** (J.B. **Fourier**) that waves, no matter how complicated, have a beautiful common description in mathematical terms. That is the concept of Fourier series. Fourier series allows us to describe tones in terms of harmonics. A tone is obtained as a superposition of pure tones of the form $A_k \sin(\text{kwt} + \theta_k)$ where A_k is the amplitude and w is the basic frequency. The most striking and interesting feature about the Fourier series is that it also allows a description of something as complex as a tone of a violin. Indeed, for any given tone, Fourier analysis allows us to determine the A_k's and w, both electrically and mathematically. The object of this section is to describe those results which are often used in the image processing, especially image enhancement, image degradation, and image compression and transmission.

Invention of the fast Fourier transform algorithm in 1965 and studies of the related topics like Walsh and Haar-Fourier series and recent studies of wavelets have aroused great interest in this field. With the help of the results of this section, one can also analyze and understand a large number of other scientific and technological problems. The main themes discussed in this chapter include:

1. amplitude, frequency and phase;
2. Bessel's inequality, Riesz-Fischer theorem and basic convergence theory including Poisson's summation formula;
3. continuous and discrete Fourier transforms and their convolution and fast Fourier transform algorithm; and
4. Fourier analysis via computer.

Our presentation here is based mainly on Weaver [1983, 1989].

5.2.1 Amplitude, frequency and phase

Fourier analysis in the simplest sense is the study of the effects of adding together sine and cosine functions. Daniel Bernoulli, while studying vibration of a string in the year 1750, first suggested that a continuous function over the interval

$(0, \pi)$ could be represented by an infinite series consisting only of sine functions. This suggestion was based on his physical intuition and it was a serious bone of contention. About 70 years later, a French soldier and a personal friend of Napoleon, J.B. Fourier reopened this debate while studying heat transfer. He argued that a function continuous on an interval $(-\pi, \pi)$ could be presented as a linear combination of both sine and cosine functions. These studies led to the development of the subject which can be found in the voluminous monographs of Antoni Zygmund [1959] and N.K. Bari [1961]. We introduce in this section the concept of frequency content, that is, the effect of a linear combination of sine and cosine functions.

Let us consider the functions $A \sin wx$ and $A \cos wx$ where A is called the **amplitude** and w is called the **(radial) frequency**. Thus the amplitude is simply a constant that scales the height of the sine and cosine functions and causes them to vary in the same prescribed way between A and $-A$. The (radial) frequency w is a measure of how often the function repeats itself. $\sin w$ repeats every 2π radians, whereas $A \sin wx$ repeats every $2\pi/w$ radians. An appropriate representation of frequency is the number of cycles or complete revolutions of the radian vector. When measured in cycles, the frequency is called circular frequency and is denoted, say, by μ. Clearly, $2\pi\mu = w$. The period T of $A \sin 2\pi\mu x$ or $A \cos 2\pi\mu x$ is defined as the number of x units required to complete one cycle or 2π radians. It is given mathematically as $T = 1/\mu = 2\pi/w$.

When the variable x represents a spatial measurement (such as the length of a vibrating string), a slightly different terminology is used. The period T is called the wavelength and is denoted by λ, whereas the radial frequency is called the wave number and is denoted by k. Let

$$f(t) = \sum_{k=1}^{n} (A_k \cos 2\pi\mu_k t + B_k \sin 2\pi\mu_k t). \qquad (5.43)$$

From a physical or intuitive point of view, the higher (larger) frequency terms help to make up the finer details of the function while the lower (smaller) ones contribute more to the overall or basic shape of the function. The **frequency content** of the function $f(t)$, given by (5.43), is a measure of all frequencies μ_k, $k = 1, 2, 3, \cdots, n$, used in the summation, in which mode they are used and how much of each is used. To be more precise, it is the set of triplex (A_k, B_k, μ_k).

A_k and B_k are called pure cosine and pure sine frequency contents of μ_k, respectively. The most comfortable way to know the frequency content of a function is to construct graphs of A_k and B_k versus μ_k. These graphs are called frequency domain plots. The graph of A_k versus μ_k is called pure cosine frequency plot while the graph B_k versus μ_k is called pure sine frequency plot.

5.2.2 Basic results

Let $f(t)$ be a periodic function with period T and Lebesgue integrable (say, in particular, continuous) over $(-T/2, T/2)$, then the Fourier series of $f(t)$ is the trigonometric series

$$\frac{1}{2}A_0 + \sum_{k=1}^{\infty}(A_k \cos kx + B_k \sin kx), \tag{5.44}$$

where

$$A_k = \frac{2}{T}\int_{-T/2}^{T/2} f(t) \cos\frac{2\pi kt}{T} dt, \ k = 1, 2, 3, \ldots \tag{5.45}$$

$$A_0 = \frac{1}{T}\int_{-T/2}^{T/2} f(t) dt$$

$$B_k = \frac{2}{T}\int_{-T/2}^{T/2} f(t) \sin\frac{2\pi kt}{T} dt, \ k = 1, 2, 3, \ldots, \tag{5.46}$$

and we write it as

$$f \sim \frac{1}{2}A_0 + \sum_{k=1}^{\infty}(A_k \cos kx + B_k \sin kx).$$

Here, we take

$$w_k = \frac{k}{T}, \ k = 0, 1, 2, 3, \ldots \tag{5.47}$$

Very often, we choose $T = 2\pi$.

A_k and B_k are called cosine Fourier coefficient and sine Fourier coefficient, respectively. The set of triplex (A_k, B_k, w_k), where A_k, B_k, w_k are given by (5.45), (5.46) and (5.47), respectively, is called the Fourier series frequency content.

The complex form of the Fourier series of $f(x)$ is

$$\sum_{-\infty}^{\infty} C_n \, e^{2\pi i n t/T},$$

where

$$C_n = \frac{A_n + iB_n}{2}, \ n < 0$$

$$C_0 = A_0$$

$$C_n = \frac{A_n - iB_n}{2}, \ n > 0$$

$$w_n = \frac{n}{T}, n = -2, -1, 0, 1, 2\,.$$

Let

$$T = 2\pi, \text{ and } S_n(f)(x) = \sum_{k=-n}^{n} C_k e^{ikx}$$

$$= \frac{1}{2}A_0 + \sum_{k=1}^{n}(A_k \cos kx + B_k \sin kx)$$

be the n-th partial sum of the Fourier series of f. Then

$$\begin{cases} S_n(f)(x) = , \frac{1}{\pi} \int_0^{2\pi} f(x-t) D_n(t) dt \\ \qquad\qquad = \frac{1}{\pi} \int_0^{2\pi} f(t) D_n(x-t) dt , \end{cases} \tag{5.48}$$

where

$$D_n(x) = \frac{1}{2} + \sum_{k=1}^{n} \cos kx = \frac{\sin(n+\frac{1}{2})x}{2 \sin \frac{x}{2}} \tag{5.49}$$

is the **"Dirichlet kernel"**, and

$$\sigma_n(f)(x) = \frac{S_0(f) + S_1(f) + \cdots + S_n(f)}{n+1} = \frac{1}{\pi} \int_0^{2\pi} f(x-t) K_n(t) dt, \tag{5.50}$$

where

$$K_n(x) = \frac{D_0(x) + D_1(x) + \cdots + D_n(x)}{n+1} = \frac{1}{n+1} \frac{\sin^2(\frac{n+1}{2})(x)}{2 \sin^2(\frac{x}{2})} \tag{5.51}$$

is called the **"Fejer kernel"**.

Theorem 5.1 (Bessel's Inequality).

$$\sum_{k=-\infty}^{\infty} |C_k|^2 \leq \|f\|^2_{L_2(0,2\pi)},$$

or

$$\frac{1}{2} A_0^2 + \sum_{n=1}^{\infty} (A_n^2 + B_n^2) \leq \|f\|^2_{L_2(0,2\pi)}.$$

This also means that $\{A_k\}$ and $\{B_k\}$ are elements of l_2.

Theorem 5.2. (Riesz-Fisher Theorem). *Let $\{C_k\} \in l_2$. Then there exists $f \in L_2(-\pi, \pi)$ such that $\{C_k\}$ is the k-th Fourier coefficient of f. Furthermore,*

$$\sum_{k=-\infty}^{\infty} |C_k|^2 = \|f\|_{L_2(-\pi,\pi)}.$$

For $\{A_k\}$, $\{B_k\}$ belonging to l_2, there exists $f \in L_2(0, 2\pi)$ such that A_k, B_k are, respectively, k-th cosine and sine Fourier coefficients of f. Furthermore,

$$\frac{1}{2} A_0^2 + \sum_{k=1}^{\infty} (A_k^2 + B_k^2) = (\int_{-\pi}^{\pi} |f|^2 dt).$$

Theorem 5.3. *Let $f \in L_2(-\pi, \pi)$, then*

$$\lim_{n\to\infty} \|f - S_n(f)\|_{L_2(-\pi,\pi)} = 0.$$

Theorem 5.4. *Let $f \in C[0,,2\pi]$ such that*

$$\int_\delta^{2\pi} \frac{\omega(f,t)}{t} dt < \infty, w(f,t) = \sup_t |f(x+t) - f(x)|.$$

Then the Fourier series of f converges uniformly to f; that is,

$$\lim_{n\to\infty} \|f - S_n(t)\|_{L^\infty(-\pi,\pi)} = 0.$$

If $\omega(f,\eta) = 0(\eta^\alpha)$, then the condition of the theorem holds.

Theorem 5.5. *If f is a function of bounded variation, then*

$$S_n(f) \longrightarrow \frac{f(x^+) + f(x^-)}{2} \text{ as } n \to \infty,$$

where

$$f(x^+) = \lim_{h\to 0^+} f(x+h)$$
$$f(x^-) = \lim_{h\to 0^-} f(x-h)$$

exists at every $x, a < x < b$.

Theorem 5.6. *Let $f \in L_1(R)$ satisfy the following two conditions:*

(i) *The series*

$$\sum_{k=-\infty}^{\infty} f(x + 2\pi k)$$

converges everywhere.

(ii) *The Fourier series*

$$\frac{1}{2\pi} \sum_{k=-\infty}^{\infty} \hat{f}(k) e^{ikx}$$

converges everywhere.

Then the following "Poisson's Summation Formula" holds:

$$\sum_{k=-\infty}^{\infty} f(x+2\pi k) = \frac{1}{2\pi}\sum_{k=-\infty}^{\infty} \hat{f}(k)e^{ikx}, x \in R. \tag{5.52}$$

In particular,

$$\sum_{k=-\infty}^{\infty} f(2\pi k) = \frac{1}{2\pi}\sum_{k=-\infty}^{\infty} \hat{f}(k),$$

where ($\hat{f}$denotes the discrete Fourier transform of f).

5.2.3 Continuous and discrete Fourier transforms

Let $f(x)$ be a continuous function of a real variable x, then the **Fourier transform** of $f(x)$ denoted by $\mathcal{F}(f)$ is defined by the following equation:

$$\mathcal{F}(f(x)) = \int_{-\infty}^{\infty} f(x)e^{-i2\pi tx}dx = F(t). \tag{5.53}$$

Given $F(t)$ the **inverse Fourier transform** denoted by $\mathcal{F}^{-1}$ is defined as

$$\mathcal{F}^{-1}\{F(t)\} = (f(x)) = \int_{-\infty}^{\infty} F(t)e^{i2\pi tx}dt. \tag{5.54}$$

Equations (5.54) and (5.55) are called the Fourier transform pair. If $F(t)$ is also continuous, then (5.55) always exists. The Fourier transform of $F(t)$ of a real variable function $f(x)$ is generally a complex-valued function, say

$$F(t) = R(t) + iJ(u), \tag{5.55}$$

where R and J are real and imaginary parts, respectively.

We can write (5.55) as

$$F(t) = |F(t)|e^{i\phi(t)},$$

where

$$|F(t)| = [R^2(t) + J^2(t)]^{\frac{1}{2}}$$
$$\phi(t) = \tan^{-1}\left[\frac{J(t)}{R(t)}\right].$$

The magnitude function $|F(t)|$ is called the **Fourier spectrum** of $f(x)$, and $\phi(t)$ its **phase angle.** $P(t) = |F(t)|^2$ is called the **power spectrum** or **spectral density** of $f(x)$. The variable t appearing in the Fourier transform may be called **frequency**

variable. The name arises from the fact that, when using Euler's formula, the exponential term $\exp[-i2\pi tx]$ can be written as

$$\exp[-i2\pi tx] = \cos 2\pi tx - i \sin 2\pi tx .$$

If we interpret the integral in (5.53) as a limit summation of discrete terms, it is clear that each value of t determines the frequency of its corresponding sine-cosine pair.

The set of triplex $(C(\mu),\ S(\mu),\ \mu)$ where

$$C(\mu) = F(\mu) + F(-\mu);$$
$$S(\mu) = i[F(\mu) - F(-\mu)],$$

is called the **Fourier transform frequency content.** It can be verified that

$$C(\mu) = 2\int_{-\infty}^{\infty} f(t) \cos 2\pi\mu t,$$

and

$$S(\mu) = 2\int_{-\infty}^{\infty} f(t) \sin 2\pi\mu t dt$$

$$C_1(\mu) = 2\int_{0}^{\infty} f(t) \cos 2\pi\mu t$$

is called the Fourier cosine transform and

$$S_1(\mu) = 2\int_{0}^{\infty} f(t) \sin 2\pi\mu t dt$$

is called the **Fourier sine transform.**

Example 5.1. Let $p_a(x) = \left\{ \begin{array}{l} 1,\ |x| \leq a, \quad a > 0 \\ 0,\ \text{otherwise} \end{array} \right\}$. $p_a(\cdot)$ is known as the pulse function. Its Fourier transform is

$$\begin{aligned} \mathcal{F}(p_a(x)) = F(t) &= \int_{-\infty}^{\infty} p_a(x) e^{-2\pi i t x} dx = \int_{-a}^{a} e^{-2\pi i t} dt \\ &= \frac{e^{2\pi t a} - e^{-2\pi t a}}{2\pi i t} \\ &= 2a \frac{\sin 2\pi t a}{2\pi t a}. \end{aligned}$$

The graph of this function $F(t)$ is given in Figure 5.7.

The Fourier transform of $f(t)$ defined by

$$f(t) = \begin{cases} e^{-at}, \ t \geq 0, & a > 0 \\ 0, & \text{otherwise}, \end{cases}$$

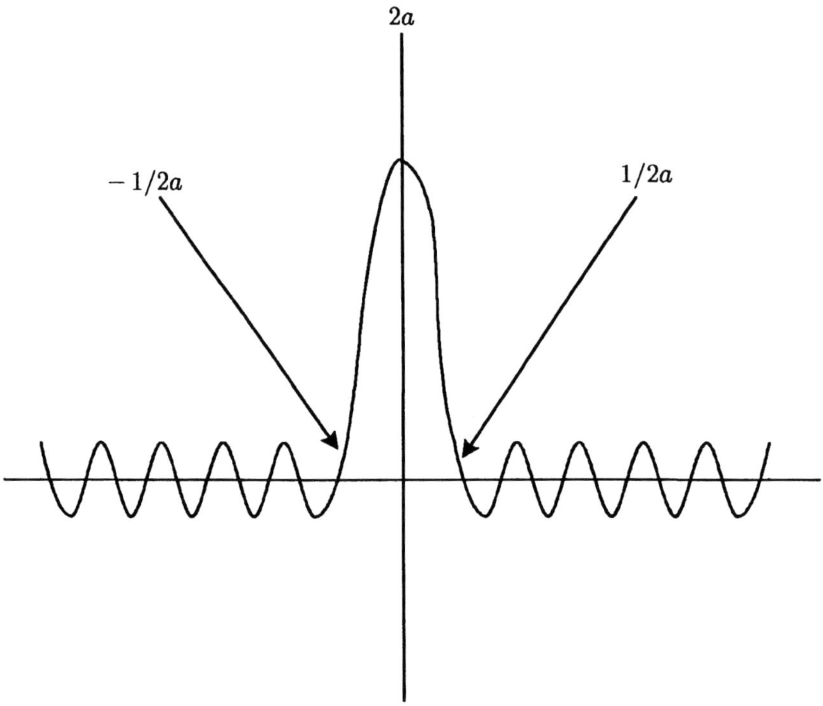

Figure 5.7. Fourier transform of the pulse functions.

is

$$F(t) = \frac{1}{a + 2\pi it} = \frac{a - 2\pi it}{a^2 + 4\pi^2\omega^2} .$$

The space of the set of points $(\mu, F(\mu))$ is called the **frequency domain** and the space of the set of the points $(t, f(t))$ is called the **temporal** or **time domain** when t represents time, where

$$\mathcal{F}(F(t)) = \int_{-\infty}^{\infty} f(x)e^{-2\pi i\mu t} = F(\mu) . \tag{5.56}$$

If t represents an optional variable, we call $[t, f(t)]$ the spatial domain.

An interesting property of the Fourier transform is that it is norm-preserving; namely,

$$\|f\|_{L_2} = \|F(f)\|_{L_2} .$$

Properties of Fourier transforms. The fundamental properties are enumerated in the form of the following theorems.

Theorem 5.7 (Linearity). *If $F(\mu)$ and $G(\mu)$ are Fourier transforms of $f(t)$ and $g(t)$, respectively, then $a\,F(\mu) + bG(\mu)$ is the Fourier transform of $h(t) = af(t) + bg(t)$; a and b are scalars.*

Theorem 5.8 (First shift theorem). *If the function $f(t)$ has a Fourier transform given by $F(\mu)$, then the Fourier transform of $f(t-a)$ is given by $F(\mu)e^{-2\pi i\mu a}$.*

Theorem 5.8 (Second shift theorem). *If $F(\mu)$ is the Fourier transform of the function $f(t)$, then $F(\mu - a)$ is the Fourier transform of the function $f(t)e^{2\pi iat}$.*

Theorem 5.9 (Scale change). *If the Fourier transform of $f(t)$ is $F(\mu)$, then the Fourier transform of $f(at)$ is given by $\frac{1}{|a|}F(\frac{\mu}{a})$, where a is any real number not equal to 0.*

Theorem 5.9 tells us that if $F(\mu)$ is the Fourier transform of $f(t)$, then $F(-\mu)$ is the Fourier transform of $f(-t)$.

Theorem 5.10 (Transform of a transform). *$\mathcal{F}(\mathcal{F}(f(t))) = f(-t)$, that is, Fourier transform of the Fourier transform of a function is equal to the function with the minus sign in the variable (rotated function of f).*

Example 5.2. The Fourier transform of the Gaussian function $f(x) = e^{-ax^2}$ is equal to

$$\sqrt{\frac{\pi}{a}}e^{-\pi^2\mu^2/a} .$$

If $a = \pi$, then we find that it is that function whose Fourier transform is the function itself.

Example 5.3. The Fourier transform of the Dirac delta function $\delta(x)$ is 1.

$$\begin{aligned}\delta(x) &= 0 \text{ if } x \neq 0\\ &= 1 \text{ if } x = 0 .\end{aligned}$$

Definition 5.1. The convolution of two continuous functions $f(x)$ and $g(x)$ is defined by the equation

$$f(x) * g(x) = \int_{-\infty}^{\infty} f(y)g(x-y)dy . \tag{5.57}$$

The concept of convolution is inherent in almost every field of the physical sciences and engineering. For example, in mechanics, it is known as the superposition or Duhamel integral. In system theory, it plays a crucial role as the impulse response integral and, in optics, as the point spread or smearing function. It has applications of vital importance in image processing.

The convolution of functions is associative, commutative and distributive; that is,

$$
\begin{aligned}
&f(x) * [g(x) * h(x)] = [f(x) * g(x)] * h(x) && \text{(Associative)}\\
&f(x) * g(x) = g(x) * f(x) && \text{(Commutative)}\\
&f(x) * [g(x) + h(x)] = f(x) * g(x) + f(x) * h(x)\,. && \text{(Distributive)}
\end{aligned}
$$

The convolution of two equal impulse functions is given by

$$
\begin{aligned}
h(x) &= \int_{-\infty}^{\infty} p_a(y)p_a(x-y)dy \\
&= \int_{-a}^{x+a} dy \;=\; x + 2a, \qquad -2a \le x \le 0,
\end{aligned}
$$

and

$$
h(x) = \int_{x-a}^{a} dy = 2a - x, \quad 0 \le x \le a,
$$

The fundamental properties of the convolution are summarized in the form of the following theorems.

Theorem 5.11 (Convolution theorem.) *If* $\mathcal{F}(f(x)) = F(x)$ *and* $\mathcal{F}(g(x)) = G(x)$, *then*

(i) $\mathcal{F}(f(x) * g(x)) = F(x)G(x)$.

(ii) $\mathcal{F}(f(x)g(x)) = F(x) * G(x)$.

Notions of the Fourier transform and convolution can be extended for functions of two variables and most of the properties carry over to the two-dimensional case. For example, if the Fourier transform pair for the two-dimensional function $f(x, y)$ is given by

$$
F(u,v) = \int_{-\infty}^{\infty}\int_{-\infty}^{\infty} f(x,y)e^{-2\pi i(ux+uy)}dxdy \tag{5.58}
$$

$$
f(x,y) = \int_{-\infty}^{\infty}\int_{-\infty}^{\infty} F(u,v)e^{2\pi i(ux+uy)}dudv, \tag{5.59}
$$

then the Scale Change Theorem takes the following form:

If the function $f(x, y)$ has a Fourier transform given by $F(u, v)$, then the Fourier transform of $f(ax, by)$ is given by

$$
\frac{1}{|a||b|}F\left(\frac{u}{a}, \frac{v}{b}\right).
$$

If $f(x,y) = h(x)g(y)$, then $F(u,v) = H(u)G(v)$ where $F(u,v)$ is the Fourier transform of $f(x,y)$ and $H(u)$ and $G(v)$ are the Fourier transforms of $h(x)$ and $g(y)$. The equation

$$f(x,y) * g(x,y) = \int_{-\infty}^{\infty}\int_{-\infty}^{\infty} f(\xi,\eta)g(x-\xi, y-\eta)d\xi d\eta\,,$$

is called the **convolution** of functions of two variables $f(x,y)$ and $g(x,y)$.

Example 5.4. (i) The convolution of a function $f(x)$ with the Dirac delta function (unit impulse function) is the function $f(x)$ itself; that is,

$$\begin{aligned} \delta(x) * f(x) &= \int_{-\infty}^{\infty} \delta(\alpha)f(x-\alpha)d\alpha \\ &= \int_{0^-}^{0^+} \delta(\alpha)f(x-\alpha)d\alpha \\ &= f(x)\int_{0^-}^{0^+} \delta(\alpha)d\alpha \\ &= f(x). \end{aligned}$$

(ii)

$$\begin{aligned} \delta(x,y) * f(x,y) &= \int_{-\infty}^{\infty}\int_{-\infty}^{\infty} \delta(\alpha-\beta)f(x-\alpha, y-\beta)d\alpha d\beta \\ &= f(x,y)\,. \end{aligned}$$

Discrete Fourier transform. We have seen in the previous section that images are transmitted through the Fourier transform and very often we face problems in evaluating the integrals involved. In such a situation, digital computers can come to our rescue provided integrals can be converted into a form amenable to computer analysis. We know that the computer recognizes sequences of numbers that represent functions. Therefore we discretize an arbitrary function to obtain a sequence of numbers that can be handled by a computer.

A **discrete Fourier transform** is an operation that maps a sequence $\{f(k)\}_0^{N-1}$ or $\{f_k\}$ to another sequence $\{F(j)\}_0^{N-1}\}$ or $\{F_j\}$ which is defined by the equation

$$F(j) = \frac{1}{N}\sum_{k=0}^{N-1} f(k)e^{-2\pi ikj/N}, \quad j \in [0, N-1]. \tag{5.60}$$

The sequence defined by the equation

$$f(k) = \frac{1}{N}\sum_{j=0}^{N-1} F(j)e^{2\pi ikj/N}, \quad k \in [0, N-1], \tag{5.61}$$

is called the **inverse discrete Fourier transform.** $\{F, t\}$ is called the **discrete Fourier transform pair.**

If we write $W_N = e^{2\pi i/N}$, called weighting kernel, then (5.60) and (5.61) take the form

$$F(j) = \frac{1}{N}\sum_{k=0}^{N-1} f(k)eW^{-kj} \tag{5.62}$$

$$f(k) = \frac{1}{N}\sum_{j=0}^{N-1} F(j)W^{kj}. \tag{5.63}$$

All theorems for the continuous Fourier transform have their counterparts in the discrete case; for example, the First Shift Theorem and the Transform of a Transform take the following form.

Theorem (5.12) (First Shift Theorem). *Let $\{F(j)\}_0^{N-1}$ be the discrete Fourier transform of $\{f(k)\}_0^{N-1}$, then the discrete Fourier transform of the shifted sequence $\{f(k-n)\}_0^{N-1}$ is equal to $\{F(j)W_N^{-jn}\}_0^{N-1}$.*

Theorem (5.13.) (Transform of a transform). *If $\{F(j)\}_0^{N-1}$ is the discrete Fourier transform of the sequence $\{f(k)\}_0^{N-1}$, then the discrete Fourier transform of the Fourier transform of the sequence $\{f(k)\}_0^{N-1}$ is equal to*

$$\frac{1}{N}\{f(N-k)\}_0^{N-1} = \frac{1}{N}\{f(-k)\}_0^{N}.$$

It can be easily checked that

$$F(-j) = F(N-j),\ j \in [0, N-1]$$
$$f(-j) = f(N-k),\ k \in [0, N-1].$$

5.2.4 The fast Fourier transforms

J.W. Tukey and J.W. Cooley published an algorithm in 1965 which tremendously reduces the number of computations required for computing the discrete Fourier transform of a sequence under certain conditions. This technique is called the **Fast Fourier Algorithm** (FFA) and is considered one of the most significant contributions of numerical analysis of this century. We present this algorithm here.

Let $F(j)$ be the discrete Fourier transform of $\{f(k)\}_0^{N-1}$. Let us assume that N is an even integer and form the two subsequences

$$\left\{\begin{array}{l} f_1(k) = f(2k) \\ f_2(k) = f(2k+1), \end{array}\right\} \; k = 0, \cdots, M-1, \text{ where } M = \frac{N}{2}. \tag{5.64}$$

It can be checked that

$$f_1(k+M) = f(2(k+M)) = f(2k+N) = f(2k) = f_1(k)$$
$$f_2(k+M) = f(2(k+M)+1) = f(2k+N+1) = f(2k+1) = f_2(k),$$

and so $\{f_1(k)\}$ and $\{f_2(k)\}$ are periodic sequences with period M.

By equation (5.63), we have

$$\left.\begin{cases} F_1(j) = \dfrac{1}{M}\displaystyle\sum_{k=0}^{M-1} f_1(k)W_M^{-kj} \\ F_2(j) = \dfrac{1}{M}\displaystyle\sum_{k=0}^{M-1} f_2(k)W_M^{-kj} \end{cases}\right\} \quad j = 0, 1, \cdots, M-1 . \tag{5.65}$$

It can be verified that both $F_1(j)$ and $F_2(j)$ are periodic with period M; that is,

$$\begin{aligned} F_1(j+M) &= F_1(j) \\ F_2(j+M) &= F_1(j). \end{aligned}$$

Now we shall show that the discrete Fourier transform of $\{f(k)\}_{k=0}^{N-1}$, namely, $F(j) = \frac{1}{N}\sum_{k=0}^{N-1} f(k)W_N^{-kj}$ can be expressed as the sum of the two Fourier transforms, one of even order and the other of odd order

$$F(j) = \frac{1}{N}\sum_{k=0}^{M-1} f(2k)W_N^{-2kj} + \frac{1}{N}\sum_{k=0}^{M-1} f(2k+1)W_N^{-(2k+1)j}.$$

Since

$$\begin{aligned} W_N^{-2kj} &= e^{-2\pi ikj/(N/2)} = W_M^{-kj}, \\ W_N^{-(2k+1)j} &= e^{-2\pi ij(2k+1)/N} = W_M^{-kj}W_N^{-j}. \end{aligned}$$

Therefore, the previous equation becomes

$$\begin{aligned} F(j) &= \frac{1}{N}\sum_{k=0}^{M-1} f_1(k)W_M^{-kj} + \frac{W_N^{-j}}{N}\sum_{k=0}^{M-1} f_2(k)W_M^{-kj}, \ j = 0, \cdots, N-1 \\ &= \frac{1}{2}F_1(j) + \frac{1}{2}W_N^{-1}F_2(j), \ j = 0, \cdots, N-1. \end{aligned}$$

Because $\{F_1(j)\}$ and $\{F_2(j)\}$ are periodic with period M, we have

$$\left.\begin{aligned} F(j) &= \tfrac{1}{2}[F_1(j) + F_2(j)W_N^{-1}] \\ F(j+M) &= \tfrac{1}{2}[F_1(j) - F_2(j)W_N^{-1}], \qquad j = 0, \cdots, M-1. \end{aligned}\right. \tag{5.66}$$

To calculate the discrete Fourier transform of $\{f(k)\}$, N^2 complex operations (additions and multiplications) are required, whereas to calculate the discrete Fourier transform of $\{f_1(k)\}$ or $\{f_2(k)\}$ requires only M^2 or $N^2/4$ complex operations.

When we use (5.66) to find $F(j)$ from $F_1(j)$ and $F_2(j)$, we need $N + 2(N^2/4)$ complex operations. In other words, we first require $2(N^2/4)$ operations to calculate the two Fourier transforms $\{F_1(j)\}$ and $\{F_2(j)\}$, and then we require the N additional operations prescribed by equation (5.66). Thus, we have reduced the number of operations from N^2 to $N + \frac{N^2}{2}$. Let us assume that N is divisible by 4 or $M = \frac{N}{2}$ is divisible by 2. Then the subsequences $\{f_1(k)\}$ and $\{f_2(k)\}$ can be further subdivided into four $\frac{M}{2}$ order sequences as per equation (5.64) as follows:

$$\begin{aligned} g_1(k) &= f_1(2k) \\ g_2(k) &= f_1(2k+1) \\ h_1(k) &= f_2(2k), \qquad k = 0, 1, \dots, \frac{M}{2} - 1 \\ h_2(k) &= f_2(2k+1). \end{aligned}$$

Thus, we can use (5.66) to obtain the discrete Fourier transforms $\{F_1(j)\}$ and $\{F_2(j)\}$ with only $M + \frac{M^2}{2}$ complex operations and then use these results to obtain $\{F(j)\}$ which requires $N + 2(M + \frac{M^2}{2}) = 2N + \frac{N^2}{4}$ operations. Thus, when we subdivide a sequence twice ($N > 4$ and N divisible by 4), we reduce the number of operations from N^2 to $2N + \frac{N^2}{4}$. The $2N$ term is the result of applying equation (5.66) (twice) whereas the $\frac{N^2}{4}$ term is the result of transforming the four reduced sequences. For the case $N = 4$, we completely reduce the sequence to four first order sequences that are their own transforms and, therefore, we do not require the additional $N^2/4$ transform operations. The formula then becomes $2N$. Continuing this process, we can show that if N is divisible by 2^p (p is a positive integer), then the number of operations required to compute the discrete Fourier transform of $\{f(k)\}_0^{N-1}$, the N-th order sequence by repeated subdivision is

$$pN + \frac{N^2}{2^p}.$$

Again, for complete reduction (i.e., $N = 2^p$), the $\frac{N^2}{2^p}$ term is not required and we obtain pN for the number of operations required. This results in a reduction factor of

$$\frac{pN}{N^2} = \frac{p}{N} = \frac{\log_2 N}{N}.$$

Thus, the essence of the Cooley-Tukey algorithm (fast Fourier transform algorithm) is to choose sequences with $N = 2^p$ and go to complete reduction. Although most sequences do not have such a convenient number of terms, we can always artificially add zeros to the end of the sequence to reach such a value. The extra number of terms in the sequence is more than compensated for by the tremendous saving of time due to the use of Cooley-Tukey algorithm. For example, a direct implementation of the transform for $N = 8192$ requires approximately 45 minutes on an IBM7094 machine while the same job can be done in 5 seconds by the same machine using the FFT algorithm (Cooley-Tukey algorithm). For details like implementation and the inverse FFT, we refer to Gonzalez and Woods [1993] and Press et al. [1990].

Forward and inverse transformation. Let T be the transform of $f(x)$ defined by the relation

$$T(u) = \sum_{x=0}^{N-1} f(x)g(x,u), \quad u = 0,1,2,\cdots,N-1,$$

where $g(x,u)$ is called the **forward transformation kernel.**

The **inverse transform** of T is given by the relation

$$f(x) = \sum_{u=0}^{N-1} T(u)h(x,u), \quad x = 0,1,2,\cdots,N-1$$

where $h(x,u)$ is called the **inverse transformation kernel.**

For two-dimensional square arrays, the forward and inverse transforms are given by the equations

$$T(u,v) = \sum_{x=0}^{N-1}\sum_{y=0}^{N-1} f(x,y)g(x,y,u,v),$$

and

$$f(x,y) = \sum_{u=0}^{N-1}\sum_{v=0}^{N-1} T(u,v)h(x,y,u,v).$$

The forward kernel is said to be **separable** if

$$g(x,y,u,v) = g_1(x,u)g_2(y,u).$$

The forward kernel is called **separable** and **symmetric** if

$$g(x,y,u,v) = g_1(x,u)g_1(y,v).$$

In a similar manner, we can define the separable and symmetric inverse transformation.

The two-dimensional Fourier transform and the inverse Fourier transform are separable and symmetric. For applications of the fast Fourier transform to different areas, we refer to van Loan [1992].

5.2.5 Fourier analysis via computer

From the discussion in the preceding sections, it is quite clear that the concepts of the Fourier series, the continuous Fourier transform and the discrete Fourier transform are basically transforms or mappings. The Fourier series maps an analytic function (continuous function) into a sequence (sequence of Fourier coefficients) while the continuous Fourier transform carries one function to another. The discrete Fourier transform is a transform from one sequence space into another sequence space. The Fourier series and the continuous Fourier transform both require the evaluation of integrals which may be very tedious and sometimes quite time-consuming and cumbersome. The discrete Fourier transform deals with bounded sequences and requires only straight-forward addition and multiplication of terms and, furthermore, by applying the FFT algorithm, it can be computed very rapidly and efficiently through the computer. We discuss here the methods through which the FFT algorithm can be used to calculate the Fourier transform and Fourier coefficients of a function. In the first place, one can convert, with the help of the **sampling theorems**, the function into a sequence and then calculate the discrete Fourier transform of this sequence, and from this we obtain the Fourier series or Fourier transform of the original function.

Sampling a function. Broadly speaking, digitizing or discretizing or sampling means conversion of a function $f(x)$ into a sequence $\{\overline{f}(k)\}$. This can be done by choosing values of that function at discrete locations of x. Let us assume that the discrete points or locations are evenly spaced in x with the distance between any two samples being Δx. The $\overline{f}(k)$ term of the sequence is equal to the value of the function $f(x)$ at $x = x_0 + k\Delta x$ [x_0 is the location of the first sample, that is, at $k = 0$]. Naturally, we want this sampled sequence to properly represent the function. $\{\overline{f}(k)\}$ is considered as an adequate representation of $f(x)$, if we can recover that function exactly from the sequence. That is to say, if we can interpolate between the sequence terms $\overline{f}(k)$ to retrieve the function $f(x)$. The theorem mentioned below is known as the **sampling theorem** which theoretically answers the question of how small our sample size Δx must be in order that the sampled values reasonably resemble the original function.

Theorem 5.14 [Sampling Theorem]. *Let $f(x)$ be a band-limited function with bandwidth $2a$, that is, $F(\omega) = 0$ for $|\omega| \geq a$. Then $f(x)$ is uniquely determined by a knowledge of its values at uniformly spaced intervals Δx apart ($\Delta x = \frac{1}{2a}$). Specially, we have*

$$f(x) = \sum_{k=-\infty}^{\infty} f(k\Delta x)\frac{\sin(2\pi a[x - k\Delta x])}{2\pi a[x - k\Delta x]}.$$

For proof, we refer to Weaver [1983].

This theorem provides a method for sampling a function in order to be able to uniquely recover the function from its sampled sequence. In other words, for this band-limited function, the highest frequency component present in a and the sampling theorem requires that we must have a sampling rate of at least $1/2a$. Physically, this means that we must have at least two samples per cycle of the highest frequency component present. This sample rate ($\Delta x = \frac{1}{2a}$) is often called the rate and the sequence obtained using this rate, the **Nyquist samples**. The sampling theorem also supplies us with an interpolation formula with which to recover the function from its Nyquist samples. For more explanation and discussion on the merits and demerits of the theorem, one may consult Weaver [83]. A function $f(x)$ is said to have bounded support if $f(x) = 0, \quad |x| \geq b$, where b is some positive constant.

In Fourier analysis, such a function is called **time-limited** or space-limited.

$f(x)$ is called almost time-limited if and only if, given any $\xi_T > 0$, there exists a positive real number b, called **time-limit**, such that

$$\int_{-\infty}^{-b} |f(x)|dx < \xi_T, \text{ and } \int_{b}^{\infty} |f(x)|dx < \xi_T.$$

A function $f(x)$ is called **almost band-limited** if and only if, given any $\xi_B > 0$, there exists a positive real number a, called the band limit, such that

$$\int_{-\infty}^{-a} |F(w)|dw < \xi_B, \text{ and } \int_{a}^{\infty} |F(w)|dw < \xi_B.$$

The Gaussian function $f(x) = e^{-ax^2}$ is both almost time-limited and almost band-limited. The band and time limits are determined, to a large extent, by the resolution, sensitivity, and/or dynamic range of the detection instruments. A more realistic theorem is as follows:

Theorem 5.15 [Real World Sampling Theorem]. *If the function $f(x)$ is almost band-limited with bandwidth $2a$, and almost time-limited, with time width $2b$, then $f(x)$ can be recovered from its sampled sequence to any desired accuracy. That is, given $\xi > 0$, we can always choose b and a such that*

$$f(x) = \sum_{k=-M}^{M} f(k\Delta x)\frac{\sin(2\pi 5a[x - k\Delta x])}{2\pi 5a[x - k\Delta x]} + \lambda_\xi(x),$$

where

$$\begin{aligned} |\lambda_\xi(x)| &< \xi \\ M\Delta x &> b \\ \Delta x &< \frac{1}{10a}. \end{aligned}$$

The following procedure is followed to calculate the Fourier transform by the computer (details can be found in Weaver [1989] and Gonzalez and Woods [1993]).

Assume that $f(x)$ is an almost time-limited function over the domain $[-c, b]$. Let us assume that it is almost band-limited over the domain $[-a, a]$. Then to digitally obtain the Fourier transform of this function, we proceed as follows:

1. If necessary ($c \neq 0$), form the new function $g(x)$ by shifting $f(x)$ to the right by an amount c; that is
$$g(x) = f(x-c).$$

2. Sample the function $g(x)$ with the sampling rate $\Delta x = \frac{1}{10a}$ and choose the number of samples N such that
$$N\Delta x > 10(b+c).$$

3. Calculate the discrete Fourier transform of this sampled sequence and multiply the resulting sequence by $N\Delta x$ to obtain the sequence $\{G(j\Delta w)\}$.

4. By the relations
$$F(-j) = F(N-j),\ j \in [0, N-1]$$
$$F(-k) = f(N-k),\ k \in [0, N-1],$$
we obtain values for the negative indices j that represent values for the negative frequencies $-j\Delta w (\Delta w = \frac{1}{N\Delta x})$.

5. Recover $G(w)$ from $\{G(j\Delta w)\}$ as per the real-world sampling theorem or by simply constructing a smooth curve between the sampled values.

6. If necessary ($c \neq 0$), recover $F(w)$ from $G(w)$ as per the formula
$$F(w) = G(w)e^{2\pi icw}.$$
It is clear that if $c = 0$, then $F = G$.

The methods in the previous section can be employed for sampling the function and then Fourier coefficients can be obtained from the terms of the discrete Fourier transform of the sampled sequence. Let $g(x)$ be a periodic function over $(-T, T)$ with period 2π and be Lebesgue integrable over $(-T, T)$. Define a function $f(x)$ as follows:
$$f(t) = g(t)\ ,\ t \in [\tfrac{-T}{2}, \tfrac{T}{2}]$$
$$f(t) = 0\ ,\ |t| > \tfrac{T}{2}.$$

It can be seen that [Weaver 1983] if C_k is k-th complex Fourier coefficient of $g(t)$, then $C_k = \frac{1}{T}F(\frac{k}{T})$, where $F(w)$ is the continuous Fourier transform of $f(t)$. Thus, we see that the Fourier coefficients of a function may be obtained from the Fourier transform of that function at equally spaced increments $w = \frac{k}{T}$.

Illustrative examples can be found in Weaver [1983]. For current developments concerning Shannon's sampling theory, we refer to Zayed [1993]. For a comprehensive discussion of the material presented in this section, we refer to Nievergelt [Chapters 4–6, 1999].

5.3 Wavelets with applications

5.3.1 Introduction

Wavelet analysis is the outcome of the synthesis of ideas that have emerged in different branches of mathematics, physics and engineering. Since the days of Fourier, scientists and engineers have made vigorous efforts to represent square integrable functions (signals having finite energy) as a linear combination of functions having some nice properties. Radamacher, Haar, Walsh, Franklin and Vilenkin constructed non-trigonometric orthogonal systems in their endeavour to accomplish this goal. The Walsh function was extensively studied and applied by electrical and electronic engineers during the seventies and eighties prior to the invention of wavelets in mid-eighties (see, for example, Siddiqi [1978, 1987] and reference therein). In 1981, Stromberg [1981] constructed an orthonormal spline system on a real line which is now termed as the first example of wavelet constructed by a mathematician. However, even without having the knowledge of this work, physicists like Grossman and geophysicists like Morlet were developing a technique to study non-stationary signals which led to the development of wavelet theory in the last decade (see, for example, Daubechies [1992] and Meyer [1993]). Meyer, Daubechies, Mallat et al. have put this theory on a firm foundation through the multiresolution analysis and establishing relationships between function spaces and wavelet coefficients. This scientific discipline of vital importance has been elegantly introduced by Meyer [1993], where he has also explained the relationship between fractals (another exciting scientific discipline) and wavelets along with future avenues of researches, especially in understanding the hierarchical organization and formation of distant galaxies. For latest interaction of fractals and wavelets, we refer to Berkner [1997], Mendivil and Vrscay [1997], and Siddiqi et al. [1997, 1999]. For current references for results on theory and applications of wavelets, we refer to Kovačevič and Daubechies [1996], Louis, Maaβ and Rieder [1997], Dahmen [1997], De Vore [1998], Kobayashi [1998], Ogden [1997], Liu and Chan [19098], and Canuto and Cravero [1997]. Since 1991, a generalization of wavelets, known as wavelet packets, has been studied by Coifman et al. [1992]. Wavelet packets were also named arborescent wavelets and they are particular linear combinations or superposition of wavelets. Discrete wavelet packets have been thoroughly studied by Wickerhauser [1994] who has also developed computer programmes and implemented them.

The aim of signal analysis is to extract relevant information from a signal by transforming it. In order to study the spectral behaviour of an analog signal from its Fourier transform, full knowledge of signal in the time-domain must be acquired. If a signal is altered in a small neighbourhood of some time instant, then the entire spectrum is affected. Indeed, in the extreme case, the Fourier transform of the delta distribution $\delta(t - t_0)$, with support at a single point t_0, is $e^{-it_0\omega}$, which certainly covers the whole frequency domain. Hence, in many applications such as analysis of **non-stationary signals and real-time signal processing**, the formula of the Fourier transform alone is quite inadequate.

A typical problem is the analysis of the sound which we hear when we blow a flute. We can observe that this sound consists of high-frequency parts as well as low-frequency parts. If we use the normal Fourier analysis, we would need extremely high frequencies to represent this jump from high to low frequencies. We can avoid this jump if we look not at the whole time interval but just at an interval where we find mainly frequencies of the same order. This means that we introduce **'time windows'**. We achieve such a time window technically by introducing a window function g. The usual approach is to introduce time-dependency in the Fourier analysis while preserving linearity. The idea is to introduce 'local frequency' parameter (local in time) so that the 'local' Fourier transform looks at the signal through a window over which the signal is approximately stationary.

The deficiency of the formula of the Fourier transform in time-frequency analysis was already observed by D. Gabor who, in his 1946 paper, introduced a time-localization 'window function' $g(t-b)$, where the parameter b is used to translate the window in order to cover the whole time-domain for extracting local information of the Fourier transform of the signal. In fact, Gabor used a Gaussian function for the window function g. Since the Fourier transform of a Gaussian function is again a Gaussian, the inverse Fourier transform is localized simultaneously. It is observed that the time-frequency window of any Gabor transform is rigid and, hence, is not very effective for detecting signals with high frequencies and investigating signals with low frequencies. This motivates the introduction of wavelet transform which windows the function (signal) and its Fourier transform directly. It allows room for a dilation (or scale) parameter that narrows and widens the time-frequency window according to high and low frequencies. In other words, **the wavelet transform is a tool** that cuts up data or function into different frequency components and then studies each component with a resolution matched to its scale.

The wavelet theory provides a unified framework for a number of techniques which had been developed independently for various signal processing applications. For example, multiresolution signal processing, used in computer vision; subband coding, developed for speech and image compression; and wavelet series expansions, developed in applied mathematics, have been recently recognized as different views of a single theory.

We present here some basic results of the wavelet theory, a fast-developing field, which have brought about tremendous improvements in computing time while solving models of real-life problems; compression ratio, and noise reduction in image processing. For a deeper insight, we refer to original sources like Beylkin, Coifman and Rokhlin [1991], Glowinski, et al. [1990], Daubechies [1992], Amartunga and Williams [1993], Walker [1997], Siddiqi [1998], Siddiqi and Ahmad [1998], and Kobayashi [1998].

5.3.2 Wavelets and multi-resolution analysis

Definition 5.3.1. *A* **wavelet** *is a function $\psi(t) \in L_2(R)$ such that the family of* *functions*

$$\psi_{j,k}(t) = 2^{j/2}\psi(2^j t - k), \tag{5.67}$$

where j and k are arbitrary integers, is an orthonormal basis in the Hilbert space $L_2(R)$.

Remark 5.3.1. This definition means that

(i) $\langle \psi_{j,k}(t), \psi_{m,n}(t)\rangle = \delta_{j,m}\delta_{k,n}$, where $\delta_{j,m}$ and $\delta_{k,n}$ are Kronecker delta $\left(\begin{array}{ll}\delta_{j,m} = 1 & \text{if } j = m \\ \quad\;\; = & \text{if } j \neq m\end{array}\right)$ and $\langle\cdot,\cdot\rangle$ denotes the inner product of $L_2(R)$.

(ii) $f \in L_2(R)$ can be written as

$$f = \sum_{j\in Z}\sum_{k\in Z} \langle f, \psi_{j,k}(t)\rangle\, \psi_{j,k}(t), \tag{5.68}$$

where Z denotes the set of integers.

Definition 5.3.2. Wavelet coefficients *of a function $f \in L_2(R)$, denoted by $c_{j,k}$ are defined as the inner product of f with $\psi_{j,k}(t)$; that is,*

$$\begin{aligned} c_{j,k} &= \langle f, \psi_{j,k}(t)\rangle = \textit{ the inner product of } f \textit{ with} \\ \psi_{j,k}(t) &= \int_R f(t)\psi_{j,k}(t)dt. \end{aligned} \tag{5.69}$$

The series

$$\sum_{j\in Z}\sum_{k\in Z} \langle f, \psi_{j,k}\rangle\, \psi_{j,k}(t), \tag{5.70}$$

is called the **wavelet series** *of $f \in L_2(R)$.*

Remark 5.3.2. (i) For a given wavelet $\psi(t)$, a scaled and translated version is obtained by

$$\psi_{a,b}(t) = \frac{1}{\sqrt{a}}\psi\left(\frac{t-b}{a}\right), \quad a \neq 0,\ b \in R. \tag{5.71}$$

The parameter a corresponds to the scale while b is the translation parameter. The wavelet $\psi_{1,0}(t) = \psi(t)$ is called the **basic wavelet** or **mother wavelet**.

(ii) Since $\psi_{j,k}(t)$ oscillates more quickly, therefore it is more suited for representing finer details in the signal $\psi_{j,k}(t)$ which is localized about the point $t = 2^{-j}k$. The wavelet coefficient $c_{j,k}$ measures the amount of fluctuation in the function about the point $t = 2^{-j}k$ with the frequency determined by the dilation index j.

(iii) (a) Given a real number h we define the **translation operator** T_h acting on functions defined on R by the formula

$$T_h(f)(x) = f(x-h).$$

(b) Given an integer s, we define the dyadic **dilation operator** J_s acting on functions defined on R by the formula

$$J_s(f)(x) = f(2^s x).$$

(iv) In order to have the existence of an inverse transform of the continuous wavelet transform defined below, a technical condition must be satisfied by $\psi \in L_2(R)$; namely

$$0 < c_\psi = 2\pi \int_R \frac{|\hat{\psi}(w)|^2}{|w|} dw < \infty, \tag{5.72}$$

where $\hat{\psi}(w)$ is the Fourier transform of ψ. Very often, this property is taken as the definition of wavelet, that is, a function ψ in $L_2(R)$ satisfying (5.72) is called a wavelet.

Definition 5.3.3. *Let ψ be a wavelet. Then the* **wavelet transform** *of $f \in L_2(R)$ is defined as*

$$T_\psi f(a,b) = \frac{1}{\sqrt{a}} \int_R f(t)\psi\left(\frac{t-b}{a}\right) dt = \langle f, \psi_{a,b}(t) \rangle . \tag{5.73}$$

Remark 5.3.3. (i) If ψ satisfies (5.72), then f can be constructed by

$$f(t) = c_\psi^{-1} \int_0^\infty \frac{da}{a^2} \int_{-\infty}^\infty T_\psi f(a,b)\psi_{a,b}(t)db, \tag{5.74}$$

that is, the truncated integral

$$\int_{1/A}^{A} \frac{da}{a^2} \int_{-B}^{B} T_\psi(t,b)\psi_{a,b}(t)db,$$

converges to $c_\psi f(t)$ in $L_2(R)$ as A and B approach $+\infty$.

(ii) condition (5.72) implies that $\hat{\psi}(0) = 0$ so that $\int_R \psi(t)dt = 0$.

(iii) More generally, one may impose the condition

$$\left[\frac{d^k}{dw}(\hat{\psi})\right](0) = 0 = \int_R t^k \psi(t)dt, \quad k = 0,1,2,\ldots,N.$$

(iv) Problem 5.13 provides a method for constructing a variety of wavelets.

(v) It can be proved that T_ψ is an isometry, that is, $||T_\psi(f)||_{L_2} = ||f||_{L_2}$.

(vi) the adjoint operator

$$\begin{cases} T_\psi^* : L_2\left(R^2, \dfrac{dadb}{a^2}\right) \to L_2(R) \\ g \to c_\psi^{-1/2} \displaystyle\int_R \int_R |a|^{-1/2} \psi\left(\dfrac{t-b}{a}\right) g(a,b) \dfrac{dadb}{a^2}, \end{cases} \tag{5.75}$$

inverts the wavelet transform on its range; that is,

$$T_\psi^* T_\psi = I \text{ and } T_\psi T_\psi^* = P_\psi, \tag{5.76}$$

where P_ψ is the orthogonal projection of T_ψ onto its range.

For the proof of these two results, we refer to Louis, Maaβ and Rieder [pp. 7–8, 1997].

(vii) It is interesting to observe that the wavelet transform for $a = 2^{-j}$ and $b = k2^{-j}$ is the wavelet coefficient $c_{j,k}$.

Definition 5.3.4. *A* **multiresolution analysis** *(MRA) is a sequence of closed subspaces of $L_2(R)$ such that*

(i) $V_{-2} \subset V_{-1} \subset V_0 \subset V_1 \subset V_2 \subset \cdots$.

(ii) span $\bigcup_{j \in Z} V_j = L_2(R)$.

(iii) $\bigcap_{j \in Z} V_j = \{0\}$.

(iv) $f(x) \in V_j$ *if and only if* $f(2^{-j}x) \in V_0$.

(v) $f \in V_0$ *if and only if* $f(x-m) \in V_0$ *for all* $m \in Z$.

(vi) There exists a function $\varphi \in V_0$*, called the* scaling function, *such that the system* $\{\varphi(x-m)\}_{m \in Z}$ *is an orthonormal basis in* V_0.

It may be observed that a scaling function φ determines the multiresolution completely. It induces a wavelet, often referred to as the **father wavelet**. The scaling equation and its equivalent forms will be discussed in the next section along with decomposition and reconstruction algorithms.

In the remaining part of this section, we would like to answer the following natural questions:

Q1. What is the general problem involving wavelets?

Q2. Are there functions satisfying conditions of Definition 5.3.1?

Q3. Is there any relationship between wavelets and multiresolution analysis?

Q4. Are there any advantages of wavelet transforms and wavelet series over such concepts in Fourier analysis?

Q5. What is the convergence theory of the wavelet series?

Q6. What are the applications of concepts introduced above in real-life problems, in particular, industrial problems?

Q7. Is it possible to extend Definitions 5.3.1 and 5.3.4 to $L_2(R^n)$, where n is any natural number?

Q1 General problem. Let $f(t)$ be a function defined for $t \in R$. Let us imagine that this function describes some real-life phenomenon. To make things mathematically simple, let us suppose that $f \in L_2(R)$. Our object is to transmit/store/analyze this function using some finite device. For example, f represents a voice signal and we want to transmit it over the telephone lines or put it on a compact disk. If we can find an orthonormal basis $\{\varphi_n\}$ in $L_2(R)$, then we can write

$$f = \sum_{n \in N} c_n \varphi_n, \tag{5.77}$$

where the series converges in $L_2(R)$, and the coefficients c_n are uniquely determined by the formulas

$$c_n = \langle f, \varphi_n \rangle \text{ for } n \in N. \tag{5.78}$$

Thus, instead of transmitting the function f, it suffices to transmit the sequence of coefficients $\{c_n\}_{n \in N}$ and let the recipient sum the series himself. It is not a finite procedure. To make it finite, we have to choose a finite set $A \subset N$ such that $\sum\limits_{n \in A} c_n \varphi_n$ will be very close to $\sum\limits_{n \in N} c_n \varphi_n$. This means that the recipient is really forming the sum $\sum\limits_{n \in A} \tilde{c}_n \varphi_n$, where $\tilde{c}_n$ and c_n are almost equal, that is, the distance between them could be ignored. This is a very general theme and there have been many ways to deal with various special instances of different aspects of this arch-type problem. Wavelets are one of the new tools to tackle this type of problem effectively and efficiently.

Q2 Existence and examples of wavelet. According to Problem 5.13, every differentiable function whose derivative belongs to $L_2(R)$ will define a wavelet. In fact, the set of all wavelets satisfying (5.72) is dense in $L_2(R)$ (Problem 5.14). We mention below the Haar, Daubechies, Shannon and Gaussian-related wavelets.

Haar wavelet [Alfred Haar, 1911].

$$\begin{cases} H(t) = 1, & \text{if } 0 \le t \le \frac{1}{2} \\ \quad\;\; = -1, & \text{if } \frac{1}{2} \le t < 1 \\ \quad\;\; = 0, & \text{otherwise} \end{cases} \tag{5.79}$$

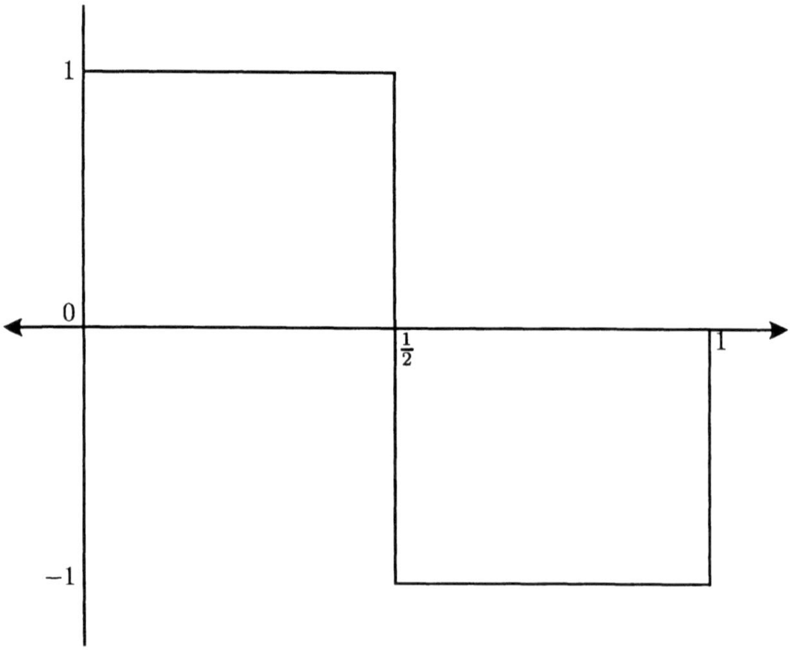

Figure 5.8a. The Haar function.

$$\{H_{j,k}(t)\} = \left\{2^{j/2}H(2^j t - k)\right\} ,$$

is an orthonormal basis of $L_2(R)$. Support of $H_{j,k} = [k2^{-j}, (k+1)2^{-j}]$. The intervals $[k2^j, (k+1)2^j]$ for $k, j \in Z$ form the family of dyadic intervals.

Daubechies wavelets Ingrid Daubechies 1988]. To design continuous wavelets amenable to a fast transform with an interval with finite length, Ingrid Daubechies introduced a basis block or scaling function by ϕ as shown in Figure 5.9.

$$\phi(r) = 0 \text{ if } r \leq 0 \text{ or } 3 \leq r, \tag{5.80}$$

ϕ does not admit any algebraic formula in terms of elementary mathematical functions. Starting from the initial values

$$\begin{aligned} h_0 &= \phi(0) = 0 \\ h_1 &= \phi(1) = \frac{1+\sqrt{3}}{2} \\ h_2 &= \phi(2) = \frac{1-\sqrt{3}}{2} \\ h_3 &= \phi(3) = 0\,. \end{aligned}$$

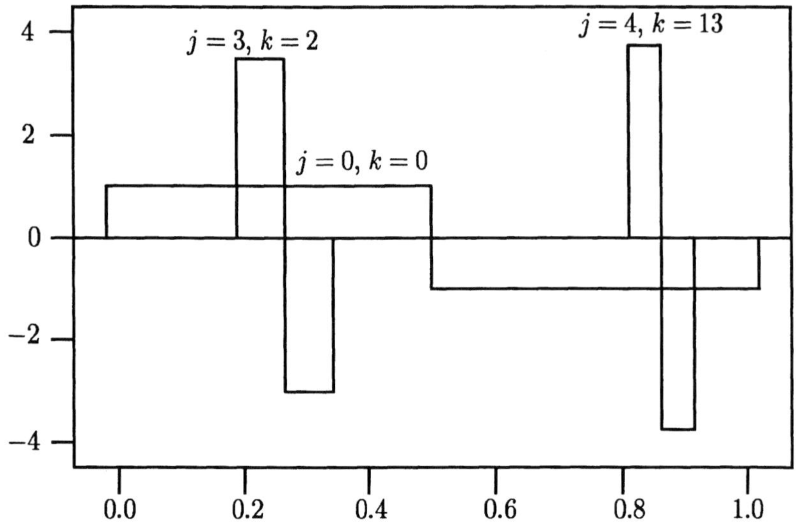

Figure 5.8b. Haar wavelet examples at different scales $H_{0,0}(t)$, $H_{3,2}(t)$, $H_{4,3}(t)$.

The building block ϕ satisfies the recurrence relation

$$\begin{cases} \phi(r) = \dfrac{1+\sqrt{3}}{4}\phi(2r) + \dfrac{3+\sqrt{3}}{4}\phi(2r-1) \\ \qquad + \dfrac{3-\sqrt{3}}{4}\phi(2r-2) + \dfrac{1-\sqrt{3}}{4}\phi(2r-3). \end{cases} \tag{5.81}$$

It is clear that $\varphi(0)+\varphi(1)+\varphi(2)+\varphi(3) = 1$. This recurrence relation takes the form of an inner product with (h_0, h_1, h_2, h_3) :

$$\varphi(r) = h_0 \cdot \varphi(2r) + h_1 \cdot \varphi(2r-1) + h_2 \cdot \varphi(2r-2) + h_3 \cdot \varphi(2r-3).$$

It can be seen that

$$\phi\left(\frac{1}{2}\right) = \frac{2+\sqrt{3}}{4} \text{ and } \phi\left(\frac{3}{2}\right) = 0.$$

Gaussian related wavelet. $\psi(t) = -\dfrac{1}{2}\dfrac{d^2}{dt^2}e^{-t^2}$.

Shannon wavelet. If $\varphi(x)$ is equal to the Shannon sampling function $\frac{\sin \pi x}{\pi x}$, then the corresponding wavelet is the Shannon wavelet which is given by the expression

$$\psi_{\text{Shannon}}(x) = \frac{\sin(2\pi x) - \sin \pi x}{\pi x}. \tag{5.82}$$

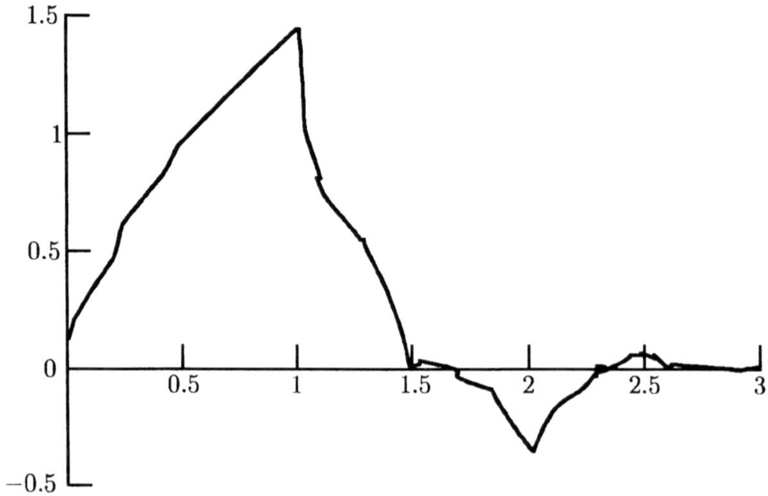

Figure 5.9. Ingrid Daubechies' basic building block φ.

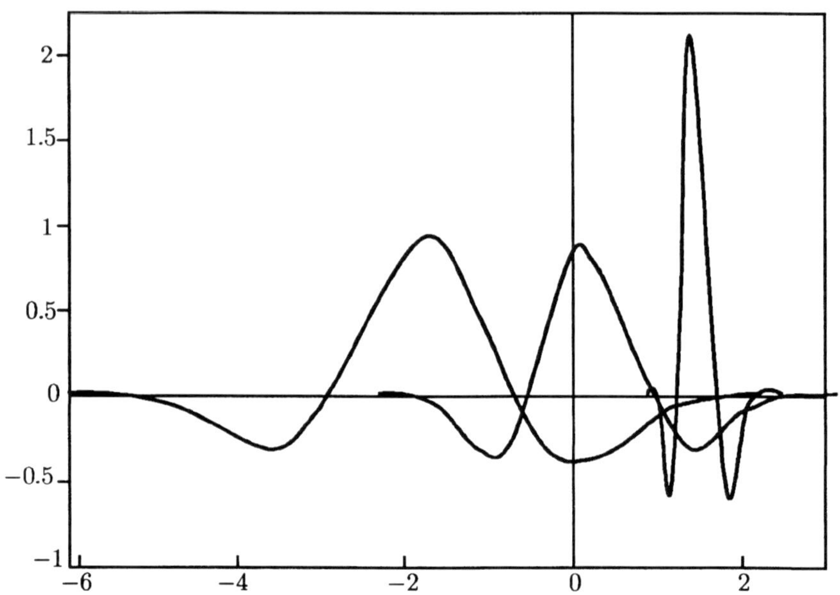

Figure 5.10. A few wavelets obtained from the mother wavelet $\psi(t) = (1 - 2t^2)^{-t^2} = \psi_{1,0}(t)$, which is the second derivative of a Gaussian. Displayed are (from left to right): $\psi_{3(/2),-2}(t)$, $\psi_{1,0}(t)$, $\psi_{(1/4),2^{1/2}}$.

Q3 Relationship between wavelets and multiresolution analysis. Corresponding to every wavelet, we can construct a multiresolution analysis; see, for example, Problem 5.16 where it is shown that there is a multiresolution analysis corresponding to the Haar wavelet. Conversely, one can construct a wavelet from a multiresolution analysis, see, for example, discussion on pp. 32–38 [Wojtaszczyzk, 1997] where it is proved that if φ is a scaling function, then the corresponding wavelet is given by the formula

$$\psi(x) = \sum_{n\in Z} \bar{a}_n(-1)^n\varphi(2x+n+1), \tag{5.83}$$

where

$$a_n = \int_{-\infty}^{\infty} \varphi\left(\frac{x}{2}\right)\overline{\varphi(x-n)}dx. \tag{5.84}$$

More details are given in the next section.

Q4 Advantages of wavelets over Fourier methods. During the last three decades, signal processing has become a discipline of vital importance. Most of the signals in nature or otherwise are transient (time-dependent/non-stationary). It was realized that the tools of Fourier analysis are incapable of studying such signals properly. This realization provided motivation for invention of wavelet analysis which removed effectively the following drawbacks of the Fourier analysis:

(i) The computation of Fourier transform for only one frequency at a time.

(ii) Exact representations cannot be computed in real time.

(iii) The Fourier transform provides information only in the frequency domain, but none in the time domain.

The main disadvantage of the Fourier transform in signal processing is its missing localization property, namely, if a signal changes at a specific time, its transform changes everywhere and a simple inspection of the transformed signal does not reveal the alteration. The main reason for this is the periodic behaviour of the trigonometric functions. In order to remove this deficiency of the Fourier method, we use small waves or wavelets and, in this case, translation and scaling allows for a frequency resolution at arbitrary positions. The Fourier transform considers phenomena in an infinite interval, and this conflicts from everyday point of view. It decomposes signals in plane waves (trigonometric functions), which oscillate infinitely with the same period, and these have no local character. The wavelet transform allows more flexibility; the wavelet which can be almost any chosen function can be shifted and dilated to analyze signals. Wavelets may be thought of as generalization of oscillations, abstractly expressed in a zero mean value.

If f in Definition 5.3.3 shows a big change in a neighbourhood $N(b)$ of the point b, it has a high frequency spectrum there. Since the set $\{\psi((\cdot - b)/a/a \in R\backslash\{0\}\}$ zooms into b for sufficiently small a, the corresponding values of the wavelet transform characterize the high frequency parts of f in $N(b)$. The wavelet transform $T_\psi f$ can be interpreted as

(i) a phase-space representation of f,

(ii) an approximation of a derivative of f, and

(iii) the splitting up of f into different frequency bands.

The first interpretation gives localization properties of the wavelet transform yielding a generalized uncertainty principle. The interpretation as the approximation of a derivative of f leads to finding jumps in derivatives, a crucial property for pattern recognition.

As we have seen wavelets are intrinsically connected to the notion of multiresolution analysis, by which objects like signals, functions, data can be examined using widely varying levels of focus. As a simple analogy, consider observing a car show-room. The observation can be made from a greater distance, at which the viewer can discern only the basic shape of the structure. As the observer moves near, various other details of the car show-room can be observed like the number of cars parked there. Moving further closer, the observer may notice the models of cars and other objects in the room. Continuing still further, it is possible to observe the details of paintings in the room. The basic framework of all these views is essentially application of the wavelet methods. This capability of multiresolution analysis is known as "zoom-in, zoom-out" property. Thus, wavelet analysis is an excellent tool to examine features of the signal of any size adjusting a scaling parameter in the analysis.

Signals are typically contaminated by random noise and a major part of signal processing is accounting for this noise. A special emphasis is on denoising, that is, extracting the true (pure) signal from the noisy version actually observed. Wavelets have performed admirably well in this field and are superior to any other tools. The wavelet method is well suited for denoising signals not only those with smooth, well behaved natures, but also those signals with abrupt jumps, sharp spikes, and other irregularities. If signal processing is to be done in real time, that is, if the signals are treated as they are observed, it is important that fast algorithms are implemented. One of the key advantages that wavelets have in signal processing is the associated fast algorithms which are faster than the fast Fourier transform. The wavelet compression technique is far superior to Fourier methods, say DCT for signals and image compression with minimum loss of originality or with maximum accuracy. This new technique has done a remarkable job in compression of 30 million sets of finger prints collected by the United States Federal Bureau of Investigation (FBI) in less than 30 kilobytes of storage space for an adequate representation of the digital data achieving compression ratio of 20:1. Recently Walker [1997] has demonstrated with the specific examples and, using Haar and Daubechie wavelets, that a compression ratio of 25:1 could be achieved with permissible distortions. Siddiqi and Ahmad [1998] have investigated the compression ratio of images available at the "Brag-zone site" of the university of Waterloo applying Fourier method (JPEG), fractal method and wavelet method. It has been observed that the performance of wavelet method is superior to the JPEG in all cases and only similar to fractal method in few cases. Details are presented in Section 5.3.4.

Q5 Convergence of wavelet series. In 1966, Carleson proved the famous Lusin conjecture that the Fourier series of an $L_2(R)$ function f converges pointwise almost everywhere to f. This result was extended by Hunt to L_p functions when $1 < p < 2$. For Fourier series of R^2, in 1971, C. Feffeman proved that spherically summed two-dimensional Fourier series of $L_p(R^2)$ functions do not converge in $L_p(R^2)$ for a certain real p. Kelly, Kon, Raphael [1994, 1994] have studied the convergence of wavelet expansions including the inter-connection between classical results in this area and their results including those concerning such expansions by Meyer and Walter. Recently, Manchanda, Mukheimer and Siddiqi [1998] have studied the pointwise convergence of two-dimensional wavelet expansions incorporating rotation. Essentially, they have proved that such an expansion of a continuous function f in $L_1(R^2) \cap L_2(R^2)$ converges uniformly on a compact subset. Relaxation of continuity and weakening of regularity condition on the wavelet require investigation on the lines of Kelly et al [1994].

Q6. Real-life and industrial applications. Wavelet analysis has outperformed all other mathematical theory known till today as far as the practical problems are concerned. Some typical examples will be discussed in Sections 5.3.4 and 5.3.5. For more case studies like wavelet multiresolution display of coastline data; wavelet analysis for a text-to-speech system, atmospheric wind, turbulent fluid, seismic data, we refer to Kobayashi [1998]. For a Spanish cement production data set analysis applying the discrete wavelet transform, we refer to Arino and Vidakovic [1999]. For applications of wavelet transforms to financial stock index event detection, one may go through Nievergelt [pp. 33–34, 1999].

Q7. Multivariable wavelet and multi-dimensional multiresolution analysis. In Section 5.3.4, we briefly introduce th'se concepts in two dimensions; for more information, we refer to Dai, Larson and Speegle [1997] and Wojtaszczyk [1997].

5.3.3 Special features of wavelets

Interpretation of multiresolution analysis and scaling equations. Let $\{V_j\}_{j\in Z}$ be an MRA with the scaling function $\varphi(x)$ with each V_j having ONB $\{\varphi_{j,k}, k \in Z\}$. Furthermore, let $P_j(f)$ denote the projection of a function f onto the space V_j (the best approximation f in V_j) and $\psi(x)$ be the wavelet associated with φ by (5.83). Then

$$P_j(f) = P_{j-1}(f) + g_{j-1}. \tag{5.85}$$

The detail function g_{i-1} (residual between two approximations $P_j(f) - P_{j-1}(f)$) can be written in terms of $\psi_{j,k}(x)$ as follows:

$$P_j(f) = P_{j-1}(f) + \sum_{k \in Z} \langle f, \psi_{j-1,k} \rangle \psi_{j-1,k} \tag{5.86}$$

or

$$P_j(f) = P_{j_0}(f) + \sum_{\ell=j_0}^{j-1} \sum_{k} \langle f, \psi_{\ell,k} \rangle \psi_{\ell,k}. \tag{5.87}$$

Let

$$W_j = \text{span}\{\psi_{j,k}, k \in Z\}, \tag{5.88}$$

often called the *wavelet subspace*;

$$V_j = \text{span}\{\varphi_{j,k}, k \in Z\}, \tag{5.89}$$

$$\text{where } \varphi_{j,k}(x) = 2^{j/2}\varphi(2^j x - k). \tag{5.90}$$

Then

$$V_j = V_{j-1} \oplus W_{j-1}, \tag{5.91}$$

where $\oplus$ represents the orthogonal sum of two subspaces of $L_2(R)$ $(W_{j-1} \perp V_{j-1})$.

It is also clear that

$$f(x) \in W_j \text{ if and only if } f(2x) \in W_{j+1}. \tag{5.92}$$

In fact, it can be verified that

$$L_2(R) = \overline{\oplus W_j}. \tag{5.93}$$

Equations (5.86) and (5.88) tell us the basic property of wavelet, namely, it is possible to construct approximations at increasing levels of resolution that are linear combinations of dilations and combinations of a scaling function φ, the differences in approximations being expressed as linear combinations of dilations and translations of a wavelet function ψ. Furthermore, the scaling function φ and the wavelet ψ (their dilates and translates) are orthogonal. The ideas of the MRA with detail spaces between successive levels of approximation are expressed in Figure 5.11, where arrows denote composition, namely, V_j is composed of V_{j-1} and W_{j-1}, etc.

From conditions (i) and (vi) of the MRA (Definition 5.3.4), we find that the scaling function φ is in V_1. By condition (iv) of Definition 5.3.4, $\varphi(x/2)$ belongs to V_0 and applying condition (vi) again, we get

$$\varphi(x/2) = \sum_{n \in Z} a_n \varphi(x - n), \tag{5.94}$$

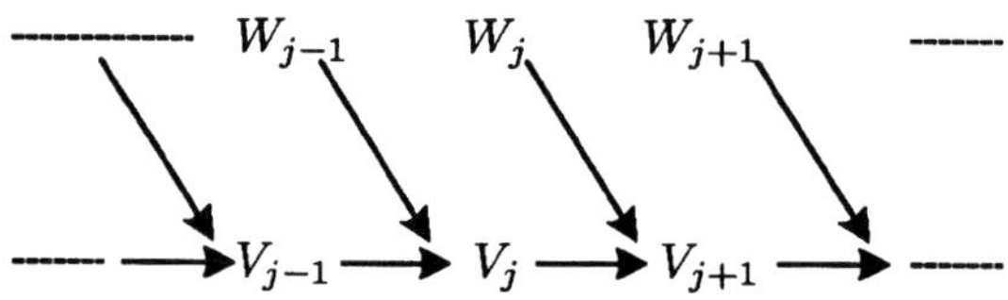

Figure 5.11. Relation of approximation and detail spaces.

or equivalently

$$\varphi(x) = \sum_{n \in Z} a_n \varphi(2x - n), \tag{5.95}$$

where $\{a_n\}$ is a sequence of scalars.

Equation (5.95) and the following equivalent equations are called **scaling equations**:

$$\hat{\varphi}(\xi) = m_\varphi(\xi/2)\hat{\varphi}(\xi/2) \tag{5.96}$$

$$\hat{\varphi}(2\xi) = m_\varphi(\xi)\hat{\varphi}(\xi), \tag{5.97}$$

where $m_\varphi(\xi)$ is a 2π-periodic function given by

$$m_\varphi(\xi) = \frac{1}{2} \sum_{n \in Z} a_n e^{-in\xi}. \tag{5.98}$$

(Here, $\hat{\varphi}$ denotes the Fourier transform of the scaling function φ).

Since $||\varphi(x/2)|| = \sqrt{2}$, we see that

$$\sum_{n \in Z} |a_n|^2 = 2, \tag{5.99}$$

so

$$\left(\frac{1}{2\pi} \int_0^{2\pi} |m_\varphi(\xi)|^2 d\xi \right)^{1/2} = \frac{1}{\sqrt{2}}. \tag{5.100}$$

Scaling equations (5.95)–(5.97) play a fundamental role in the wavelet analysis.

It can be checked that

$$|m_\varphi(\xi)|^2 + |m_\varphi(\xi + \pi)|^2 = 1 \text{ for almost all } \xi \in R. \tag{5.101}$$

For results concerning the generation (formation/creation) of an MRA from a function of L_2, in particular a wavelet, and vice-versa, that is, creation of a wavelet from an MRA, we refer to Theorems 2.13 and 2.20 in Wojtaszczyk [1997].

Wavelet decomposition and reconstruction. It is clear that if $\varphi \in V_0$, then $\varphi \in V_1$ as $V_0 \subset V_1$. Since $\{\varphi_{1,k}, k \in Z\}$ is an orthonormal basis for V_1, there exist two sequences $\{p_k\}$ and $\{q_k\}$ in $\ell_2(Z)$, where ℓ_2 is a space of square summable sequences over $Z(\{a_n\} \in \ell_2(Z)$ if $\sum_{n \in Z} |a_n|^2 < \infty)$ such that

$$\varphi(x) = \sum_{k \in Z} p_k \varphi_{1,k}(x) = 2^{1/2} \sum_{k \in Z} p_k \varphi(2x - k), \tag{5.102}$$

$$\psi(x) = \sum_{k \in Z} q_k \varphi_{1,k}(x) = 2^{1/2} \sum_{k \in Z} q_k \varphi(2x - k) \tag{5.103}$$

Equations (5.102) and (5.103) are called the *two-scale relations.* $q_k = (-1)^k p_{1-k}$, if φ and ψ are orthogonal.

Equation (5.102) is also known as the **dilation equation** or **refinement equation**. We may write $p_k = \langle \varphi, \varphi_{1,k} \rangle$.

Let $\psi(x)$ be equal to the Haar function and $\varphi(x)$ the characteristic function of $[0, 1]$, that is

$$\begin{aligned} \varphi(x) &= 1 \quad \text{if } x \in [0,1] \\ &= 0 \quad \text{if } x \notin [0,1]. \end{aligned}$$

Then it may be verified that for each $j \in Z$, $V_j = \{\varphi_{j,k}, k \in Z\}$ is an MRA with the scaling function $\varphi(x)$ with each V_j having ONB. In this case

$$p_k = \begin{matrix} \dfrac{1}{\sqrt{2}}, & k = 0, 1 \\ 0, & \text{otherwise} \end{matrix} . \tag{5.104}$$

Equations (5.102) and (5.103) take the form

$$\varphi(x) = \frac{1}{\sqrt{2}} \varphi_{1,0}(x) + \frac{1}{\sqrt{2}} \varphi_{1,1}(x) \tag{5.105}$$

$$\psi(x) = \frac{1}{\sqrt{2}} \varphi_{1,0}(x) - \frac{1}{\sqrt{2}} \varphi_{1,1}(x). \tag{5.106}$$

The decomposition algorithm Let $\{c_{j,k}\}, j, k \in Z$ and $\{d_{j,k}\}, j, k \in Z$ represent the wavelet and scaling function coefficients of a function $f \in L_2(R)$; that is,

$$c_{j,k} = \int_R f(x) \psi_{j,k}(x) dx, \tag{5.107}$$

and

$$d_{j,k} = \int_R f(x) \varphi_{j,k}(x) dx. \tag{5.108}$$

Then we may prove that

$$d_{j,k} = \sum_{\ell \in Z} p_{\ell - 2k} d_{j+1,\ell} \tag{5.109}$$

and

$$c_{j,k} = \sum_{\ell \in z} (-1)^{\ell} p_{-\ell+2k+1} d_{j+1,\ell}. \tag{5.110}$$

Thus, given scaling coefficients at any level m, all lower-level scaling function coefficients for $j < m$ can be computed recursively applying (5.105) and all lower-level wavelet coefficients ($j < m$) can be computed from the scaling function coefficients using (5.110). This decomposition algorithm is represented schematically in Figure 5.12. In Figure 5.12, arrows represent the decomposition computations, namely,

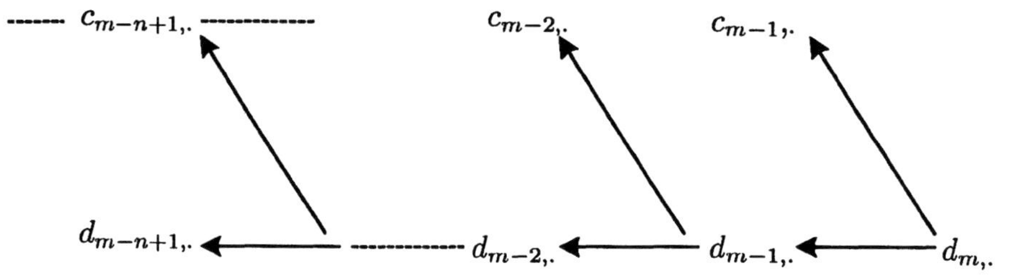

Figure 5.12. Schematic representation of the decomposition algorithm.

$d_{m-2,\cdot}$; $c_{m-2,\cdot}$ can be computed using only the coefficients d_{m-1}.

It can be noted that in either equation, (5.109) and (5.110), if the dilation index k is increased by one, the indices of the $\{p_\ell\}$ sequence are all offset by two. Consequently, if there are only finitely many non-zero elements in the sequence $\{p_\ell\}$, then applying the decomposition algorithm to a set of non-zero scaling function coefficients at level $j+1$ will yield only half as many non-zero scaling function coefficients at level j. Thus, computing the decomposition algorithm recursively yields fewer coefficients at each level. This is the basis of the *fast wavelet transform* to be discussed after the reconstruction algorithm.

The reconstruction algorithm. Let us begin with the MRA $\{V_j, j \in Z\}$ with $\{\varphi_{j,k}, k \in Z\}$ and $\{\psi_{j,k}, k \in Z\}$ forming orthonormal bases for V_j and W_j, respectively. Since $\varphi_{1,0} \in V_1$ and $V_1 = V_0 \oplus W_0$, $\varphi_{1,0}$ can be written as a linear combination of the $\varphi_{0,k}$'s (basis of V_0) and the $\psi_{0,k}$'s (basis of W_0). Let for $k \in Z$,

$$a_{2k} = \langle \varphi_{1,0}, \varphi_{0,k} \rangle \text{ and } a_{2k-1} = \langle \varphi_{1,1}, \varphi_{0,k} \rangle$$
$$b_{2k} = \langle \varphi_{1,0}, \psi_{0,k} \rangle \text{ and } b_{2k-1} = \langle \varphi_{1,1}, \psi_{0,k} \rangle.$$

Therefore

$$\varphi_{1,0}(x) = \sum_{k \in Z} (a_{2k} \varphi_{0,k}(x) + b_{2k} \psi_{0,k}(x)) \tag{5.111}$$

and

$$\varphi_{1,1}(x) = \sum_{k \in Z} (a_{2k-1} \varphi_{0,k}(x) + b_{2k-1} \psi_{0,k}(x)). \tag{5.112}$$

By using (5.111) and (5.112), we can write a similar expression for any $\varphi_{1,k}$. Let k be even. Then

$$\begin{aligned}\varphi_{1,k}(x) &= \varphi_{1,0}\left(x-\frac{k}{2}\right)\\ &= \sum_{\ell\in Z} a_{2\ell}\varphi_{0,\ell}\left(x-\frac{k}{2}\right) + b_{2\ell}\psi_{0,\ell}\left(x-\frac{k}{2}\right)\\ &= \sum_{\ell\in Z} a_{2\ell}\psi_{0,\frac{k}{2}+\ell}(x) + b_{2\ell}\varphi_{0,\frac{k}{2},\ell}(x)\end{aligned}$$

or

$$\varphi_{1,k}(x) = \sum_{\ell\in Z} a_{2\ell-k}\varphi_{0,\ell}(x) + b_{2\ell-k}\psi_{0,\ell}(x). \tag{5.113}$$

Assuming k to be odd, we get the same formula. For odd (even) k, only the odd indexed (even indexed) elements of the sequences $\{a_\ell\}$ and $\{b_\ell\}$ are accessed.

Proceeding on similar lines, a formula relating each scaling function $\varphi_{j,k}$ to scaling functions and wavelets at level $j-1$ can be obtained; namely,

$$\varphi_{j,k}(x) = \sum_{\ell\in Z} a_{2\ell-k}\varphi_{j-1,\ell}b_{2\ell-k}\psi_{j-1,\ell}. \tag{5.114}$$

We can apply (5.114) to obtain the following **reconstruction algorithm** for the scaling function and wavelet coefficients:

$$d_{j,k} = \sum_{\ell\in Z} a_{2\ell-k}d_{j-1,\ell} + b_{2\ell-k}c_{j-1,\ell}. \tag{5.115}$$

We can show that

$$\begin{aligned}b_{2k-1} &= \langle\varphi_{1,1},\psi_{0,k}\rangle\\ &= \int_{-\infty}^{\infty}\sqrt{2}\varphi(2x-(1-2k))\psi(x)dx\\ &= \sum_{\ell\in Z}(-1)^{\ell}_{p-\ell+1}\langle\varphi_{1,1-2k},\varphi_{1,\ell}\rangle\\ &= -p_{2k}.\end{aligned}$$

Similarly it can be seen that

$$a_k = p_{-k} \text{ and } b_k = (-1)^k p_{k+1} \text{ for all } k \in Z.$$

Thus the reconstruction algorithm (5.115) can be written as

$$d_{j,k} = \sum_{\ell\in Z} p_{k-2\ell}d_{j-1,\ell} + (-1)^k p_{2\ell-k+1}c_{j-1,\ell}. \tag{5.116}$$

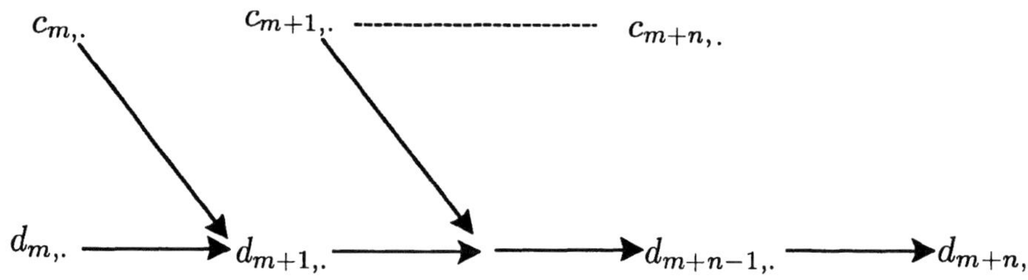

Figure 5.13. Schematic representation of the reconstruction algorithm.

The scaling function coefficients at any level can be computed from only one set of low-level scaling function coefficients and all the intermediate wavelet coefficients by applying (5.115) or (5.116) recursively. This concept is shown schematically in Figure 5.13.

The filter representation. The decomposition and reconstruction algorithms can be treated as examples of signal processing filters. As we know in every-day life, filter is used to purify certain things; for example, in a laboratory it is used to purify a liquid from solid impurities or to remove a solid from solid impurities. A filter in the signal processing may attempt either to isolate the pure signal from its noise contamination or to extract the noise, depending on which (signal or noise) is of primary consideration. A discrete signal f is represented by a sequence $\{f_k\}_{k\in Z} \in \ell_2(Z)$. A filter may also be represented by a sequence $\{a_k\}_{k\in Z} \in \ell_2(Z)$, and is denoted by A. Applying a filter to a signal yields another signal. The filtering process comprises a discrete convolution of the filter sequence with the signal. By applying the filter A to the signal f, we get

$$(Af)_k = \sum_{\ell\in Z} a_{\ell-2k} f_\ell, \tag{5.117}$$

where $(Af)_k$ is a new signal indexed by k ranging over Z. $A^2 f$ will mean that A has been applied twice. If A and B are two filters then ABf will mean that, first B has been applied to f and then A has been applied to Bf. Our goal now is to express the decomposition algorithm given by equation (5.110) in terms of filters. Let H be a filter represented by the sequence $\{p_k\}_{k\in Z}$ given as

$$p_k = \langle \varphi, \varphi_{1,k} \rangle. \tag{5.118}$$

Let $\{d_{j,k}\}_{k\in Z}$, scaling function coefficients at a particular level, be a signal. The scaling function coefficient at the next lower level signal is obtained by applying the filter H to the signal $d_{j,\cdot} = \{d_{j,k}\}_{k\in Z}$:

$$d_{j-1,\cdot} = H d_{j,\cdot}\,, \tag{5.119}$$

(5.119) corresponds to (5.109).

The scaling function coefficients at any level can be obtained by repeatedly applying the filter H:

$$d_{j-m,.} = H^m d_{j,.} \,. \tag{5.120}$$

Now we define a new filter G by

$$q_k = (-1)^k p_{1-k}, \quad k \in Z, \tag{5.121}$$

where p_k's are defined by (5.118).

Wavelet coefficients at level $j-1$ can be obtained from scaling function coefficients at level j via the filter

$$c_{j-1,.} = G d_{j,.} \,. \tag{5.122}$$

(5.122) corresponds to (5.110).

By combining filters H and G, wavelet coefficients at any lower level can be computed from scaling function coefficients at level j

$$c_{j-m,.} = G H^{m-1} d_{j,.} \,, \tag{5.123}$$

where H and G are examples of quadrature mirror filters well known in the engineering literature. H is known as a **low-pass filter** and G is an example of a **high-pass filter**. Broadly speaking, low-pass filters are related to averaging operations, and high-pass filters correspond to differencing.

It may be observed that for compactly supported wavelets, that is, those having zero values outside a compact interval, the sequence $\{c_{i,j}\}$ and $\{d_{i,j}\}$ will have only finite number of non-zero elements. Furthermore, the two-scale sequence $\{p_k\}$ completely characterizes each wavelet basis. For wavelets with support on all of R, each sequence element p_k is, in general, non-zero but elements decay exponentially as $|k|$ becomes large.

Let H be the low-pass filter associated with the Haar system, then $p_0 = p_1 = \frac{1}{\sqrt{2}}$; other p_k's are zero. For a signal $f = \{f_k\}_{k \in Z}$

$$f' = Hf \quad \text{where } f'_k = \frac{1}{\sqrt{2}} (f_{2k} + f_{2k-1}) \,, \tag{5.124}$$

which is proportional to the average of adjacent elements.

For a signal $f = \{f_k\}_{k \in Z}$, G is defined as

$$f^* = Gf \text{ where } f^*_k = \frac{1}{\sqrt{2}} (f_{2k} - f_{2k-1}) \,, \tag{5.125}$$

which is proportional to the difference between adjacent elements.

In general, one may say that p_k's for low-pass filter satisfy the relation

$$\sum_{k \in Z} p_k = \sqrt{2} \,, \tag{5.126}$$

and for high-pass filters, they satisfy the relation

$$\sum_{k \in Z} q_k = 0. \tag{5.127}$$

The fast wavelet transform. We have seen in Section 5.2 that the Cooley and Tukey fast Fourier transform (algorithm) of data has a reduced computational cost of only $O(n \log n)$, a very significant achievement. The main idea is to reduce the number of computations made by recursively computing the discrete Fourier transform of subsets of the data. This is achieved by reordering the data subsets so as to take advantage of some redundancies in the usual discrete Fourier transform algorithm. Along similar lines, we could take approximate wavelet transforms by replacing the function f in the definition of a wavelet coefficient by an estimate such as

$$c_{j,k} = \int_{-\infty}^{\infty} \tilde{f}(x)\psi_{j,k}(x)dx,$$

where

$$\begin{aligned} \tilde{f}(t) &= X_k, \quad k-1 < t \leq k \\ &= 0, \quad \text{otherwise} . \end{aligned} \tag{5.128}$$

This idea could be used to estimate the following scaling function coefficients as well:

$$\begin{aligned} \tilde{d}_{j,k} &= \int_{-\infty}^{\infty} \tilde{f}(x)\varphi_{j,k}(x)dx \\ &\simeq \sum_{\ell=1}^{n} X_\ell \varphi_{j,k}(\ell) . \end{aligned} \tag{5.129}$$

Let us start with a set of high-level scaling function coefficients, and assume that we have only a finite number of coefficients. This assumption is reasonable, as any signal $f \in L_2(R)$ must have rapid decay in both directions, so $d_{j,k}$ can be neglected for large $|k|$. Rescale the original function, if necessary, so that the scaling function coefficients at level m are $d_{m,0,\dots}, d_{m,n-1}$. Computing scaling function and wavelet coefficients at level $m-1$ is accomplished via equations (5.109) and (5.110). As observed earlier, the $\{p_k\}$ sequence used in these calculations has only finite number of non-zero values if the wavelets are compactly supported; otherwise p_k values decay exponentially; so it can be approximated by finitely number of terms. In either case, let K denote the number of non-zero terms used in the sequence, possibly truncated. Computing a single coefficient at level $m-1$ according to either (5.109) or (5.110) would take at most K operations. If scaling function coefficients $d_{m,k}$ for $k \notin \{0, \dots, n-1\}$ are set to zero, giving exactly n non-zero coefficients at level m, then the number of non-zero scaling function coefficients at level $m-1$ would be at most

$$\left[\frac{n}{2}\right] + \left[\frac{K}{2}\right] + 2, \tag{5.130}$$

where $[x]$ denotes the greatest integer function of x. If M is even large, the number of non-zero wavelet coefficients is no more than the number in (5.130). Therefore, the total number of non-zero scaling function coefficients at level $m-1$ is approximately $\frac{n}{2}$, and the total number of operations required to compute the one-level-down wavelet and scaling function coefficients is approximately $2K \cdot \frac{n}{2}$.

Let n_1 be the number of non-zero scaling function coefficients at level $m-1$. Applying the decomposition again requires no more than

$$\left[\frac{n_1}{2}\right] + \left[\frac{K}{2}\right] + 2,$$

operations to compute the wavelet coefficients and the same number of operations for the scaling function coefficients. There will be approximately $\frac{n_1}{2} \simeq \frac{m}{4}$ non-zero scaling function coefficients at level $m-2$, and the computation will require approximately $2K \cdot \frac{n_1}{2} \simeq 2K \cdot \frac{n}{2}$ operations.

Continuing this process, the total number of operations required for all decompositions is approximately

$$2K\left(\frac{n}{2} + \frac{n}{4} + \frac{n}{8} + \cdots\right) = O(n)\,.$$

If we do the similar computation using the fast Fourier transform, the required number operations is of the $O(n \log n)$. Therefore the fast wavelet transform is faster than the fast Fourier transform.

5.3.4 Performance of Fourier, fractal and wavelet methods in image compression

In many fast-growing industries, image compression plays very vital role. Therefore, intensive researches are being carried out in this area and different methods are tested for different class of images to improve compression ratio with permissible distortion. Fourier method (discrete Fourier transform (DCT) or JPEG), fractal digital image compression (IFS method, etc.) and wavelet method or WIC are most popular and compete with each other.

Walker [1997] has compared the performance of Fourier and wavelet methods (using Haar and Daubechies transforms) for one-dimensional signal. In this section, we shall compare the performance of Fourier, fractal and wavelet methods for some of the test images from the "Bragzone site" of the University of Waterloo, Canada available for research purposes. It is mainly based on Siddiqi and Ahmad [1998]. We shall present here a brief account of test images, JPEG, multiresolution analysis in

$L_2(R^2)$ and the fast wavelet transform for image compression, encoding and decoding in fractal method along with compression results.

Description of a set of test images. Most of our test images have come from 'BragZone site" of the University of Waterloo. This site provides diverse set of test images for research purposes. The aim of providing such images is to bring uniformity to the image processing community as far as test images are concerned. The images of selected compression ratios for various compression algorithms are also available on this site.

We explain the details of these images. The size of the image and number of bits required to store one pixel is mentioned in the bracket.

- Bridge [$256 \times 256 \times 8$]: This is the smaller version of the classical 512×512 bridge image. Being originally derived from a 6-bit spanning process, it exhibits low resolution and contrast, as evident in the regions of foliage.

- Circles [$256 \times 256 \times 8$]: A set of nested circles with the outside circle darkest and the innermost circle lightest. This image tests edges at various orientations. Also, the frame truncates the larger circles, creating two horns that narrow to a point.

- Crosses [$256 \times 256 \times 8$]: A pattern of diagonal and horizontal lines. The lines are narrow, varying from one to three pixels wide. Fine lines are essential to architectural drawings, flow charts, graphic designs, etc. Yet, many coders have great trouble with this type of content, especially along the diagonals.

- Lena [$512 \times 512 \times 8$]: The image contains a nice mixture of details, flat regions, shading and texture that do a good job of testing various image processing algorithms.

- Peppers [$512 \times 512 \times 8$]: The absence of fine detail makes it comparatively easy to compress.

- Squares [$256 \times 256 \times 8$]: A series of concentric squares of decreasing width from the outside in, and of increasing luminance. The plateaus are completely flat with grey level values of 50, 100, 150 and 200. This image tests the preservation of flat regions and step edges. The rectangular boundaries are at even pixel locations, helping to make this image the easiest to compress the entire site.

- Washat [$512 \times 512 \times 8$]: This is a satellite photograph of the Washington DC area. The contrast is low but there is much fine detail that must be preserved in order to facilitate interpretation by earth scientists and intelligence analysts.

- Wood [$512 \times 512 \times 8$]: The wood image is taken from ITWM, University of Kaiserslautern, Germany. Due to the presence of enough fine details in image, it is often difficult for the coders to compress it.

- Zelda [512 × 512 × 8]: From the USC database. Like Lena, the absence of much fine detail makes it comparatively easy to compress. However, humans are very sensitive to faces. In some places, such as around the eyes and mouth, slight mathematical distortions can be perceptually disconcerting.

JPEG: A brief description. The JPEG, Joint Photographic Experts Group, is a working group of technical experts in image encoding. The goal of JPEG is to develop an international standard for compression of grey scale or color images. Here we describe the basic coding machinery of the JPEG baseline system encoder. It provides a simple and efficient algorithm that is adequate for most image encoding applications. The image is partitioned into 8 × 8 blocks and each block is independently transformed using the 2-dimensional discrete cosine transform (DCT). The forward 2-D DCT of any block of pixels is defined as (see Wallace [1990] for more details.)

$$F(u,v) = \frac{1}{4}C(u)C(v)\sum_{x=0}^{7}\sum_{y=0}^{7} f(x,y)\cos\frac{(2x+1)u\pi}{16}\cos\frac{(2y+1)v\pi}{16},$$

and the inverse 2-D DCT is defined as

$$f(x,y) = \frac{1}{4}\sum_{u=0}^{7}\sum_{v=0}^{7} C(u)C(v)F(u,v)\cos\frac{(2x+1)u\pi}{16}\cos\frac{(2y+1)v\pi}{16},$$

where

$$C(w) = \begin{cases} \frac{1}{\sqrt{2}} & \text{for } w = 0 \\ 1 & \text{otherwise}. \end{cases}$$

All transformed coefficients are normalized by applying a user-defined normalization array that is fixed for all blocks. The normalized coefficients are then uniformly quantized by rounding to the nearest integer. Thus, the quantization function is defined as

$$FQ(u,v) = \text{Integer Round}\left(\frac{F(u,v)}{Q(u,v)}\right),$$

where $Q(u,v)$ is called quantization matrix.

The top-left coefficients in the 2-D DCT array are referred to as the DC value and are proportional to the average brightness of the spatial block. After quantization, this coefficient is encoded with a lossless differential pulse code modulation (DPCM) scheme. The remaining 63 coefficients of 8 × 8 block are called AC coefficients. The quantization of AC coefficients produces many zeros, especially at higher frequencies. To take advantage of these zeros, the 2-D array of DCT coefficients is formatted into a 1-D vector using a zig-zag reordering. This rearranges the coefficients in approximately decreasing order of their average energy with the aim of creating large number of zero values. If the first non-zero AC value is separated from the DC value with a (one) zero, then it is called a run of length of 1 zero.

The DC value and AC coefficients in the 1-D sequence are encoded using separate Huffman code tables. At the receiver, the sequence is decoded using the Huffman decoding table and then denormalized. The denormalized block is then inverse transformed using the inverse 2-D DCT.

Although the JPEG image standard has been widely used but it has some drawback. The error caused due to the quantization process is isolated in each local block. At the high compression ratio, the block effect becomes obvious. The second, energy compactness is only achieved within each local block and as a result, redundancy between blocks can not be utilized.

Fast wavelet transform for image compression

Wavelets and multiresolution analysis in R^2. Let $f^1(x)$ and $f^2(x)$ be two functions of one variable. We can form a function of two variables as follows:

$$f = f^1 \otimes f^2 = \bigotimes_{j=1}^{2} f^j ,$$

where

$$\begin{aligned} \bigotimes_{j=1}^{2} f^j(x_1, x_2) &= \prod_{j=1}^{2} f^j(x_j) \\ &= f^1(x_1) f^2(x_2) . \end{aligned}$$

Let X_1 and X_2 be two closed subspaces of $L_2(R)$, then we can form a closed subspace of $L_2(R^2)$ denoted by $\bigotimes_{j=1}^{2} X_j$ or $X_1 \otimes X_2$ and defined as the closed linear span in $L_2(R^2)$ of all functions of the form $f^1(x_1) \cdot f^2(x_2)$. The symbol $\otimes$ is known as the tensor product. It can be checked that if the system $\{f^j_s\}_{s \in A_j}$ are orthonormal bases in subspaces $X_1, X_2 \subset L_2(R)$, then the system

$$\left(\bigotimes_{j=1}^{2} f^j_{s_j} \right)_{(s_1, s_2) \in A_1 \times A_2}$$

is an orthonormal basis in $\bigotimes_{j=1}^{2} X_j$. It can also be verified that

$$\bigotimes_{j=1}^{2} L_2(R) = L_2(R^2) .$$

From the above results it is quite clear that (i) if ψ_1 and ψ_2 are two wavelets in $L_2(R)$, then $\psi(x_1, x_2) = \psi_1(x_1)\,\psi_2(x_2)$ is a wavelet in $L_2(R^2)$;

(ii) if $\{V_j\}$ and $\{\tilde{V}_j\}$ are two multiresolution analyses in $L_2(R)$, then $V_j \otimes \tilde{V}_j$ is a multiresolution analysis in $L_2(R^2)$.

Mallat transform. We now describe the fast wavelet transform of Mallat [1989] for two-dimension signals, such as images.

$\tilde{W}_j$ is defined as the orthogonal complement of $\tilde{V}_j$ into $\tilde{V}_{j+1}$. Thus we have the following orthogonal bases:

- basis for $\tilde{V}_j$:
  $(\phi^j_{m,n}) = \left(\phi^j_m(x)\phi^j_n(y)\right)_{m,n}, \quad m, n \in \mathbb{Z},$

- three bases for $\tilde{W}_j$:
  $(\psi^{j,1}_{m,n}) = \left(\phi^j_m(x)\psi^j_n(y)\right)_{m,n}, (\psi^{j,2}_{m,n}) = \left(\psi^j_m)(x)\phi^j_m(y)\right)_{m,n},$
  $(\psi^{j,3}_{m,n}) = \left(\psi^j_m(x)\psi^j_n(y)\right)_{m,n}.$

For an image $f \in L_2(\mathbb{R}^2)$ its approximation at scale 2^j is given by its orthogonal projection onto $\tilde{V}_j$

$$f_{\tilde{V}_j}(x, y) = \sum_{m,n \in \mathbf{Z}} \langle f, \phi^j_{m,n} \rangle \phi^j_{m,n}(x, y).$$

This approximation is thus characterized by the sequence $S^j_{m,n} = 2^j \langle f, \phi^j_{m,n} \rangle$. The sequence $S_j = (S^j_{m,n})_{m,n \in \mathbf{Z}}$ is called discrete approximation of f at the resolution 2^j. The additional details from the scale 2^j to 2^{j+1} are given by its orthogonal projection onto $\tilde{W}_j$.

$$f_{\tilde{W}_j}(x, y) = \sum_{m,n \in \mathbf{Z}} \langle f, \psi^{j,d}_{m,n} \rangle \psi^{j,d}_{m,n}(x, y), \quad d = 1, 2, 3.$$

This component is thus characterized by the sequences,

- $D^{j,1}_{m,n} = 2^j \, \langle f, \phi^j_m \psi^j_n \rangle$,
- $D^{j,2}_{m,n} = 2^j \, \langle f, \psi^j_m \phi^j_n \rangle$, and
- $D^{j,3}_{m,n} = 2^j \, \langle f, \psi^j_m \psi^j_n \rangle$.

These sequences

$$D^d_j = (D^{j,d}_{m,n})_{m,n \in \mathbf{Z}}, d = 1, 2, 3,$$

are called details of the resolution 2^{j+1}.

In image compression, we get a discretized image at resolution, say, 2^j and the goal is to decompose it into lower resolutions. According to the Mallat algorithm S_j could be computed from S_{j+1} with the action of low-pass filter $\tilde{h}(n)$ followed by a decimation.

$$S^j_{m,n} = \sum_{k,\ell\in\mathbb{Z}} S^{j+1}_{k,\ell}\tilde{h}(2m-k, 2n-\ell).$$

The bidimensional filter $\tilde{h}$ is the tensor product of the same 1-D low-pass filter h defined as, $\tilde{h}(m,n) = h(-m)h(-n)$ with $h(n) = 2^{-1/2}\langle\phi_0^{-1}, \phi_n^0\rangle$.

Similarly, D^d_j, $d = 1,2,3$, can be computed from S_{j+1} by the action of high-pass filter $\tilde{g}_d$ ($\tilde{g}_d$ is the tensor product of 1-D high- and low-pass filters g and h) followed by the same decimation.

$$D^{j,d}_{m,n} = \sum_{k,\ell\in\mathbb{Z}} S^{j+1}_{k,\ell}\tilde{g}_d(2m-k, 2n-\ell),$$

where $\tilde{g}_1(m,n) = h(-m)g(-n), \tilde{g}_2(m,n) = g(-m)h(-n)$, $\tilde{g}_3(m,n) = g(-m)g(-n)$ and $g(n) = 2^{-1/2}\langle\psi_0^{-1}, \psi_n^0\rangle$.

The filters $H = \{h(n) : n \in \mathbb{Z}\}$ and $G = \{g(n) : n \in \mathbb{Z}\}$ are called quadrature mirror filters (QMF). H and G correspond to a low- and high-pass filters, respectively. The rows of the image are filtered by computing their correlation with the low- and high-pass filters H and G followed by 2:1 decimation. The same procedure is then applied to columns. Thus we get a four-channel orthogonal subband decomposition using separable QMF. That is, from the discretized image S_j, we get the four-channel decomposition $S_{j-1}, D^1_{j-1}, D^2_{j-1}$ and D^3_{j-1}. The same is repeated on channel S_{j-1} and so on. Figures 5.14 and 5.15 represent a general scheme of wavelet decomposition. The transformed wavelet coefficients are then quantized followed by entropy encoding.

Encoding and decoding process

Digital fractal image processing. Suppose that we are given an image f that we wish to encode. This means that we want to find a collection of maps $w_1, w_2, \dots, w_N$ with $W = \bigcup_{i=1}^{N} w_i$ and $f = x_W$. That is, we want f to be the fixed point of the map W. We seek a partition of f into pieces to which we apply transforms w_i and get back f. Being able to exactly cover f with parts of itself is not likely, so the least possible cover is sought with the hope that x_W and f will not look too different, i.e, $d(f, x_W)$ is small. The hope for this comes from the Collage theorem. Thus, the encoding process is to partition I^2 by a set of ranges B_i. For each B_i, an $A_i \subset I^2$ and $w_i : A_i \times I \to I^3$ are sought such that $w_i(f)$ is as close to $f \cap (B_i \times I)$ as possible; i.e.

$$d\Big(f \cap (B_i \times I), w_i(f)\Big)$$

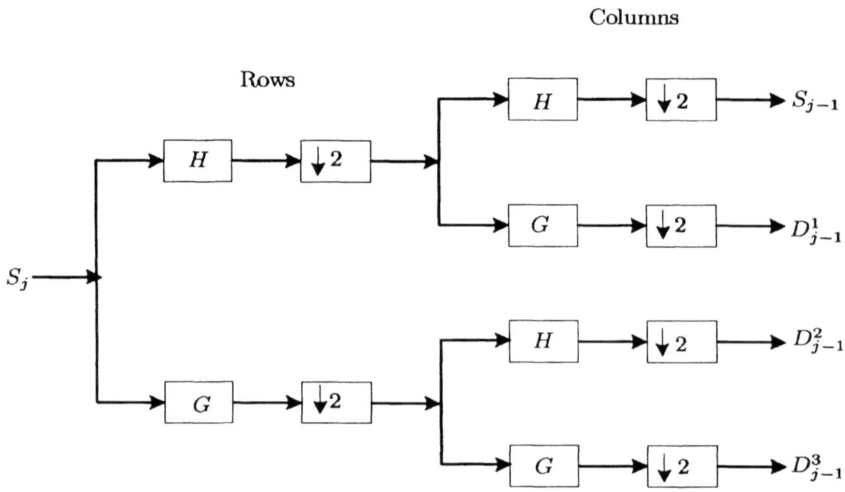

Figure 5.14. One stage in wavelet decomposition.

is minimized.

Example. Let us start with a 512×512 pixel image with 256 grey levels. We partition the image recursively until the image size is $d \times d$ (a typical consideration is 32×32) and call these square blocks B_i range blocks. For each range block of the image, we find an overlapping domain block A_i, twice the size of the range block, so that when we apply a suitable transformation w_i to A_i, we get some thing very close (in the sense of the root mean square metric) to the part of the image over B_i, i.e., we seek to minimize the expression

$$d_{\text{rms}}\Big(f \cap (B_i \times I), w_i(f)\Big), \quad i = 1, \dots, N.$$

The d_{rms}, in case of two images f and g, is defined as

$$d_{\text{rms}}(f, g) : \sqrt{\int\!\!\int_{I^2} \Big(f(x,y) - g(x,y)\Big)^2 \, dx\, dy}.$$

The pixels in a domain block are arranged in groups of four so that the domain is reduced to the size of range. Let $a_1, a_2, \dots, a_n$ and $b_1, b_2, \dots, b_n$ be the pixel intensities of the blocks A_i and B_i, respectively. We want to determine s and o (contrast and brightness adjustments, respectively) such that we minimize the quantity,

$$S = (sa_1 + o - b_1)^2 + (sa_2 + o - b_2)^2 + \cdots + (sa_n + o - b_n)^2.$$

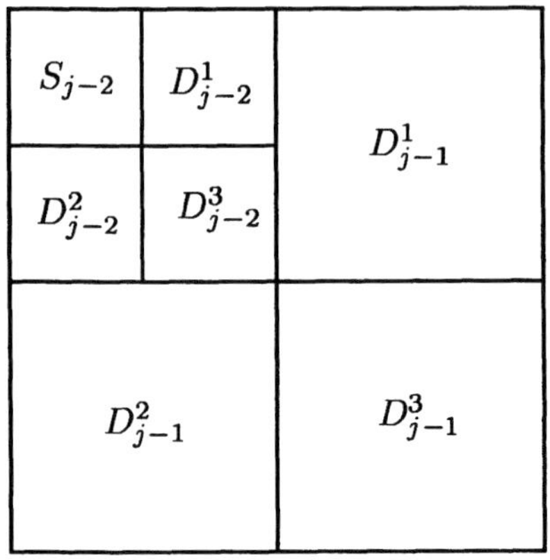

Figure 5.15. Wavelet decomposition scheme.

After solving, we get

$$s = \frac{\left[n \sum_{i=1}^{n} a_i b_i - \sum_{i=1}^{n} a_i \sum_{i=1}^{n} b_i \right]}{\left[n \sum_{i=1}^{n} a_i^2 - \left(\sum_{i=1}^{n} a_i \right)^2 \right]},$$

and

$$o = \frac{1}{n} \left[\sum_{i=1}^{n} b_i - s \sum_{i=1}^{n} a_i \right] .$$

For more details, we refer to Fisher [1995].

Computational results and conclusion. Original image and compressed images of Zelda, Holz and Peppers are given in Figures 5.16 to 5.18, compressed in different compression ratios with wavelet, fractal and JPEG. The distortion measures are given in Table 1. Peppers image is compressed in different CRs ranging from 16:1 to 50:1. The plots of CRs versus distortion error of Peppers images are given in Figure 5.19. The CRs and the corresponding errors (mean square error, Sobolev and total variation error) are shown on X and Y axes, respectively.

It has long been accepted that distortion measures like mean square error (MSE) are inaccurate in predicting distortion, Teo and Heeger [1994], though MSE is the widely used distortion measure in image processing literature. We have studied various types of images ranging from natural to artificial and classified them into three classes (A, B and C); see Siddiqi and Ahmad [1998]. The Peppers image (Figure 5.16) belongs to Class A of our classification and FIC is the best compression technique for this class of images, Siddiqi and Ahmad [1998].

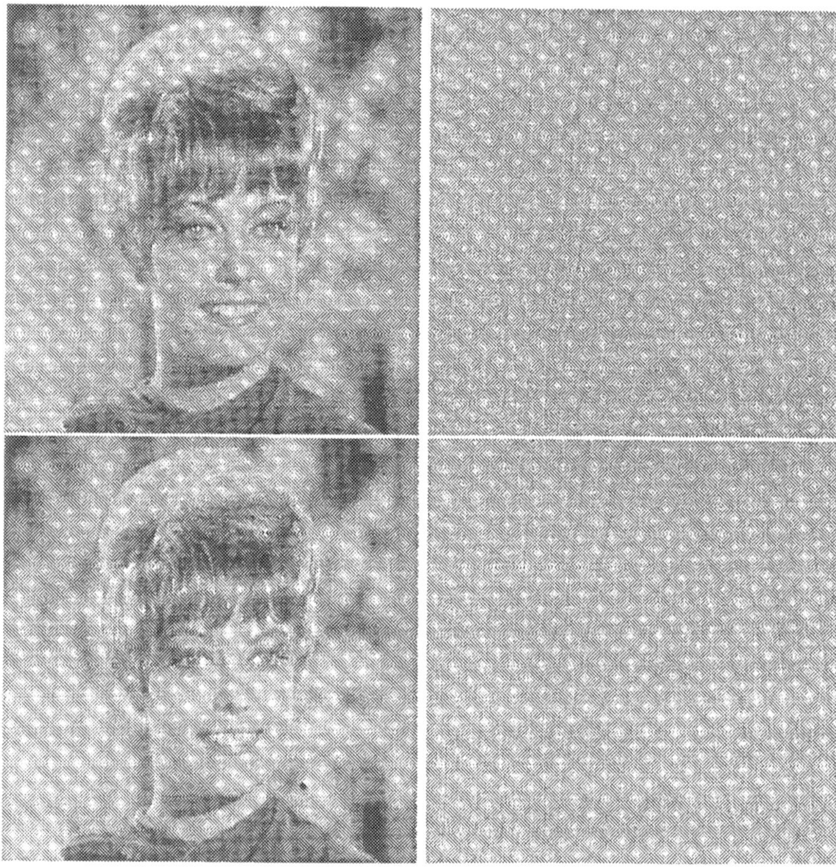

Figure 5.16 Top: Original image of Zelda and Holz8, Bottom: Compressed images with wavelet, CR 16:1 and 20:1, respectively.

In the present study and in case of Peppers image (true in general for the images of Class A), we observe that up to CR 40:1, all the distortion measures behave alike (all match human perception) but if we increase CR above 40:1, only Sobolev or TVE match human perception. MSE fails to measure it. The performance of the fractal method is better than wavelet and JPEG, in case of Peppers image for CR above 40:1, as clearly seen from the plot of Sobolev or TVE. A similar observation is made in case of other images of this class at high compression ratios. Therefore, we conclude that for a particular type images and for higher CRs, Sobolev or TVE are better measures than MSE. For the rest of the images, Sobolev and TVE are as good as MSE.

Our computational results for different distortion measures are given in the following table:

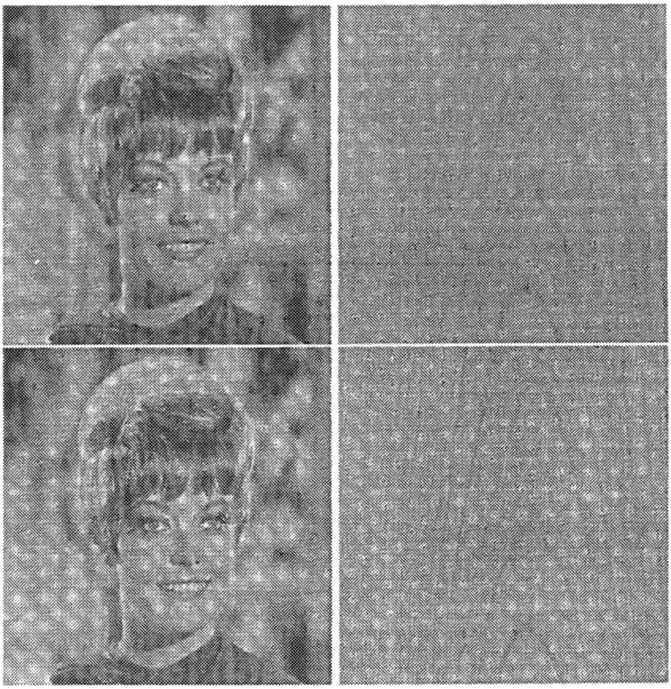

Figure 5.17 Top: Compressed images with FIC, CR 16:1 and 20:1, respectively. Bottom: Compressed images with JPEG, CR 16:1 and 20:1, respectively.

Table 1. Distortion measure of Peppers, Zelda and Holz images.

Image: Peppers, CR 32:1				
Algorithm	MSE	PSNR	TVE	Sobolev
Wavelet	39.7012	32.1428	10.5462	12.2851
Fractal	43.3606	31.7599	10.6643	12.5335
JPEG	62.4071	30.1785	12.4391	14.7356
Image: Zelda, CR 16:1				
Algorithm	MSE	PSNR	TVE	Sobolev
Wavelet	8.80429	38.6839	5.54797	6.2916
Fractal	17.5769	35.6814	6.80572	7.99342
JPEG	11.7603	37.4266	6.25757	7.13565
Image: Holtz, CR 20:1				
Algorithm	MSE	PSNR	TVE	Sobolev
Wavelet	92.021	28.4919	17.8883	20.2981
Fractal	158.822	26.1217	19.8303	23.496
JPEG	98.5858	28.1927	18.194	20.727

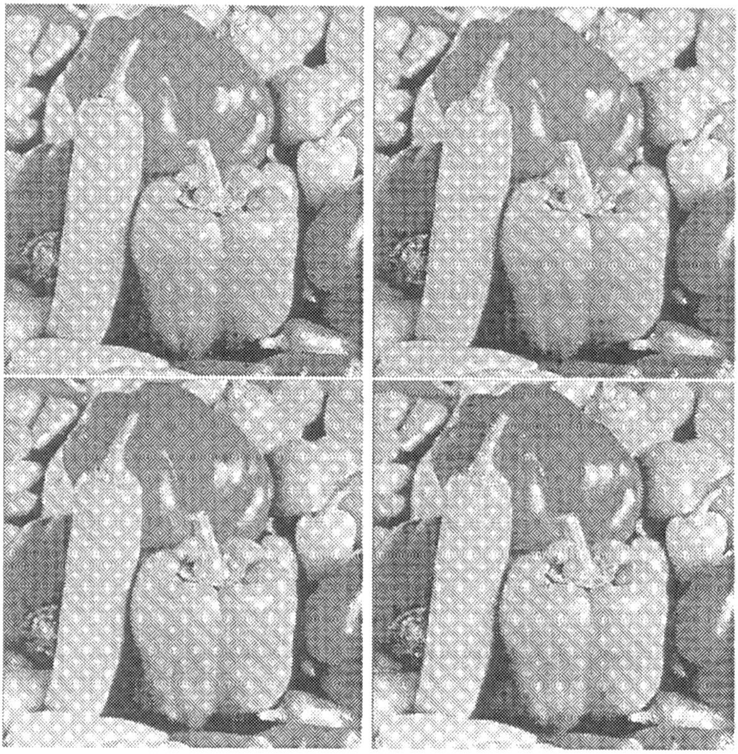

Figure 5.18 Top left: Original image of peppers. Top right: Compressed image with wavelet, CR 32:1. Below left: Compressed image with FIC, CR 32:1. Below right: Compressed image with JPEG, CR 32:1.

5.3.5 Differential equation and wavelets

Wavelets are currently being investigated for the numerical solution of differential and integral equations. We shall consider here a simple example to illustrate the application of wavelets in solving it. It is clear from the MRA that any function in $L_2(R)$ can be approximated arbitrarily well (in the $L_2(R)$ norm) by the piece-wise constant functions from V_j provided j is large enough. V_j is the space of piece-wise constant $L_2(R)$ functions with breaking at the dyadic integers $k \cdot 2^{-j}$, $j, k \in Z$. We take the following boundary value problem:

$$\begin{cases} -u''(x) + C\,u(x) = f(x), \quad x \in \Omega = (0,1) \\ u(0) = u(1) = 0\,, \end{cases} \tag{5.99}$$

with $C > 0$, a constant, $f \in H^1(\Omega)$ and solve it for $u = u(x)$.

We apply the variational method of approximation (Galerkin method) for solving equation (5.99). As the existence of the solution of variational problem is guaranteed only in a complete space, so we take the Sobolev space $H^1(\Omega)$. The idea is to solve variational equation on the finite-dimensional subspace of $H^1(\Omega)$.

In variational form, the solution $u \in H^1(\Omega)$ of the above equation satisfies

$$\int_\Omega (u'v' + uv) = \int_\Omega fv, \tag{5.100}$$

for all $v \in H^1(\Omega)$.

To approximate u by the Galerkin's method, we choose a finite-dimensional subspace of $H^1(\Omega)$ which is a space spanned by wavelets defined on the interval [0,1]. We have already discussed the wavelet bases $\phi_{j,k}$, $\psi_{j,k}$ and spaces V_j, W_j generated by them respectively in previous sections. For getting numerical solution of (5.99), we choose a positive value m and approximate u by an element $u_m \in V_m$ that satisfies

$$\int_\Omega (u_m'v' + u_m v)dx = \int_\Omega f(x)\, v\, dx, \quad v \in V_m, \tag{5.101}$$

where u_m and v can be written as

$$u_m = P_0 u_m + \sum_{k=0}^{m-1} Q_k u_m. \tag{5.102}$$

P_k is the projection from $H^1(\Omega)$ onto V_k and $Q_{k-1} = P_k - P_{k-1}$. Therefore,

$$u_m = C_0^0\, \phi_{0,0} + \sum_{j=0}^{m-1} \sum_{k=0}^{2^j-1} d_k^j\, \psi_{j,k}. \tag{5.103}$$

Here, $\phi_{0,0}$ is identically 'one' on Ω, C_0^0 is the average of u_m on Ω. We have used a "multiresolution" approach, i.e., we have written a successive coarser and coarser approximation to u_m.

Therefore, for all practical purposes, u_m can be approximated to an arbitrary precision by a linear combination of wavelets. We need now to determine d_k^j. Equation (5.103) together with (5.101) gives the following:

$$\sum_{j=0}^{m-1} \sum_{k=0}^{2^j-1} d_k^j \int_\Omega (\psi_{j,k}\psi_{j',k'} + \psi_{j,k}\psi_{j',k'})dx = \int_\Omega f\psi_{j',k'}dx, \tag{5.104}$$

for $k' = 0, \cdots, 2^{j'} - 1$ and $j' = 0, \cdots, m-1$ or, more precisely,

$$LA = f.$$

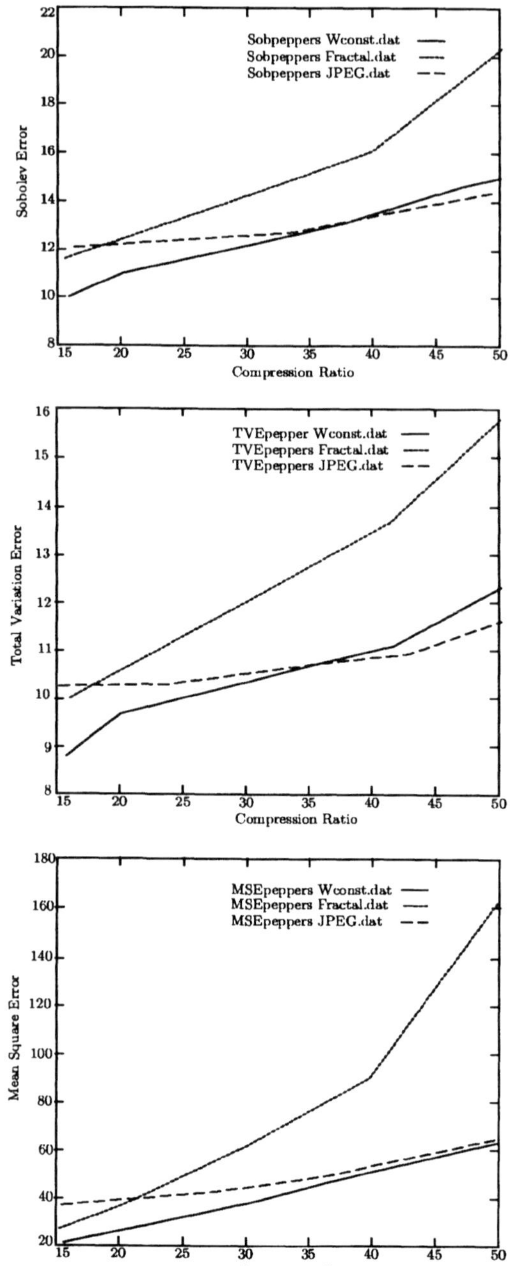

Figure 5.19 Different distortion measures: Sobolev, TVE and MSE.

Let us work out the details for the boundary value problem considered in Chapter 3

$$-u'' = f \text{ on } (0,1),\ u(0) = u(1) = 0 .$$

The standard weak formulation

$$\langle u', v' \rangle = \langle f, v \rangle, v \in H_0^1([0,1]) . \tag{5.105}$$

Let h_j be a finite-dimensional subspace of $H_0^1([0,1])$. The simplest conforming choice of the trial space H_h are the spans of scaled test functions

$$\phi_{j,k}(x) = 2^{j/2}\phi(2^j . - k),\ k = 0, \cdots, 2^j, \tag{5.106}$$

where

$$\phi(x) = \begin{cases} 1+x & -1 \le x < 0, \\ 1-x & 0 \le x < 1, \\ 0 & \text{otherwise.} \end{cases} \tag{5.107}$$

Choosing the $\phi_{j,k}$ as a basis function for H_h, the Galerkin condition

$$\langle u_h, v \rangle = \langle f, v \rangle, v \in H_h \tag{5.108}$$

gives rise to a linear system of equations

$$A_h u = F, \tag{5.109}$$

where A_h is the stiffness matrix relative to the basis functions $\phi_{j,k}$ and u, F are corresponding vectors with $f_k = \langle f, \phi_{j,k} \rangle$. Clearly, A_h is tridiagonal. Hence (5.109) is very efficiently solvable. However, for higher-dimensional analogues, the matrix would no longer have such a narrow bandwidth and one has to resort to iterative methods to preserve sparseness.

On the other hand, recalling the min-max characterization of the smallest and largest eigenvalue of a symmetric positive definite matrix, one can see that the condition numbers of A_h grow like 2^{2^h}, which renders iterative methods prohibitively inefficient. To remedy this, one has to precondition the linear systems. One way is to exploit suitable multiscale decompositions of the trial spaces H_h. First, note that since

$$\phi(x) = \frac{1}{2}\phi(2x+1) + \phi(2x) + \frac{1}{2}\phi(2x-1); \tag{5.110}$$

that is,

$$\phi_{j,k} = \frac{1}{2\sqrt{2}}\phi_{j+1,2k-1} + \frac{1}{\sqrt{2}}\phi_{j+1,2k} + \frac{1}{2\sqrt{2}}\phi_{j+1,2k+1}, \tag{5.111}$$

therefore, H_h are nested and, of course, their union is dense in $L_2([0,1])$.

In order to successively update solutions from coarser grids, one can consider the following hierarchical decomposition of the trial spaces. We use here the Lagrange projectors

$$L_j f = \sum_{k=0}^{2^j} 2^{-j/2} f(2^{-j}k)\phi_{j,k}, \tag{5.112}$$

and note that the complements

$$W_j = (L_{j+1} - L_j)H_{j+1} \tag{5.113}$$

are simply spanned by the tent functions on new grid points on the next higher scale

$$\Psi_j = \{\psi_{j,k} = \phi_{j+1,2k+1} : k = 0, \cdots, 2^j - 1\}. \tag{5.114}$$

Note that neither the $\phi_{j,k}$ nor the $\psi_{j,k}$ is orthogonal but it is not hard to show that they satisfy the stability condition

$$c_1 \left(\sum_{k=0}^{2^j} |c_k|^2 \right)^{1/2} \leq \| \sum_{k=0}^{2^j} c_k \phi_{j,k} \|_{L_2([0,1])} \leq c_2 \left(\sum_{k=0}^{2^j} |c_k|^2 \right)^{1/2}, \tag{5.114}$$

for some constants c_1, c_2 independent of the sequence $\{c_k\}_{k=0}^{2^j}$. Keeping this in mind, we now consider stiffness matrices relative to the hierarchical bases composed of the collections Ψ_j and note that

$$\frac{d}{dx}\psi_{j,k}(x) = \frac{d}{dx}\phi_{j+1,2k+1}(x) = 2^{j+\frac{3}{2}}\psi^H_{j,k}(x), \tag{5.116}$$

where $\psi^H_{j,k}(x)$ are the Haar wavelets. Therefore, we get

$$\left\langle \frac{d}{dx}\psi_{j,k}, \frac{d}{dx}\psi_{n,l} \right\rangle = 2^{n+j+3}\langle \psi^H_{j,k}, \psi^H_{n,l} \rangle = 2^{-2j+3}\delta_{j,n}\delta_{k,l}.$$

Hence A_h is, up to a 2×2 upper left block stemming from the coarse grid space H_0, a diagonal matrix, which is trivially preconditioned by symmetric diagonal scalings.

Now, one has to be somewhat careful when extrapolating from this observation. The fact that the hierarchical basis functions $\psi_{j,k}$ are actually orthogonal to the energy inner product is an artifact. In two dimensions, this is no longer the case but it turns out that the hierarchical stiffness matrices can still be preconditioned by diagonal scaling to efficiently reduce the growth of the condition numbers to logarithmic behaviour. Moreover, it has suggested similar strategies involving other multiscale bases which do even better. For interested readers, we refer Canuto and Cravero [1997] and Dahmaen [1997].

5.4 Fractal image compression

5.4.1 Introduction

Image compression techniques and methods are of vital importance in high technologies, especially in the information technology. These techniques and methods are related to communicating and storing images and data (information) in the shortest possible time (in minimum space on hard disk or CD-ROM) and retrieval with permissible distortion. Tools and techniques of the Fourier analysis like Fourier transform, convolution, Shannon sampling theorems and Walsh-Fourier methods have been successfully used for a long time. Matrix transform, optimization and variational techniques were also employed to study image and data compression. Wavelet theory entered in this area in the late eighties and started replacing the role of DCT (JPEG) which was dominating the scene until that time in view of its high compression ratio and the quality of information after retrieval.

The concept of a fractal and the discipline of fractal geometry was introduced by Benôit Mandelbrot in the early eighties to study irregular shapes like coastal lines, mountains, clouds or rain fall. Fractals are complicated looking sets like Cantor set, Sierpinski gasket, Sierpinski carpet, von-Koch curve, Julia set, but they arise out of simple algorithms. By now it is a well-established discipline and a comprehensive and updated bibliography can be found in Barnsley [1988], Barnsley and Hurd [1993], Fisher [1995], Saupe [1996], Lu [1997] and references therein.

Barnsley [1988,1996] established a close connection between functional analysis, fractals and multimedia by demonstrating that fractals can be defined in terms of fixed points of mappings defined on an appropriate metric space into itself and image compression can be studied through this methodology achieving marvelous results. He has already commercialized his achievements in the form of the top-selling multimedia encyclopedia **Encarta**, published by the Microsoft Corporation including, on one CD-ROM, seven thousand colour photographs which may be viewed interactively on a computer screen. There are diverse images like those of buildings, musical instruments, people's faces, baseball bats, ferns. This development is known as the IFS theory (Iterated Function System theory) for image compression. Stewart [1995] has explained in a simple language the far-reaching consequences of this theory which will drastically reduce the expenses on communication through fax machines manufactured using this technique and methodology. It has been established that for a fairly large class of images, the IFS theory provides a better compression ratio and quality of images after retrieval (see, for example, Fisher, Lu and recent papers on web resources) compared to the most popular method until now, DCT(JPEG). It is also expected that a combination of fractal and wavelet techniques may still yield better results. This is a fast-growing field in which, besides mathematicians, computer and information scientists, physicists, chemists and engineers are actively involved.

5.4.2 IFS theory

Let (X, d) be a complete metric space and let $T : X \longrightarrow X$ be a mapping of X into itself. The iterates of T are mappings $T^{0n} : X \longrightarrow X$, defined by $T^{00}(x) = x, T^{01}(x) = T(T^{00}(x)) = T(x), T^{02}(x) = T(T^{01}(x))$ or $T^{02} = T^{00}oT^{01}, \cdots, T^{0n}(x) = T(T^{0(n-1)}(x))$ or $T^{0n} = ToT^{0(n-1)}$. A mapping F on the set of real numbers R into itself is called an affine transformation if it is of the form $F(x) = ax + b$ for all $x \in R$, where a and b are constants. A mapping G on R^2 into itself is called an

affine transformation if it can be written as

$$G(x,y) = (ax + by + e, cx + dy + f),$$

where a, b, c, d, e and f are real numbers. G can also be written as

$$G\begin{pmatrix} x \\ y \end{pmatrix} = \begin{pmatrix} a & b \\ c & d \end{pmatrix}\begin{pmatrix} x \\ y \end{pmatrix} + \begin{pmatrix} e \\ f \end{pmatrix} = Ax + B,$$

where $A = \begin{pmatrix} a & b \\ c & d \end{pmatrix}$ is a 2×2 matrix and B is the column vector $\begin{pmatrix} e \\ f \end{pmatrix}$.

The mapping T is called a Lipschitz mapping if there exists a constant $\alpha \geq 0$ such that $d(T(x), T(y)) \leq \alpha d(x,y)$ for all $x, y \in X$. T is called a **contraction mapping** if $0 \leq \alpha < 1$. α is called the contractivity factor of T. A Lipschitz continuous function T is called **eventually contractive** if there is a number n such that T^{0n} is a contraction map. Let Y be a non-empty subset of a compact metric space (X, d), then a mapping S on Y into X is called **a local (partitioned) contraction mapping** on (X, d) if there is a number $s, 0 \leq s < 1$, such that $d(S(x), S(y)) \leq sd(x,y)$ for all $x, y \in Y$. A complete metric space (X, d) equipped with n contraction mappings $w_i : X \longrightarrow X, i = 1, 2, \cdots, n$, denoted by $\{X, d, w_i, i = 1, 2, \cdots\}$ is called an **iterated function system (IFS)**. A complete metric space (X, d) equipped with n eventually contractive mappings $w_i : X \longrightarrow X, i = 1, 2, \cdots, n$, denoted by $\{X, d, w_i, i = 1, 2, \cdots, n\}$ is called an **eventually iterated function system**. A local (partitioned) iterated function system (LIFS) is a compact metric space (X, d) equipped with n local contractive mappings $w_i : Y \longrightarrow X$, $Y \subseteq X$. A **recurrent iterated function system** (RIFS) is a collection $w_1, w_2, \cdots w_n$ of n Lipschitz maps in a complete metric space X and $n \times n$ matrix (a_{ij}) satisfying $\sum_j a_{ij} = 1$ for all i.

The iterated function system (IFS) theorem states that:

For an IFS $\{X, d, w_i\}$ with contractivity factors $\alpha_i, i = 1, 2, \cdots n$, the mapping W defined on $H(X)$ = space of all compact subsets of X, into itself by $W(B) = \cup_{i=1}^{n} w_i(B)$ is a contraction mapping on the complete metric space $(H(X), h(.,.))$, where

$$\begin{aligned} h(A,B) &:= \max\{d(A,B), d(B,A)\}\ , \\ d(A,B) &:= \max\{d(x,B) \mid x \in A\}\ , \\ d(x,B) &:= \min\{d(x,y) \mid y \in B\}, \end{aligned}$$

with the contractivity factor $\alpha = \max_{1 \leq i \leq n} \alpha_i$, that is,

$$h(W(B), (W(C)) \leq \alpha h(B, C),$$

for all $B, C \in H(X)$.

Consequently, W has a unique fixed point, say, $A \in H(X)$ which satisfies the relation $A = W(A) = \cup_{i=1}^{n} w_i(A)$, and is given by

$$A = \lim_{n \to \infty} W^{0n}(B) \quad \text{for any} \quad B \in H(X)\,.$$

This fixed point A is called the **attractor** or **deterministic fractal** or **fractal.** $h(.,.)$ is known as the Hausdorff metric.

Let $\{R^2, d, w_i, i = 1, 2, \cdots n\}$ be an IFS where w_i's are given by

$$w_i \begin{pmatrix} x \\ y \end{pmatrix} = \begin{pmatrix} a_i & b_i \\ c_i & d_i \end{pmatrix} \begin{pmatrix} x \\ y \end{pmatrix} + \begin{pmatrix} e_i \\ f_i \end{pmatrix}, i = 1, 2, \cdots n .$$

Then the following table is known as the IFS code:

Table 5.2. IFS code.

w	a	b	c	d	e	f	p
w_1	a_1	b_1	c_1	d_1	e_1	f_1	p_1
w_2	a_2	b_2	c_2	d_2	e_2	f_2	p_2
$\vdots$	$\vdots$	$\vdots$	$\vdots$	$\vdots$	$\vdots$	$\vdots$	$\vdots$
w_n	a_n	b_n	c_n	d_n	e_n	f_n	p_n

where

$$p_i \simeq \frac{|detA|}{\sum_{i=1}^{N} |detA_i|} = \frac{|a_i d_i - b_i c_i|}{\sum_{i=1}^{N} |a_i d_i - b_i c_i|},$$

for $i = 1, 2, \cdots n$; the symbol $\simeq$ means "approximately equal to". The numbers p_i's can be interpreted as probabilities for finding the attractor of an IFS using the Chaos Game Algorithm. An image can be treated as a closed bounded (compact) subset of R^2. The following result, known as the **Collage Theorem,** is very important for designing IFS whose attractors or fractals are close to a given image. Let (X, d) be a complete metric space and S be a given image, that is, $S \in H(X)$ and let $\epsilon \geq 0$ be also given. Choose an IFS $\{X, d, w_i\}, i = 1, 2, \cdots n$, with contractivity factors $\alpha_i, i = 1, 2, \cdots n$, such that

$$h(S, \cup_{i=1}^{n} w_i(S)) \leq \epsilon .$$

Then

$$h(S, A) \leq \frac{1}{1-s} h(S, W(S)) \leq \frac{\epsilon}{1-s},$$

where

$$W(s) = \cup_{i=1}^{n} w_i(S)) \quad \text{and} \quad s = \max_{1 \leq i \leq n} \alpha_i,$$

and A is the attractor of the IFS. This theorem precisely tells us that if we can find an IFS code so that the Hausdorff distance between S and $W(S)$ is very small, the attractor of W will be very close to the target image S. There are algorithms for finding attractors of the given IFS like the Chaos Game Algorithm, the Photocopy Machine Algorithm (see Barnsley and Hurd [1993] and Lu [1993]).

The basic technique of image compression through IFS is to find out appropriate affine contraction mappings $w_1, w_2, \cdots w_n$ such that the condition of the **Collage Theorem** is satisfied, namely, S is very close to $W(S)$ and so, instead of communicating/storing the image, we can communicate/store the fractal or attractor of IFS, that is, coefficients in the IFS code. For details of this method, we refer to Barnsley [1988], Barnsley and Hurd [1993], Lu [1993], Jacquin [1993], Fisher [1995], and Ning Lu [1997].

3. Inverse problem for images.

The problem of representing a given image (or a function) by the IFSs or their variations is a typical inverse problem. This involves finding the IFS parameters of an image that is exactly generated via an IFS. In recent years, the **iterated function system with probabilities** (IFSP), **iterated fuzzy set systems** (IFZS) and IFS **with grey level maps** (IFSM) have been introduced and the corresponding inverse problems have been investigated. Such an inverse problem is related to the problem of finding the image/function as the fixed point element of a given iteration algorithm of the types IFS, IFSP and IFSM on function spaces like $L_p, H^{m,p}$ (the Sobolev space of order m), the weighted Sobolev space and the Besov space. In purely abstract mathematical terms, it comprises the following steps:

(i) Finding a suitable metric space X in which to represent the image (function).

(ii) Finding an appropriate metric $d(\cdot, \cdot)$ on X.

(iii) Finding an appropriate contraction map T on X into itself.

The fact that such problems have more than one solution motivated the search for different kinds of optimality. This problem has been studied in recent years by Forte and Vrscay (see, for example, Siddiqi, Ahmad, and Mukheimer [1997] and Forte and Vrscay [1995], Manchanda, Mukheimer and Siddiqi [1998]).

We present here a summary of these results. Let $F^*(X)$ denote the set of all functions $u : X \longrightarrow [0,1]$ which are (i) upper semi-continuous on (X, d) (ii) normalized, that is, for each $u \in F^*(X)$, there exists an $x_0 \in X$ such that $u(x_0) = 1$. For $u \in F^*(X)$ and for each $\alpha \in [0,1]$, the α-level sets of u are defined by

$$[u]^\alpha = \{x \in X \mid u(x) \geq \alpha\}, \alpha \in (0,1]$$

$$[u]^0 = \overline{\{x \in X \mid u(x) > 0\}}\,.$$

It is clear that $[u]^\alpha \in H(X)$. For each $u, v \in F^*(X)$, we can define a metric

$$d_\infty(u,v) = \sup_{0 \leq \alpha \leq 1} h([u]^\alpha, [v]^\alpha) \quad (F^*(X), w_i, \Phi),$$

where $w_i : X \longrightarrow X$ are contraction maps, $\Phi = \{\phi_1, \phi_2, \cdots \phi_n\}$ and ϕ_i's $i = 1, 2, \cdots n$ are mappings of [0,1] into itself ($\phi_i : [0,1] \longrightarrow [0,1]$), each of which being (a) non-decreasing, (b) right continuous, (c) $\phi_i(0) = 0$ for all i, and (d) for at least one $i, \phi_i(1) = 1$, is called the **iterated fuzzy set system** (IFZS).

Let contraction mappings $w_1, w_2, \cdots, w_n$ be associated with probabilities $p_1, p_2, \cdots, p_n$ with $\sum_{i=1}^{n} p_i = 1$. Furthermore, let $\mathcal{B}(X)$ denote the σ−algebra of Borel subsets of X and $\mathcal{M}(X)$ denote the set of all probability measures on $\mathcal{B}(X)$. $(\mathcal{M}(X), w_i, T)$, where T is defined by the relation

$$(T\nu)(S) = (M\nu)(S) = \sum_{i=1}^{n} p_i(\nu(w_i^{-1}(S))),$$

for a $\nu \in \mathcal{M}(X)$ and each $S \in H(X)$ is called the **iterated function system with probabilities** (IFSP). $\mathcal{M}(X)$ is a metric space with respect to the metric

$$d_H(\mu, \nu) = \sup_{f \in Lip_1(X)} \left| \int_X f d\mu - \int_X f d\nu \right|,$$

where

$$\text{Lip}_1(X) = \{f : X \longrightarrow R \mid |f(x_1) - f(x_2)| \leq d(x_1, x_2) \;\; \forall \;\; x_1, x_2 \in X\}.$$

T defined above is called the Markov operator on $(\mathcal{M}(X), d_H(.,.))$. For a measure μ on $\mathcal{B}(X)$ and for any integer $p \geq 1$, let $L_p(X, \mu)$ denote the vector space of all real-valued functions u such that u^p is integrable on $(\mathcal{B}(X), \mu)$. $L_p(X, \mu)$ is a complete metric space with respect to the metric induced by the L_p norm; that is,

$$d(u, v) = \|u - v\|_{L_p} = \left(\int_X |u(x) - v(x)|^p d\mu(x) \right)^{1/p}, \quad (L_p(X, \mu), w_i, \Phi),$$

where $\Phi = \{\phi_1, \phi_2, \cdots \phi_n\}$ with $\phi_i : R \longrightarrow R$, known as the grey level maps, is called the **iterated function system with grey level maps** (IFSM).

An **operator** T can be defined on IFSM as

$$(Tu)(x) = \sum_{i=1}^{n}{}' \phi_i(u(w_i^{-1}(x))).$$

The prime ($\prime$) signifies that the sum operates on all those terms for which $w_i^{-1}(x)$ is defined. If $w_i^{-1}(x) = 0$ for all $i = 1, 2, \cdots, n$ then $(Tu)(x) = 0$. For $X \subset R^n$, let $m^{(n)} \in \mathcal{M}(X)$ denote the Lebesgue measure on $\mathcal{B}(X)$. The indicator function of a subset A of X denoted by $I_A(x)$ is defined by

$$I_A(x) = \begin{cases} 1, & x \in A \\ 0, & \text{otherwise} \end{cases}$$

$$\text{Lip}(R) = \{\phi : R \longrightarrow R \mid |\phi(t_1) - \phi(t_2)| \leq \beta |t_1 - t_2|\},$$

$\forall\, t_1, t_2 \in R$ and for some $\beta \in [0, \infty)$. It can be verified that for any $u \in L_p(X, \mu)$, $1 \leq p < \infty$, and $\phi \in \text{Lip}(R)$, $1 \leq i \leq n$, T is a mapping on $L_p(X, \mu)$ into itself. In fact, T

becomes a contraction mapping under certain assumptions and hence has a unique fixed point as $L_p(X,\mu)$ is a complete metric space.

Affine IFSM on $L_p(X,\mu)$ is that IFSM on $L_p(X,\mu)$ where ϕ_i are given by $\phi_i(t) = \alpha_i t + \beta_i, \quad t \in R, \quad i = 1, 2, \cdots n$. Let $X = [0,1]$ and $\mu = m^{(1)}$ with $w_i(x) = s_i x + a_i$ and $\alpha_i = |s_i| < 1, \quad 1 \leq i \leq n$. If T is contractive with fixed point $\bar{u}$, then

$$\bar{u}(x) = \sum_{i=1}^{n}{}' \alpha_i \bar{u}\left(\frac{x - a_i}{s_i}\right) + \beta_i I_{w_i(X)}(x) = \sum_{i=1}^{n}{}' \left[\alpha_i \psi_i(x) + \beta_i \chi_i(x)\right].$$

This means that $\bar{u}$ may be expressed as a linear combination of both piece-wise constant functions $\chi_i(x)$ as well as functions $\psi_k(x)$ which are obtained by dilations and translations of $\bar{u}(x)$ and $I_X(x) = 1$, respectively. This reminds us of the role of scaling functions in the wavelet theory.

The Collage Theorem mentioned earlier can be rephrased as follows:
Let (X,d) be a complete metric space, and for a given $x \in X$ there exists a contraction map $W : X \longrightarrow X$ with the contractivity factor α such that $d(x, W(x)) < \epsilon$. Then

$$d(x,\bar{x}) < \frac{\epsilon}{1-\alpha},$$

where $\bar{x}$ is the fixed point of W $(W(\bar{x}) = \bar{x})$.

In view of this result, the inverse problem for approximation of functions in $L_p(X,\mu)$ by IFSM may be stated as follows: Given a target function $v \in L_p(X,\mu)$ and a $\delta > 0$, find an IFSM $(L_p(X,\mu), w_i, \phi_i)$ with the associated operator T such that $\|v - Tv\|_p \leq \delta$.

For $\mu \in M(X)$, a family $\mathcal{A}$ of subsets $A = \{A_i\}$ of X is called μ-dense in a family $\mathcal{N}$ of subsets B of X if for every $\epsilon > 0$ and any $B \in \mathcal{N}$, there exists a collection $A \in \mathcal{A}$ such that $A \subseteq B$ and $\mu(B \setminus A) < \epsilon$. Let $\{w_i\}$ be an infinite sequence of contraction maps on X into itself. We say that $\{w_i\}$ generates a "μ-dense and non-overlapping" (abbreviated as "$\mu - d - n$") family $\mathcal{A}$ of subsets of X if, for every $\epsilon > 0$ and every $B \subseteq X$, there exists a finite set of integers $i_k \geq 1, \quad 1 \leq k \leq n$, such that

(i) $A = \cup_{k=1}^{n} w_{i_k}(X) \subseteq B$,

(ii) $\mu(B \setminus A) < \epsilon$, and

(iii) $\mu(w_{i_k}(X) \cap w_{i_l}(X)) = 0$ if $k \neq l$.

If $\{w_i\}$ satisfies the above conditions on (X,d), then $\inf\limits_{1 \leq i < \infty} \{\alpha_i\} = 0$, where α_i's are the contractivity factors of w_i's, independent of μ. If $X = [0,1]$ and μ is the Lebesgue measure, then the wavelet type functions

$$w_{ij}(x) = 2^{-i}(x + j - 1), \quad i = 1, 2, \cdots, j = 1, 2, \cdots, 2^i,$$

can form a "μ-d-n" family.

For each $i^* \geq 1$, the set of maps $\{w_{i^*j}, \quad j = 1, 2, \cdots, 2^{i^*}\}$ provides a set of 2^{-i^*} contractions of $[0, 1]$ which tile $[0, 1]$. In 1995, Forte and Vrscay obtained the following result which provided the solution of the inverse problem:

Theorem A[Forte-Vrscay, 1995]. *Let $v \in L_p(X, \mu), \quad 1 \leq p < \infty$, then*

$$\liminf_{n \to \infty} \|v - T^n v\|_p = 0,$$

provided the sequence of contraction maps w_i generates a "μ-d-n" a family $\mathcal{A}$ of subsets of X and w_i's are also one-to-one, where

$$(T^n v)(x) = \sum_{i=1}^{n}{}' \phi_i(u(w_i^{-1}(x))) .$$

This theorem has been studied for local IFSM and special cases like $p = 2$ and ϕ_i's affine maps. Forte and Vrscay have also carried out an approximation of the target image "Lena", a 512×512-pixel grey scale image, with each fixed pixel having 256 possible values (8 bits, with values from 0 to 255, which are rescaled to values in [0,1]). This type of approximation has been studied by Siddiqi et al. [1997] for singer and bride. The correspondence between fractal-wavelet transforms and iterated function systems with grey-level maps has been systematically studied by Mendivil and Vrscay [1997]. A wavelet-based solution to the inverse problem for fractal interpolation functions has been investigated by Berkner [1997]. Manchanda, Mukheimer and Siddiqi [1998] have extended Theorem A to the Besov space.

4. Comparison of JPEG, wavelet (EPIC) and IFS techniques for image compression.

A comparison of performance in terms of compression ratio (CR), and visual appearance (PSNR) of IFS (fractal), wavelet (EPIC) and JPEG techniques for some wood blocks and a singer and bride of Indian sub-continent is given in the following table:

Table 5.3. CR and PSNR.

Image	Compression techniques					
	Fractal		EPIC		JPEG	
	CR	PSNR	CR	PSNR	CR	PSNR
Wood 1	4.36:1	31.59	4.83:1	34.33	5:1	28.31
Wood 1	18.78:1	27.52	18.41:1	28.73	20:1	25.83
Wood 2	4.69:1	29.15	4.78:1	33.98	5:1	26.95
Wood 2	18.74:1	25.03	17.97:1	27.12	20:1	23.77
Singer	4.36:1	33.27	5.10:1	36.99	5:1	30.49
Singer	18.81:1	27.30	18.22:1	29.56	20:1	25.62
Bride	4.36:1	32.33	5.16 :1	40.35	5:1	30.64
Bride	18.79:1	27.02	18.68:1	30.65	20:1	26.06

The wavelet technique is slightly better in terms of the visual quality of images after compression followed by fractal and JPEG. The following references also provide interesting information about wavelets and their applications to image processing; Bertoluzza [1992], Daubechies et al. [1992], Daubechies [1992], DeVore et al. [1992], Kaiser [1994], Kelly, Kon and Raphael [1994], Mallat [1989, 1996], Strang [1989, 1993], Sweldons [1994] and Walter [1992, 1995].

5.5 Problems

Problem 5.1. Consider a linear position invariant image degradation system with impulse response $h(x-\alpha \cdot y-\beta) = e^{[(x-\alpha)^2+(y-\beta)^2]}$. Suppose that the input to the system is an image consisting of a line infinitesimal width located at $x = a$, and modelled by $f(x,y) = \delta(x-a)$. Assuming no noise, what is the output image $g(x,y)$?

Problem 5.2. (i) Show that the discrete Fourier transform and its inverse are periodic functions, (ii) obtain the Fourier transforms of: (a) $\frac{df(x)}{dx}$, and (b) $\frac{f(x,y)}{x}+\frac{f(x,y)}{y}$.

Problem 5.3. Discuss the variational formulation of the Perona and Malik model.

Problem 5.4. Find solutions of the dilation equation

$$\phi(x) = \sum_n c_n\phi(2x-n), \sum_n c_n = 2,$$

for the following values of c_n

(i) $c_0 = 2$ and $c_n = 0$ for $n \neq 0$, and

(ii) $c_0 = c_1 = 1$ and $c_n = 0$ for $c_n \neq 0$ and 1.

Problem 5.5. Show that $\phi(x) = \sin c(\pi x)$ is a scaling function.

Problem 5.6. Let

$$\psi \in L_2(R),$$

and

$$c_\psi = \int_R \frac{|\hat{\psi}|(w)|}{|w|} dw < \infty,$$

and

$$\psi(x) = \frac{d\phi}{dx},$$

then show that the wavelet transform of ϕ is given by

$$L_\phi(T^{\text{wav}} f)(a,b) = \left| \frac{|a|^{1/2}}{\sqrt{c_\phi}} \int_R \phi\left(\frac{(x-b)}{a}\right) f' cx\, dx \right|$$

Problem 5.7. State and prove the Parseval's identity for wavelet transforms.

Problem 5.8. Establish a relationship between wavelet transforms and wavelet Fourier coefficients.

Problem 5.9. Write a short note introducing the silent features of wavelets.

Problem 5.10. Explain the advantages of wavelet techniques in scientific computing.

Problem 5.11. Discuss the relationship between the fractal wavelet transform and the iterated function systems with grey-level maps.

Problem 5.12. How can one apply the scaling properties of continuous wavelet transforms to find a solution to the inverse problem for fractal interpolation functions?

Problem 5.13. Let φ be a given k-times, $k \geq 1$, differentiable function such that $\varphi^{(k)} \in L_2(R)$ and $\varphi \neq 0$. Then $\psi(t) = \varphi^k(t)$ is a wavelet.

Problem 5.14. The set of wavelets $\overline{\Psi} = \{\psi \in L_2(R)$ such that ψ satisfies (5.72)$\}$ is a dense subset in $L_2(R)$.

Problem 5.15. Let $H(t)$ be the Haar function. Show that $\{H_{j,k}(t)\} = \{2^{j/2}H(2^jt-k)\}$ is an orthonormal basis in $L_2(R)$.

Problem 5.16. Show that there is a multiresolution analysis corresponding to the Haar wavelet.

Solution of Problem 5.13. We know that $|\hat{\psi}(w)| = |w|^k\, |\hat{\varphi}(w)|$ by the properties of the Fourier transform. We need to show that

$$\begin{aligned}
c_\psi &= 2\pi \int_R \frac{|\hat{\psi}(w)|^2}{|w|}\, dw < \infty \\
c_\psi &= 2\pi \int_R \frac{|w|^{2k}\,|\hat{\varphi}(w)|^2}{|w|}\, dw \\
&= 2\pi \int_{-1}^{1} |w|^{2k-1}\,|\hat{\varphi}(w)|^2 dw + 2\pi \int_{|w|>1} \frac{|w|^{2k}|\hat{\varphi}(w)|^2}{|w|}\, dw \\
&\leq 2\pi \left(\|\varphi\|_{L_2}^2 + \|\varphi^{(k)}\|_{L_2}^2\right) < \infty\,.
\end{aligned}$$

Solution of Problem 5.14. Let $f \in L_2(R)$, then we get $\hat{f} \in L_2(R)$. Let us define f_ϵ as

$$\hat{f}_\epsilon(w) = \begin{cases} \hat{f}(w)\,, & |w| \geq \epsilon \\ 0\,, & |w| < \epsilon\,. \end{cases}$$

Then, for every ϵ, f_ϵ is a wavelet as it satisfies the admissibility condition (5.72). Since $\|f\|_{L_2} = \|\hat{f}\|_{L_2}$, we have

$$\|f - f_\epsilon\|_{L_2} = \int_{-\epsilon}^{\epsilon} |\hat{f}(w)|^2 dw \to 0 \text{ as } \epsilon \to 0\,.$$

Thus every function f in $L_2(R)$ can be considered as a limit of a sequence of wavelets.

Solution of Problem 5.15. We need to show that

$$\begin{aligned}
\langle H_{j,k}, H_{m,n}\rangle &= 0 \text{ if } j \neq m \text{ or } k \neq n \\
&= 1 \text{ if } j = m \text{ and } k = n\,.
\end{aligned}$$

Since

$$\langle H_{j,k}, H_{m,n}\rangle = \int_{-\infty}^{\infty} 2^{s/2} H(t) H(2^s t - r) dt,$$

where $s = m-j$ and $r = 2^{m-j}k-n$, $\langle H_{j,k}, H_{m,n}\rangle = 1$ for $j = m$ and $k = n$. If $j = m$ but $k \neq n$, then $r \neq 0$; so $\text{supp}\, H(2^s t - r) \cap \text{supp}\, H(t) = \phi$, where $\text{supp}\, H_{j,k} = \left[k2^{-j}, (k+1)\,2^{-j}\right]$. Consequently, the integral is 0 or $\text{supp}\, H(2^s t - r) \not\subseteq \text{supp}\, H(t)$. But in this case $H(t)$ is constant on $\text{supp}\, H(2^s t - r)$, so the integral is also 0 because $\int_{-\infty}^{\infty} H(t) dt = 0$.

In order to show that $\{2^{j/2} H(2^j t - k)\}_{j \in Z, k \in Z}$ is an orthonormal basis in $L_2(R)$, it is sufficient to show that

$$\begin{aligned}
S_n &= \text{span}\{H_{j,k}\}_{j<n, k \in Z} \\
&= L_n = \left\{f \in L_2(R)/f \text{ is constant on } \left[k2^{-n}, (k+1)2^{-n}\right],\ \text{for } k \in Z\right\}\,.
\end{aligned}$$

$\{S_n\}$ and $\{L_n\}$ have the following properties:

(a) $\{S_n\}$ is a family of closed subspaces of $L_2(R)$ such that

(i) $\cdots S_{-2} \subset S_{-1} \subset S_0 \subset S_1 \subset \cdots$

(ii) $f(t) \in S_n$ if and only if $f(2t) \in S_{n+1}$.

(iii) $f(t) \in S_0$ if and only if $f(t+k) \in S_0$ for $k \in Z$.

(b) $\{L_n\}$ is also a family of closed subspaces of $L_2(R)$ satisfying (i) to (iii).

(c) For all $\ell \in Z$ we have $L_\ell = S_\ell$.

In view of these results $\{H_{j,k}\}$ is an orthonormal basis of $L_2(R)$. We verify (c); other parts can be easily checked.

In view of (ii), it is sufficient to show that $L_0 = S_0$. Since each $H_{j,k}$ for $j < 0$ is constant on any interval $[r, r+1]$, we see that $S_0 \subset L_0$. On the other hand, each function of L_0 can be written as $\sum_{r\in Z} a_r \chi_{[r,r+1]}$ so by (iii), it suffices to show that $\chi_{[0,1]} \in S_0$. To show this, let us consider the series

$$\sum_{j<0} 2^{j/2} H_{j,0} = \sum_{j<0} 2^j H(2^j t).$$

Since $\|2^j H(2^j t)\|_2 = 2^{j/2}$ and $j < 0$, this series is absolutely convergent in $L_2(R)$. It is clear from the definition of the Haar function $H(t)$ that

$$\sum_{j<0} 2^{j/2} H_{j,0} = 0 \text{ for } t \le 0,$$

$$\sum_{j<0} 2^{j/2} H_{j,0}(t) = \sum_{j<0} 2^j = 1 \text{ for } 0 < t < 1,$$

and for $2^r < t < 2^{r+1}$ where $r = 0, 1, 2, \ldots$, one has

$$\sum_{j<0} 2^{j/2} H_{j,0}(t) = -2^{-r-1} + \sum_{j=r+2}^{\infty} 2^{-j} = 0.$$

This implies that $S_0 = L_0$ which, in turn, implies that $L_\ell = S_\ell$ for all $\ell \in Z$.

Solution of Problem 5.16. $\{L_n\}_{n=-\infty}^{\infty}$, where L_n is defined in the solution of Problem 5.15, is a multiresolution analysis associated with the Haar wavelet $H_{j,k}$. $\chi_{[0,1]}$ can be taken as the scaling function.

Chapter 6

Models of Hysteresis and Applications

The phenomenon of hysteresis occurs in a large number of practical situations and plays a very important role in areas like ferro-magnetism, ferro-electricity, phase transitions from solid to liquid or liquid to solid. In spite of the fact that this concept has varieties of significant applications in scientific problems, its systematic study started quite late. However, in the last couple of years, two proceedings of the conferences, edited by Visintin [1993, 1994] and four monographs by Brokate and Sprekels [1996]; Krasnosel'skii, and Pokrovskii [1989], Mayergoyz [1991] and Visintin [1994] have appeared, besides a large number of research papers on applications of the concept, especially in the estimation of life-time of machines. These publications are attracting the attention of fairly large number of mathematicians and engineers. The main goal of this chapter is to familiarize the readers with the basic tools and techniques of this field which may prove vital for industrial applications. It may be observed that this area still requires intensive study and there is no dearth of challenging problems.

6.1 Introduction to hysteresis

The word hysteresis literally means the lag exhibited by a body in reacting to changes in the forces, especially magnetic forces effecting it. More precisely, this phenomenon is exhibited, often by a system comprising a ferromagnetic or imperfectly elastic material, in which the reaction of the system is dependent upon its past reaction to change.

Let us concentrate on Figure 6.1. This figure shows an iron ring on which a magnetising winding has been wound which can be supplied with a current. There is also a secondary winding connected to a flux meter which can measure the change of flux (the integral of voltage with time) when the magnetizing current is changed.

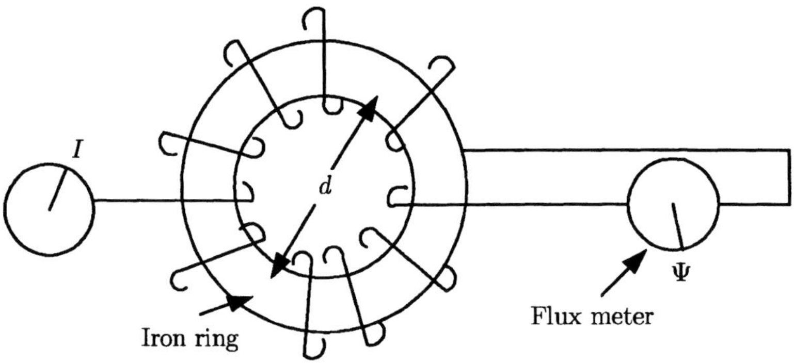

Figure 6.1. Ring used in the determination of a BH curve (hysteresis loop).

One can observe the relationship between the magnetic field strength H applied to the iron and the resultant flux density B in the iron. By carrying out an experiment, it has been found that the change in B associated with particular change in H depends not only on the values of H but also on the magnetic history of the iron. The curve shown in Figure 6.2 gives a relationship between B and H.

If the field applied to a specimen is increased to saturation and then decreased, the flux density B decreases but not as rapidly as it increased along the initial magnetization curve. Thus, there is a residual flux density or reminiscence, B_r, as shown in the figure even when $H = 0$. In order to reduce B to zero, a negative field $-H_c$ is required. This can be created by reversing the battery polarity. This force is called coercive force and is marked on the curve by H_c. The phenomenon which causes B to lag behind H so that the magnetization curve for increasing and decreasing fields is not the same is called **hysteresis** and the loop traced out by the magnetization curve is called a **hysteresis loop**. In soft (easily magnetized) materials, the hysteresis loop is thin, with a small area enclosed. For hard magnetic materials, the area enclosed by the loop is larger. In calculations on soft magnetic materials, it is often possible to neglect the width of the hysteresis loop, although the energy loss due to hysteresis is still taken into account. In such cases, engineers use the curve shown in Figure 6.3 which is called a **magnetic reversal curve** or **magnetic saturation curve.**

The saturation curve is quite helpful in predicting the flux resulting from a certain applied magnetizing current.

Materials such as iron, nickel, cobalt which exhibit a strong magnetic effect, are called ferromagnetic. The permeability of these materials is not constant but is a function of the applied field and the previous magnetic history of the specimen.

Hysteresis can be defined as a rate independent of memory effect. This is a property of some constitutive laws which relate an input variable u and output variable w. Here, memory means that at any instant t, $w(t)$ is determined by the previous evolution of u, and not just by $u(t)$, while rate-independence means that the curve described in R^2 by the pair (u, w), namely, loops are invariant for changes

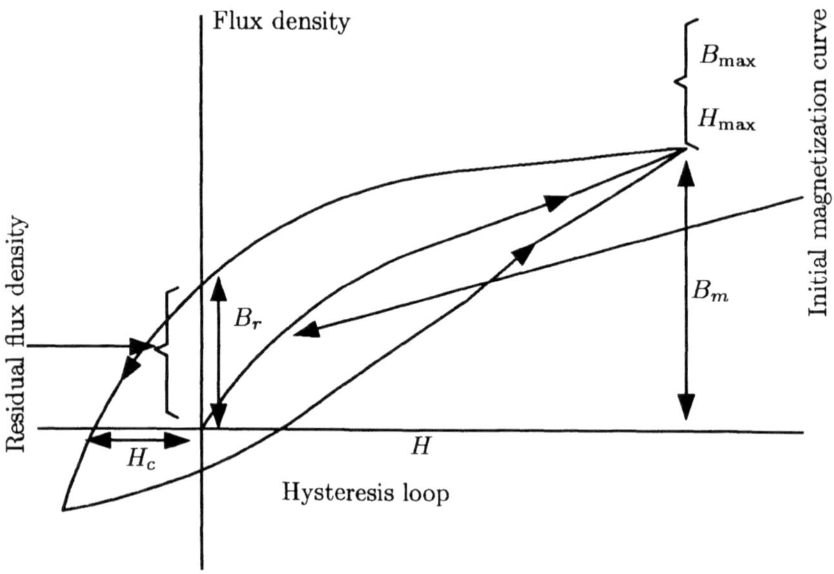

Figure 6.2. Hysteresis diagram.

of the input rate, for example, the frequency. As indicated in Chapter 1 (see Dressler and Hack [1996], Dressler, Hack and Krüger [1997]), the rainflow counting method is a well-established method for fatigue analysis and damage estimation.

Hysteresis loss (loss of energy in the process of hysteresis). We want to know the loss of energy in the work done to take the magnetic state of the material around the hysteresis loop indicated in Figure 6.2. With reference to the situation in Figure 6.1, let the magnetizing current I be supplied by a generator and let it be changed from $+I_{\max}$ to $-I_{\max}$ and back to $+I_{\max}$ and let this process take time T. Then the work done by the generator will be

$$W = \int_0^T - eIdt,$$

where e is the e.m.f. induced by the change of flux in the iron. Thus

$$e = N\,\frac{d\Phi}{dt} = -N\,S\,\frac{dB}{dt},$$

where S is the cross-section area of the ring and N the number of turns of the binding. By the circuit law, we have

$$\oint Hdl = Hl = NI.$$

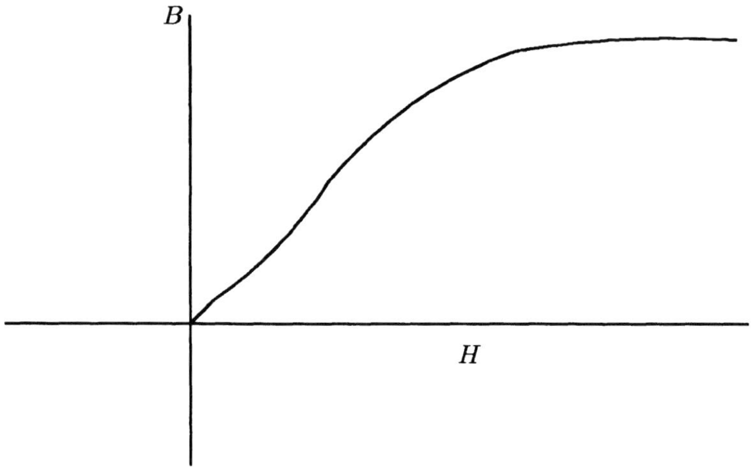

Figure 6.3. Saturation or reversal curve.

Hence

$$W = \int_0^T SIH\frac{dB}{dt} = SI\oint HdB\,. \qquad (6.1)(a)$$

where the integral is taken around complete magnetic cycle. This integral is the area of the hysteresis loop. Equation (6.1)(a) states that the hysteresis loss per cycle is given by the volume of iron multiplied by the area of the hysteresis loop. The energy is supplied by the source of the current and reappears as heat in the iron. The energy loss per cycle is constant, so the total loss depends on the number of cycles. The loss of energy due to hysteresis is therefore proportional to frequency. It is also important to know how the loss is related to the change of the maximum flux. The study of hysteresis operators provides mathematical modelling of the hysteresis loops and loss of energy which was not available earlier (see Remark on page 149 [Hammond [1986])and Section 6.3 for current research concerning dissipation of energy. In Section 2, we introduce hysteresis operators while rainflow counting method will be discussed in Section 3. Section 4 will be devoted to relations between the rainflow counting method, hysteresis operators of different kinds (Prandtl, Ishlinskii, Preisach mh-hystrons, moving model etc.) and the dissipated energy indicating the techniques for fatigue analysis. This chapter is mainly based on Brokate's work (See Brokate [1994], Brokate et al. [1996]).

6.2 Hysteresis operators

A hysteresis operator results from a translation of a hysteresis diagram into a mathematical object. If in Figure 6.2, we treat H and B axes as v and w, respectively, then for given $v : [0,T] \longrightarrow R, T > 0$, we get an output function $w : [0,T] \longrightarrow R$

such that $(v(t), w(t))$ moves along the loop in the diagram. In the sequel, Maps$[0,T]$, $C_{pm}[0,T]$, and $C_{pc}[0,T]$ denote, respectively, the set of functions, the set of piecewise monotone functions and the set of piecewise continuous functions, on $[0,T]$ into R. An operator H on $C_{pm}[0,T]$ into $Map[0,T]$ is called a **hysteresis operator** if it is rate-independent and has the Volterra property; namely,

$$\text{(i)} \quad H[v] \circ \phi = H[v \circ \phi], \text{ for all } v \in C_{pm}[0,T] \tag{6.1}$$

for all continuous monotone increasing transformations ϕ defined on $[0,T]$ into itself satisfying $\phi(0) = 0, \phi(T) = T$ (rate-independence property).

(ii) For $v, \overline{v} \in$ Map$[0,T]$ and $t \in [0,T], v(t) = \overline{v}(t)$ implies that $(H[v])(t) = (H[\overline{v}])(t)$ (Volterra property).

Remark 6.1. (i) The rate-independence implies that only the locally extremal values of v (local minima or local maxima) can have an influence on the memory of the process; we may replace input functions v by input strings $(v_0, v_1, v_2 \cdots v_n), v_i \in R$. Let us assume that

$$S = \{(v_0, v_1, v_2 \cdots v_n) | n \in N_0, v_i \in R, 0 \le i \le n\}, N_0 = N \cup \{0\} \tag{6.2}$$

is the set of all finite strings and

$$\begin{cases} S_H = \{(v_0, v_1, v_2 \cdots v_n) \in S | v_0 \neq v_1 \text{ and } n \ge 1, \\ \qquad (v_{i+1} - v_i)(v_i - v_{i-1}) < 0, \text{ for } 0 < i < n\}. \end{cases} \tag{6.3}$$

S_H is called the subset of alternating strings.

$H_f : S_H \longrightarrow R$ is defined by

$$H_f(v_0, v_1 \cdots v_n) = H[v](T), \tag{6.4}$$

where $v \in C_{pm}[0,T]$ is any input function having a monotonicity partition $0 = t_0 < t_1 < t_2 < \cdots t_n = T$ such that $v(t_i) = v_i, 0 \le i \le n$ is called the **final value mapping**.

(ii) Any mapping $H_f : S_H \to R$ defines a hysteresis operator H if we set

$$H[v](t) = H_f(v(t_0), v(t_1) \cdots, v(t_k)), \tag{6.5}$$

where $0 = t_0 < t_1 < \cdots < t_k = t$ is a monotonicity partition of $v : [0,T] \to R$ such that the string $(v(t_0), v(t_1), \cdots v(t_k))$ is alternating.

(iii) Equations (6.4) and (6.5) provide a bijective correspondence between the set of all hysteresis operators and the set of all real-valued mappings on S_H. Any hysteresis operator H can be interpreted as the mapping on S into itself if we set

$$\begin{cases} H(v_0, v_1, \cdots, v_n) = (H_f(v_0), H_f(v_0, v_1), H_f(v_0, v_1, v_2) \ldots, \\ \qquad H_f(v_0, v_1, v_2, \cdots v_n)). \end{cases} \tag{6.6}$$

In the sequel, $H[v]$ and $H[s]$ will denote the value of H at the function v and the value of H at the string s, respectively.

Example 6.1. Let us consider longitudinal vibrations of a rod. Newton's law coupled with the constitutive stress-strain relation; that is,

$$\frac{\partial^2 u}{\partial t^2} = \frac{\partial \sigma}{\partial x}, \quad \sigma = H[\epsilon], \quad \epsilon = \frac{\partial u}{\partial x}, \tag{6.7}$$

together with certain initial and boundary conditions determines the displacement u, the stress σ and the strain ϵ as functions of time t and space x. Within the elastic limit, Hooke's law

$$H[\epsilon] = E\epsilon \tag{6.8}$$

holds where E denotes the modulus of elasticity. Beyond the elastic limit, many materials exhibit plastic behaviour. The simplest description, namely, the elastic-perfectly plastic model of Figure 6.4 with a fixed yield stress $|\sigma| = r$ and pure plastic flow, admits a continuum of possible stable states $\{(\sigma, \epsilon)/\sigma \in [-r, r]\}$ for every value of ϵ. Figure 6.4 will give rise to a hysteresis operator $\mathcal{E}_r$ acting on a space of functions and its constitutive equation

$$\sigma = \mathcal{E}_r[\epsilon] \tag{6.9}$$

represents an equation between functions rather than between certain values of stress and strain. For $r \geq 0$, the hysteresis operator $\mathcal{E}_r[\cdot, w_{-1}]$, $w_{-1} \in R$ (the initial

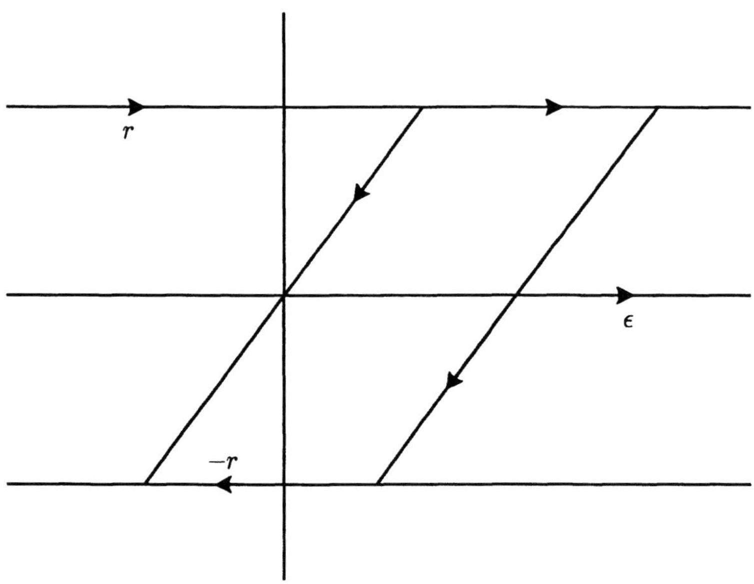

Figure 6.4. The elastic-perfectly plastic element.

value) is defined by its final value mapping $S_H \to R$ given recursively by $\mathcal{E}_{r,f}(v_0) = e_r(v_0 - w_{-1})$

$$\mathcal{E}_{r,f}(v_0, v_1, \cdots, v_n) = e_r(v_n - v_{n-1} + \mathcal{E}_{r,f}(v_0 \cdots, v_{n-1})), \tag{6.10}$$

where

$$e_r(v) = \min\{r, \max\{-r, v\}\}. \tag{6.11}$$

We take $w_{-1} = 0$ unless otherwise stated, and we write $\mathcal{E}_r[v]$ instead of $\mathcal{E}_r[v, 0]$.

Remark 6.2. (i) Let $r \geq 0$, we define $\mathcal{F}_r\ [\cdot, w_{-1}]$ for the initial value w_{-1} (w_{-1} represents the internal state before $v(0)$ is applied at time $t = 0$)

$$w(t) = \mathcal{F}_r[v, w_{-1}](t), \tag{6.11}$$

by its corresponding final value mapping $F_r : S_H \to R$ given recursively by

$$F_r(v_0, v_1, v_2 \cdots v_n) = f_r(v_n, F_r(v_0, v_1 \cdots v_{n-1})), \tag{6.12}$$

where

$$f_r(v, w) = \max\{v - r, \min\{v + r, w\}\}. \tag{6.13}$$

In case the choice of initial value is not mentioned explicitly, we assume it to be zero. Accordingly, we write $F_r(v)$ instead of $F_r[v; 0]$. $F_r(\cdot, \cdot)$ is called the **play operator**.

(ii) Let $R_{x,y}$ be defined as, for $x < y$

$$\begin{aligned} R_{x,y}[v] &= 1 \text{ if the input switches to } y \text{ from below} \\ &= -1 \text{ if the input switches to } x \text{ from above.} \end{aligned} \tag{6.14}$$

$R_{x,y}$ is called **relay with thresholds** $x < y$. The final value mapping $R_{x,y}$ with thresholds $x < y$ and the initial value $w_{-1}(x, y)$ denoted by $R_{x,y,f}$ has the following form:

$$R_{x,y,f}(v_0, v_1, v_2 \cdots, v_n) = \begin{cases} 1, v_n \geq y \\ -1, v_n \leq x \\ R_{x,y,f}(v_0, v_1 \cdots v_{n-1}), x < v_n < y, n \geq 1 \\ w_{-1}(x, y), x < v_n < y, n = 0\,. \end{cases} \tag{6.15}$$

In case the choice of initial value is not stated explicitly, we assume that $w_{-1}(x, y) = 1$ if $x + y \leq 0$ and $w_{-1}(x, y) = -1$ otherwise.

(iii) The play operator F_r incorporates the memory of all relays $R_{x,y}$ with $|x - y| = 2r$.

Lemma 6.1. [Brokate 94]. (i) *For each $r > 0$ and each $s \in R$, we have*

$$R_{s-r,s+r,f}(v_0, v_1 \cdots, v_n) = \begin{cases} 1, & \text{if } F_{r,f}(v_0, v_1, \cdots v_n) > s \\ \\ -1, & \text{if } F_{r,f}(v_0, v_1, \cdots v_n) < s\,, \end{cases} \tag{6.16)(a}$$

for every $n \geq 0$ and every string $(v_0, v_1 \cdots v_n) \in S$.

(iv) It can be seen [Brokate 1994] that $F_r + \mathcal{E}_r = id$; that is

$$F_r[v; w_{-1}] + \mathcal{E}_r[v; w_{-1}] = v \text{ for every } v \in C_{pm}[0, T]\,, \tag{6.16)(b}$$

and every $w_{-1} \in R$.

In 1928, Prandtl constructed a model as the continuous parallel combination of elastic-perfectly plastic elements, which yields a reasonable approximation in cyclic plasticity for investigation of stabilized elastic-plastic behaviour, that is, the behaviour after a large number of load cycles. We can write Prandtl's model in terms of hysteresis operators as

$$H[\epsilon] = \int_0^\infty p(r)\mathcal{E}_r[\epsilon]dr \qquad \text{(Prandtl model)}, \tag{6.17}$$

where $p(\cdot)$ is related to the stabilized σ-ϵ-curve for initial loading that is starting from $\sigma = \epsilon = 0$, as well as the shape of the hysteresis loops.

In 1935, Preisach developed a model for describing the hysteresis loops traced out by magnetic field and magnetization in ferromagnetic materials (see Figure 6.2). The operator form of the Preisach model is as follows:

$$H[v] = \int_0^\infty \int_{-\infty}^\infty \omega(r,s)R_{s-r,s+r}[v]dsdr \qquad \text{(Preisach model)}, \tag{6.18}$$

where $\omega(\cdot,\cdot)$ denotes some density function analogous to $p(\cdot)$ in (6.17) and $R_{x,y}$ denotes the relay function with thresholds, $x < y$.

Remark 6.3. (i) By putting the values of $R_{s-r,s+r}$ in (6.18), we can write the Preisach model in terms of the play operator as

$$H[v](t) = \int_0^\infty q(r, F_r[v](t))dr + q_{00}, \tag{6.19}$$

where

$$q(r,s) = 2\int_0^s \omega(r,\sigma)d\sigma \tag{6.20}$$

$$q_{00} = \int_0^\infty \left(\int_{-\infty}^0 \omega(r,\sigma)d\sigma - \int_0^\infty \omega(r,\sigma)d\,\sigma\right) dr. \tag{6.21}$$

We observe that $q_{00} = 0$ if $\omega(r,s) = \omega(r,-s)$ for all r and s.

(ii) Hysteresis loop corresponding to the play operator is given in Figure 6.5.

(iii) By 6.16(b), and keeping in mind $F_0 = id$, the Prandtl operator H can be written in terms of the play operator; that is,

$$H[v](t) = \int_0^\infty p(r)dr \cdot F_0[v](t) - \int_0^\infty p(r)F_r[v](t)dr. \tag{6.22}$$

(iv) It may be observed that the values $(F_r[v](t))$, $r \geq 0$ play a crucial role in determining the output $H[v](t)$ for all hysteresis operators discussed here. In fact, these values contain the entire memory information at time t, required to determine the future, that is, to determine $H[v]_{[t,T]}$ from $[v]_{[t,T]}$.

Preisach type operator. Let $(v_0, v_1, \cdots v_n) \in S$ be an arbitrary input string. This string successively generates the memory curves $\psi_k : R_+ \to R$,

$$\psi_k(r) = F_{r,f}(v_0, v_1 \cdots v_k), 0 \leq k \leq n, \tag{6.23}$$

through the update

$$\begin{cases} \psi_k(r) &= f_r(v_k, \psi_{k-1}(r)) \\ &= \max\left\{v_k - r, \min\left\{v_k + r, \psi_{k-1}(r)\right\}\right\}, 0 \leq k \leq n \,. \end{cases} \tag{6.24}$$

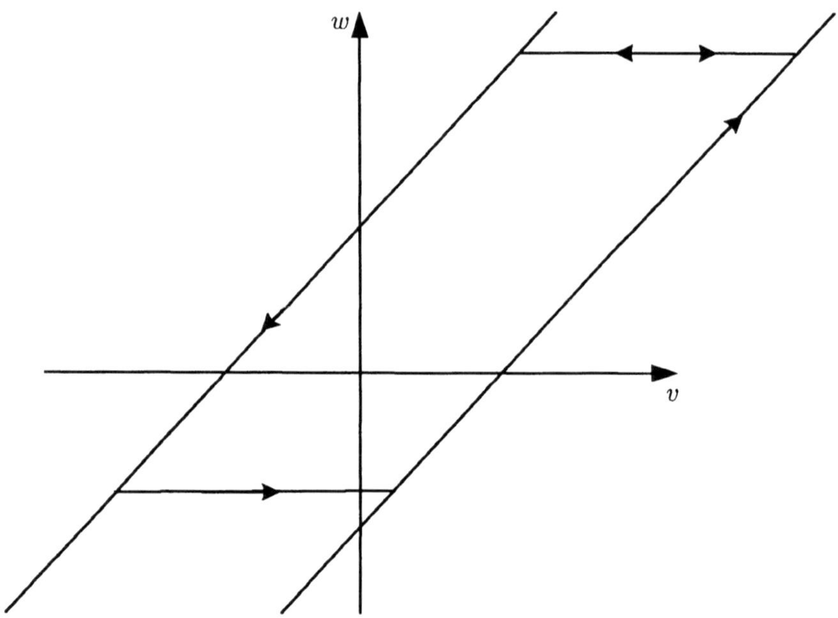

Figure 6.5. Hysteresis loop of the play operator.

Let ψ_{-1} be an element of the set Φ_0 defined by

$$\begin{cases} \Phi_0 &= \{\phi : R_+ \to R\ 1 \mid \phi(r) - \phi(\bar{r}) \mid \leq |r - \bar{r}|, \text{ for all } r, \bar{r} \geq 0\} \\ \phi_{(\rho,\infty)} &= 0 \text{ for some } \rho \geq 0\,, \end{cases} \tag{6.25}$$

where Φ_0 is known as the family of **Preisach memory curves.**

It can be checked that $\psi_k \in \Phi_0$ for all k whenever $\psi_{-1} \in \Phi_0$.

An operator H on $C_{pm}([0, t_E])$ into $\mathrm{Map}([0, t_E])$ is called a **Preisach type operator** if

(i) H is a hysteresis operator.

(ii) The final value map $H_f : S \to R$ has the form

$$\begin{cases} H_f(v_0, v_1 \cdots, v_n) = Q(\psi_f(v_0, \cdots, v_n)), \; (\psi_f(v_0, \cdots v_n))(r) \\ \qquad = F_{r,f}(v_0, v_1, \ldots v_n), \end{cases} \tag{6.26}$$

for some mapping $Q : \Phi_0 \to R$, called the output mapping and some initial condition $\psi_{-1} \in \Phi_0$.

Equation (6.26) can also be written as

$$H[v](t) = Q(\psi(t)), \quad \psi(t)(r) = F_r\left[v; \psi_{-1}(r)\right](t), t \in [0, T], r \geq 0. \tag{6.27}$$

A Preisach type operator H is called a P_o**-operator** if $\psi_{-1} = 0$.

Remark 6.4 (i) The Prandtl model is a Preisach type operator where

$$Q(\phi) = p_0\phi(0) - \int_0^\infty p(r)\phi(r)dr, p_0 = \int_0^\infty p(r)dr. \tag{6.28}$$

(ii) The Preisach model is also the Preisach type operator where

$$Q(\phi) = \int_0^\infty q(r, \phi(r))dr + q_{00}, \tag{6.29}$$

and

$$q(r, s) = 2\int_0^s \omega(r, \sigma)d\sigma. \tag{6.30}$$

(iii) It may be remarked that the class of Prandtl (Preisach) operators is not quite unique, as we have to specify the class of allowed density function $p(\cdot)$ and $\omega(\cdot, \cdot)$. However, no such ambiguity is involved in the definition of a P_o-operator.

6.3 Rainflow counting method in fatigue analysis and damage estimation

We have discussed the application of the rainflow counting method in fatigue analysis in Chapter 1. As mentioned there, the **rainflow counting method** due to Endo is widely used to decompose an arbitrary sequence of loads or deformations into cycles and to count the cycles. In combination with the Palmgren-Miner rule, one can obtain for every sequence of loads a real number which estimates the damage inflicted upon the work piece by this loading sequence. Mechanical engineers have developed several refinements and modifications of this procedure; in any case, the single cycle, respectively, the corresponding hysteresis loop in the stress-strain plane certainly constitutes a basic event for damage assessment. It is also known that the memory structure of the elastoplastic constitutive law due to Prandtl-Ishlinskii directly corresponds to the decomposition performed by the rainflow method. The role of the rainflow method to fatigue analysis has been discussed in detail in Krüger et al. [1985] and Dreßler et al. [1996, 1997]. In this section, we briefly present the description of (i) the accumulated damage equating the variation of the output of a certain Preisach hysteresis operator, (ii) the accumulated damage depending continuously on the input, (iii) the relation between the rainflow method and the hysteresis memory operator, and (iv) hysteretic constitutive laws, dissipation formulas and rainflow count formulas for nonlinear rate-independent rheological models and derivation of the accumulated damage as the output of a certain Preisach operator. For more details, we refer to Brokate et al. [1996].

A string $s \in S$ is called irreducible if neither deletion rule (monotone or Madelung) applies to it. We introduce a partial ordering on S by saying that $s' \leq s$ for $s,\ s \in S$, if s' can be obtained from s by a finite sequence of (arbitrary mixed) monotone and Madelung deletions. For any such finite sequence of deletions, we define its unsymmetric rainflow count as a function $a_u : R \times R \to N_0$, where $a_u(x,y)$ specifies how often the pair (x,y) is removed by a Madelung deletion. If we ignore the order within each pair, we obtain the symmetric rainflow count $a(x,y) = a_u(x,y) + a_u(y,x)$. It is clear that $a_u(x,x) = a(x,x) = 0$ for all $x \in R$ and that $a_u(x,y) \neq 0$ for a finite number of pairs only. In actual applications, the input values are classified in a preprocessing step, so that afterwards only finitely many, say, K different values $x_1 < \cdots < x_K$ can occur. In this case, the rainflow count reduces to the $K \times K$ rainflow matrix A_u with $a_{u,jl} = a_u(x_j, x_l)$. Accordingly, $A = A_u + A_u^T$, defines the symmetric rainflow matrix. In practice, one keeps both the rainflow counts and the residual, since the latter includes information on the order of the original sequence which has been found relevant for purposes of reconstruction and extrapolation of loading sequences from rainflow matrices. However, if we are interested in a complete cycle count, we need to count the residual, too. For any $s \in S$, there is a unique irreducible string $s_R \in S$ with $s_R \leq s$. This irreducible string is called the **rainflow residual** of s. The symmetric rainflow count of (s_R, s_R) appears as the natural way to count the residual s_R (See Brokate et al [Lemma 2.5, 1996]). Since this means that we have to compare counts of different strings, we will write $a(s)$ and $a_s(x,y)$ instead of a and $a(x,y)$ to specify the string whose count is obtained. For any $s \in S$, its periodic rainflow count $a_{\text{per}}(s) : R \times R \to N_0$ is defined as

$$a_{\text{per}}(s) \qquad = a(s) + a(s_R, s_R).$$

It can be checked that

$$a_{\text{per}}(\text{wrap}_i(s)) = a_{\text{per}}(s), \quad 0 \leq i \leq n,$$

where

$$\text{wrap}_i(s) = (v_i, \ldots, v_n, v_0, v_1, \ldots v_{i-1}). \tag{6.31}$$

The count of a single cycle of maximum amplitude is denoted by

$$a_0 = a(v_0, v_{min}, v_0, v_{min}), v_0 = \max_i v_i, v_{min} = \min_i v_i, \quad 0 \leq i \leq n. \tag{6.32}$$

For $x, y \in R$ with $x < y, w_{-1} \in \{0, 1\}$, the relay hysteresis operator $R_{x,y} : S \to S$ is defined by

$$R_{x,y}(v_0 \cdots, v_n) = (w_0, \cdots, w_n), \tag{6.33}$$

with

$$w_i = \begin{cases} 1, v_i \geq y \\ -1, v_i \leq x \\ w_{i-1}, x < v_i < y \, . \end{cases} \tag{6.34}$$

For any string $s = (v_0, v_1 \cdots v_n) \in S$, we denote its variation by

$$\text{Var}(s) = \sum_{i=0}^{n-1} |v_{i+1} - v_i|. \tag{6.35}$$

The removal of a Madelung pair (x, y) from a string decreases its variation by the amount $2|y - x|$, whereas monotone deletions do not change the variation; so

$$\text{Var}(s) = 2 \sum_{x<y} (y - x) a(s)(x, y) + \text{Var}(s_R)\, , \tag{6.36}$$

holds for any string $s \in S$.

The number $\text{Var}(R_{x,y}(s))$ represents the number of oscillations of s between the thresholds x and y. The following results can be verified

$$\text{Var}(R_{x,y}(s)) = 2 \sum_{\xi \leq x < y \leq \eta} a(s)(\xi, \eta) + \text{Var}(R_{x,y}(s_R)) \tag{6.37}$$

$$\text{Var}(R_{x,y}(s, s)) = 2\text{Var}(R_{x,y}(s)) \tag{6.38)(i}$$

$$\text{Var}(R_{x,y}(s)) = 2 \sum_{\xi \leq x < y \leq \eta} a_{\text{per}}(s)(\xi, \eta) \tag{6.38)(ii}$$

$$\text{Var}(R^{\text{per}}_{x,y}(s)) = 2 \sum_{\xi \leq x < y \leq \eta} a_{\text{per}}(s)(\xi, \eta), \tag{6.39}$$

if $s = (v_0, v_1, \cdots, v_n)$ with $v_0 = v_n$, and if $w_{-1} = w_0 = w_N$.

Let $N(x, y)$ denote the number of times a repetition of the input cycle (x, y) leads to failure. Then on a unit scale,

$$\Delta(x, y) = \frac{1}{N(x, y)} \tag{6.40}$$

represents the contribution of the single cycle (x, y) to the damage of the structure. The total damage $D(s)$ due to an input string $s \in S$ is then estimated as

$$D(s) = \sum_{x<y} a_{\text{per}}(x, y) \Delta(x, y). \tag{6.41}$$

Within the context of fatigue analysis, there is no reason to consider arbitrary large input values. In order to facilitate the exposition from now on, we fix on a priori bound M for admissible input values. The relevant threshold values for the relays $R_{x,y}$, then lie within the triangle $P = \{(x,y) \in R^2 \mid -M \leq x < y \leq M\}$. For the sake of simplicity, we also assume that

$$w_{-1}(x,y) = \begin{cases} 1, x+y \leq 0; \\ 0, x+y \geq 0. \end{cases} \tag{6.42}$$

Under these assumptions, the Preisach operator $H : S \to S$ takes the form

$$H(s) = \int_{x<y} \varphi(x,y) R_{x,y}(s) dx\ dy, \quad \varphi \in L_1(P), \tag{6.43}$$

and its periodic version $H^{\text{per}} : S \to S$ is defined by

$$H^{\text{per}}(s) = \int_{x<y} \varphi(x,y) R^{\text{per}}_{x,y}(s) dx\ dy. \tag{6.44}$$

Here, the integral is understood to be componentwise with respect to the elements of the string $R_{x,y}(s)$, and the function φ is set to zero outside the triangle P. An operator $H : S \to S$ of the form $s = (v_0, \cdots, v_n) \to H(s) = (w_0, \cdots, w_n)$ is called **piecewise monotone** if $(w_k - w_{k-1})(v_k - v_{k-1}) \geq 0, 1 \leq k \leq n$ holds for every $s = (v_0, \cdots, v_n) \in S$.

Theorem 6.1. *Let $H : S \to S$ be a Preisach operator for some given density function $\varphi \in L_1(P)$. Then for each $s = (v_0, v_1 \cdots, v_n) \in S$ with $v_0 = v_n$ we have*

$$Var(H^{per}(s)) = \sum_{x<y} a_{per}(x,y)\phi(x,y), \tag{6.45}$$

where

$$\phi(x,y) = 2\int_x^y \int_x^\eta \varphi(\xi,\eta) d\ \xi\ d\ \eta. \tag{6.46}$$

Corollary 6.1. *Let $\Delta \in H^2(P)$, $\partial_x\Delta(x,y), \partial_y\Delta(x,y) \geq 0$ and $\partial_x\Delta(x,y) \leq 0$ for all $(x,y) \in P$ and $\Delta(x,x) = \partial_x\Delta(x,x) = \partial_y\Delta(x,x) = 0$ for each $x \in R$. Furthermore, let the density function of the Preisach operator be $\varphi(x,y) = -\frac{1}{2}\partial_{xy}\Delta(x,y)$. Then for each string $s = (v_0, v_1, \cdots, v_n) \in S$ with $\|s\|_\infty \leq M$ and $v_0 = v_n$, the total damage $D(s)$ associated to s can be represented as*

$$D(s) = \sum_{x<y} a_{per}(x,y)\Delta(x,y) = Var(H^{per}(s)). \tag{6.47}$$

Remark 6.5. (i) It may be remarked that the corollary indicates that the mathematical theory developed for the Preisach operator can be used for the analysis of the damage functional D arising from the combination of rainflow counting and the Palmgren-Miner-rule.

(ii) It has been shown that the total damage depends continuously on the input (See Theorem 3.1 [Brokate et al. 1996]) or [Brokate et al. 1996, Section 2.6]) and, as a consequence, the total damage is stable with respect to small variations of the input and, in particular, with respect to range discretization in data reduction algorithms.

Proof of Theorem 6.1. We have

$$\left\{\begin{aligned}&[(H^{\mathrm{per}}(s))_{i+1}-(H^{\mathrm{per}}(s))_i]\,(\mathrm{sign}\,(v_{i+1})-\mathrm{sign}\,(v_i))\\&\quad=\int_{x<y}\varphi(x,y)\left[(R^{\mathrm{per}}_{x,y}(s))_{i+1}-\left(R^{\mathrm{per}}_{x,y}(s)\right)_i\right]\\&\quad\times\ (\mathrm{sign}\,(v_{i+1})-\mathrm{sign}\,(v_i))\,dx\,dy\,.\end{aligned}\right. \tag{6.48}$$

Since $H(s,s)=(H(s),H^{\mathrm{per}}(s))$, therefore

$$(H^{\mathrm{per}}(s))_{i+1}-(H^{\mathrm{per}}(s))_i=(H(s,s))_{n+i+2}-(H(s,s))_{n+i+1}\,, \tag{6.49}$$

and the same formula holds true if we replace H by $R_{x,y}$. From the piecewise monotonicity of H and of $R_{x,y}$, it follows that if we sum over i in (6.48), then

$$\mathrm{Var}\,(H^{\mathrm{per}}(s))=\int_{x<y}\varphi(x,y)\mathrm{Var}\left(R^{\mathrm{per}}_{x,y}(s)\right)dx\,dy\,. \tag{6.50}$$

We obtain (6.45) by integrating (6.39) with the density function φ over the domain where $x<y$ and keeping in mind the relation

$$\left\{\begin{aligned}&\int_{-\infty}^{\infty}\int_{-\infty}^{y}\varphi(x,y)2\textstyle\sum_{y\le\eta}\sum_{\xi\le x}a_{\mathrm{per}}(\xi,\eta)dx\,dy=\sum_{\eta\in R}\sum_{\xi<\eta}a_{\mathrm{per}}(\xi,\eta)2\\&\qquad\times\int_{\xi}^{\eta}\int_{\xi}^{y}\varphi(x,y)dx\,dy\,.\end{aligned}\right. \tag{6.51}$$

Proof of Corollary 6.1. The assertion will follow from the theorem if we show that H with conditions given in the corollary is piecewise monotone. Let $s=(v_0,\cdots,v_n)$ with $v_n>v_{n-1}$ and $v_n\ne v_j$ for all $j<n$. For the Preisach operator (6.43), there holds

$$\partial_n H(v_0,\cdots,v_n)=2\int_0^{r^*}\varphi(v_n-2r,v_n)dr, \tag{6.52}$$

where $r^*>0$ is a certain number depending on $(v_0,v_1,\ldots,v_n)$ and $\partial_n H(v_0,v_1,\ldots,v_n)$ denotes the partial derivatives of the last component of the output string with respect to the last input value. From (6.46) and the assumptions on Δ, we get

$$\partial_n H(v_0,v_1,\cdots,v_n)=\frac{1}{2}\partial_y\Delta(v_n-2r^*,v_n)-\frac{1}{2}\partial_y\Delta(v_n,v_n)\ge 0. \tag{6.53}$$

An analogous argument in the case $v_n < v_{n-1}$ yields

$$\partial_n H(v_0, v_1, \cdots, v_n) = -\left(\frac{1}{2}\partial_x \Delta(v_n, v_n + 2r^*) - \frac{1}{2}\partial_x \Delta(v_n, v_n)\right) \geq 0. \tag{6.54}$$

Therefore, H is piecewise monotone.

Rainflow and hysteresis memory. The rainflow method is linked intimately to the memory structure of scalar rate-independent hysteresis. One may establish the connection for every hysteresis model whose memory is based upon a continuous family of play (or equivalently, relay) operators. Their collective memory evolution generated by a particular input string $s = (v_0, \cdots, v_n)$ is completely described by a string of functions

$$\begin{cases} \psi_k : R_+ \to R \\ (\psi_0(r_1), \cdots, \psi_n(r)) = F_r(s), r \geq 0 \,. \end{cases} \tag{6.55}$$

The function ψ_k represents the memory after v_k has been processed. For a systematic study of various properties of the hysteresis memory operator which maps input strings to memory curves, we refer to Brokate and Visintin [1989] and Visintin [1994]. We indicate here how it is applied to calculate the partial derivative $\partial_{v_n} H(v_0, \cdots, v_n)$ in the damage analysis.

By (6.24), we have

$$\psi_k(r) = \min\left\{v_k + r, \max\left\{v_k - r, \psi_{k-1}(r)\right\}\right\}, \quad \psi_{-1}(r) = 0, r \geq 0. \tag{6.56}$$

We call the function ψ_n or, more precisely, its graph $\{(r, \psi_n(r)) \,/ r \geq 0\}$ the **hysteresis memory curve** belonging to the string $s = (v_0, v_1, \cdots, v_n)$ and denote it as

$$\Phi_r(s) = \psi_n(r). \tag{6.57}$$

If s is alternating and satisfies

$$|v_0| > |v_1|, |v_{i+1} - v_i| < |v_i - v_{i-1}|, 0 < i < n, \tag{6.58}$$

then its hysteresis memory curve ψ_n consists of $n + 1$ pieces of straight lines of the slope $\pm$ 1 starting at $(0, v_n)$ and ending at $(r_0, 0) = (|v_0|, 0)$ with corners $(r_i, \psi_n(r_i))$,

$$r_i = \frac{|v_i - v_{i-1}|}{2}, \ \psi_n(r_i) = \frac{v_i + v_{i-1}}{2}, 1 \leq i \leq n \tag{6.59}$$

The connection to the rainflow method is as follows: Let v_n tend to v_{n-2}, then r_n tends to r_{n-1}. When v_n becomes equal to v_{n-2}, the corner $(r_n, \psi_n(r_n))$ merges with $(r_{n-1}, \psi_n(r_{n-1}))$ and both vanish. At that moment, the rainflow method counts and deletes the Madelung pair (v_{n-1}, v_n) and, in the input-output plane, a hysteresis loop is closed. For an arbitrary string s, one may check that

$$\Phi_r(s) = \Phi_r(s_R) = \Phi_r(s_{B_R}), \quad r \geq 0, \tag{6.60}$$

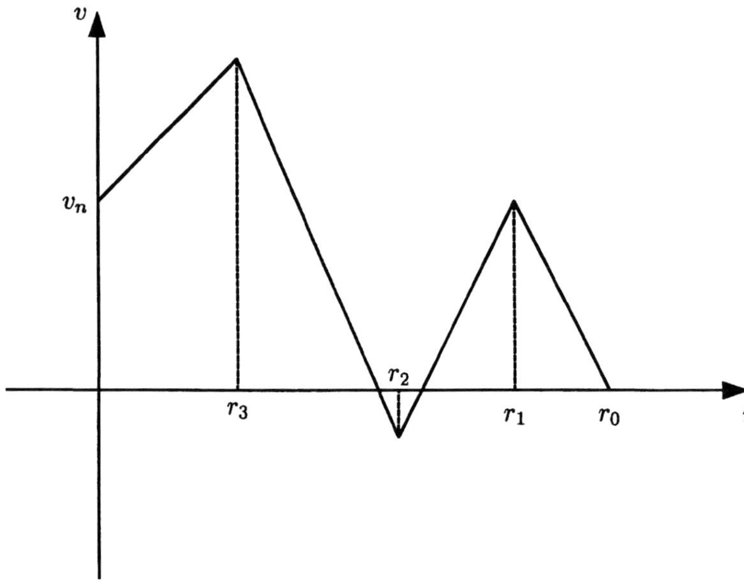

Figure 6.6. The hysteresis memory curve.

where we obtain the so-called backward residual s_{B_R} from the residual $s_R = (v_0', v_1', \ldots, v_n')$ by deleting successively $v_0', v_1' \cdots$ until the remaining string satisfies (6.58).

Let $s = (v_0, v_1, \ldots, v_n)$ be alternating and furthermore assume that (6.58) holds. Consider the case $v_n > v_{n-1}$, or the case $n = 0$ and $v_0 > 0$. From Figure 6.6, we see that

$$\partial_n F_r(v_0, v_1, \ldots, v_n) = \partial_{v_n} \Phi_r(v_0, v_1, \cdots, v_n) = \begin{cases} 1, r < r_n \\ 0, r > r_n \end{cases}, \tag{6.61}$$

as well as

$$\Phi_r(v_0, v_1, \cdots, v_n) = v_n - r, r < r_n. \tag{6.62}$$

By (A6) and (A7), in Brokate et al. [1996a].

$$\begin{cases} \partial_n H(v_0 \cdots, v_n) = \displaystyle\int_0^\infty \partial_z p\,(r, \Phi_r(v_0, \cdots, v_n)\,\partial_n F_r(v_0 \cdots, v_n) dr \\ \qquad = \displaystyle\int_0^r \partial_z(r, v_n - r) dr = 2\int_0^{r_n} \varphi(v_n - 2r, v_n) dr\,. \end{cases} \tag{6.63}$$

If the string $s = (v_0, \cdots, v_n)$ is arbitrary, the computation above applies to its backward residual S_{BR}; so we have

$$\partial_n H(v_0, \cdots, v_n) = 2\int_0^{r^*} \varphi(v_n - 2r, v_n) dr, \tag{6.64}$$

with a certain number $r^* > 0$. Similarly, we deduce that in case $v_n < v_{n-1}$.

$$\partial_n H(v_0, v_1, \cdots, v_n) = 2\int_0^{r^*} \varphi(v_n, v_n + 2r)dr, \tag{6.65}$$

for some r^*.

6.4 Energy dissipation in rate-independent rheological models

We mention here a formal analogy between energy dissipation and damage accumulation for scalar rate-independent constitutive laws which are established by the calculus of hysteresis operators. This kind of technique has been developed and widely used in material science and engineering by the name rheological models. The concept of a rheological element consists of a constitutive relation between the sress σ and the strain ϵ, and an internal energy U (one may consider the free energy or the potential energy in isothermal situations; these concepts coincide).

Basic rheological elements. (i) $\mathcal{E}$ denotes the linear elastic element with the constitutive equation and internal energy

$$\epsilon = \frac{1}{E}\sigma, \quad U = \frac{1}{2}\epsilon\sigma, \tag{6.66}$$

where E denotes the modulus of elasticity.

(ii) $\mathcal{N}$ denotes the nonlinear elastic element with the constitutive equation and internal energy

$$\epsilon = g(\sigma), U = G(\sigma) = \sigma\ g(\sigma) - \int_0^\sigma g(\xi)d\ \xi, \tag{6.67}$$

where $g : R \to R$ is a non-decreasing function with $g(0) = 0$.

(iii) $\mathcal{P}$ denotes the rigid plastic element whose constitutive relation is described by the variational inequality

$$\dot{\epsilon}(t)(\sigma(t) - \tilde{\sigma}) \geq 0 \quad \text{for all } \tilde{\sigma} \in [-r, r] \tag{6.68}$$

$$\sigma(t) \in [-r, r], \tag{6.69}$$

with the yield stress $r > 0$.

(iv) $\mathcal{B}$ denotes the brittle element described by

$$\epsilon(t) = 0 \text{ if } \|\sigma\|_{[0,t]} < h, \sigma(t) = 0 \text{ if } \|\sigma\|_{[0,t]} \geq h, \tag{6.70}$$

with the fracture stress $h > 0$. Here, $\|\cdot\|_{[0,t]}$ denote the sup norm over the time interval $[0, t]$.

For the elements $\mathcal{P}$ and $\mathcal{B}$, the internal energy U is set to 0 to express the fact that no reversible power can be stored by these elements.

The definition $\mathcal{B}$ via (6.69) makes sense if $|\sigma(t)| < h$ for all $t \geq 0$. The material remains rigid as long as $|\sigma(t)|$ stays bounded away from the value h; as soon as $|\sigma(t-)| = h$, the material breaks, σ jumps to zero and we lose any control on ϵ. Condition (6.70) can be equivalently written in terms of the Heaviside function H, defined as $H(x) = 1$ if $x > 0$, and $H(x) = 0$ otherwise, as

$$\epsilon(t)H(h - \|\sigma\|_{[0,t]}) = 0, \sigma(t)\left(1 - H(h - \|\sigma\|_{[0,t]}\right) = 0. \tag{6.71}$$

From the two given rheological elements R_1 and R_2, we may form a new element R as the

(a) series combination $R = R_1 - R_2$ or $R = \sum_{i\, \epsilon\{1,2\}} R_i$, setting

$$\epsilon = \epsilon_1 + \epsilon_2, \sigma = \sigma_1 = \sigma_2, U = U_1 + U_2, \tag{6.72}$$

(b) parallel combination $R = R_1/R_2$ or $R = \Pi_{i\in\{1,2\}} R_i$, setting

$$\epsilon = \epsilon_1 = \epsilon_2,\ \sigma = \sigma_1 + \sigma_2,\ U = U_1 + U_2. \tag{6.73}$$

Here ϵ, σ, and U are treated as functions of time t within the interval $[0, t]$. According to the second law of thermodynamics, the dissipation rate $\dot{q}(t)$ given by

$$\dot{q} = \dot{\epsilon}\sigma - \dot{U} \tag{6.74}$$

has to satisfy

$$\dot{q}(t) \geq 0. \tag{6.75}$$

A weak formulation of (6.75) which we consider here is as follows:

$$\begin{cases} q(t_2) - q(t_1) = [\epsilon(t)\sigma(t) - U(t)]_{t_1}^{t_2} - \displaystyle\int_{t_1}^{t_2} \epsilon(t)\dot{\sigma}(t)dt \geq 0 \\ 0 \qquad\qquad \leq t_1 \leq t_2 \leq T\,. \end{cases} \tag{6.76}$$

A rheological element is called thermodynamically consistent if (6.75) and (6.76) hold for all functions satisfying the constitutive relation and belonging to a suitable function space. The elements $\mathcal{E}, \mathcal{N}$ and $\mathcal{P}$ are thermodynamically consistent. In fact, $\mathcal{E}$ and $\mathcal{N}$ are conservative, that is, $\dot{q} = 0$. It is clear that any parallel or series combination of thermodynamically consistent elements is again thermodynamically consistent. It can be proved that the parallel combination $\mathcal{N}/\mathcal{B}$ is thermodynamically consistent for $\sigma \in H^1(0, T)$ if g is a real-valued continuous function satisfying $x\, g(x) > 0$ for $x \neq 0$ (for proof, see Brokate et al. [1996]).

Example 6.2 (Basic elastoplastic elements). The series combination $\mathcal{E}$-P is described by

$$\epsilon = \epsilon^e + \epsilon^p, \sigma = E\, \epsilon^e, \quad U = \frac{1}{2E}\sigma^2 \tag{6.77}$$

$$\sigma \in [-r, r], \dot{\epsilon}^p(\sigma - \tilde{\sigma}) \geq 0 \quad \text{for all } \tilde{\sigma} \in [-r, r]. \tag{6.78}$$

We observe that both $\mathcal{E}/P$ and $\mathcal{E}$-P are governed by an evolution variational inequality; namely

$$(E\dot{\epsilon} - \dot{\sigma})(\sigma - \tilde{\sigma}) \geq 0 \;\forall\; \tilde{\sigma} \in [-r, r], \tag{6.79}$$

for $\mathcal{E}$-P, and

$$(\dot{\sigma} - \dot{\sigma}^p)(\sigma^p - \tilde{\sigma}) \geq 0 \;\forall\; \tilde{\sigma} \in [-r, r], \tag{6.80}$$

for $\mathcal{E}/P$. In (6.79), σ is determined from the strain ϵ, whereas in (6.80), the plastic stress σ^p, and hence $\epsilon = \dfrac{1}{E}(\sigma - \sigma^p)$ are determined from the stress σ. Both variational inequalities have the same form, i.e., for a given function $v : [0,T] \to R$, we look for $w : [0,T] \to R$ such that

$$(\dot{v}(t) - \dot{w}(t))(w(t) - x) \geq 0 \text{ a.e. in } (0,T) \text{ for all } x \in [-r, r] \qquad (6.81)(\mathrm{i})$$

$$w(t) \in [-r, r] \quad \text{a.e. in } [0,T],\; w(0) = w_0\,. \qquad (6.81)(\mathrm{ii})$$

It is well known in the theory of variational inequalities (see, e.g., Duvaut-Lions [1976] and Brezis[1973]) that (6.81) has a unique solution $w \in H^1(0,T)$. The correspondence $v \to w$ defines a hysteresis operator S_r, called the **stop operator** by Krasnoselskii and Pokrovskii [1989] which, for the initial value

$$w_0 = \min\{r, \max\{-r, v(0)\}\},$$

is related to the play operator F_r by the relation (6.16(b)).

We can rewrite the constitutive relations for the two elastoplastic elements in terms of the operator F_r and S_r as

$$\mathcal{E} - P : \sigma = S_r[E\epsilon],\; U = \frac{1}{2E}(S_r[E\epsilon])^2 \tag{6.82}$$

$$\mathcal{E} \mid P : E\epsilon = F_r[\sigma], U = \frac{1}{2E}(F_r[\sigma])^2\,. \tag{6.83}$$

Example 6.3 (The Prandtl-Preisach-Ishlinskii model). We consider a infinite series combination

$$\mathcal{N}_0 - \sum_{r>0} \mathcal{N}_r/\mathcal{P}_r. \tag{6.84}$$

We denote by $g(r,\cdot)$ the constitutive function of the nonlinear elastic element N_r; that is,

$$\epsilon_r = g(r, \sigma_r^e), r \geq 0. \tag{6.85}$$

In the composition formula (6.72), we replace the sum by an integral so that (6.84) is described by

$$\epsilon = \epsilon_0 + \int_0^\infty \epsilon_r\, dr, \sigma = \sigma_r^p + \sigma_r^e \;\forall\; r > 0 \tag{6.86}$$

$$\sigma_r^p \in [-r, r], \dot{\epsilon}(\sigma_r^p - \tilde{\sigma}) \geq 0 \;\forall\; \tilde{\sigma} \in [-r, r] \;\forall\; r > 0, \tag{6.87}$$

$$U = G(0, \sigma_0^e) + \int_0^\infty G(r, \sigma_r^e) dr, \tag{6.88}$$

where we define G in analogy to (6.67) as

$$G(x,\sigma) = \sigma g(x,\sigma) - \int_0^{\sigma} g(x,\xi)d\,\xi. \tag{6.89}$$

Variational inequality (6.87) implies that

$$\dot{\sigma}_r^e(\sigma_r^p - \tilde{\sigma}) \geq 0 \text{ for all } \tilde{\sigma} \in [-r,r]. \tag{6.90}$$

We, therefore, have

$$\sigma_r^p = S_r[\sigma], \sigma_r^e = F_r[\sigma]. \tag{6.91}$$

Since $\sigma_0^e = \sigma$, (6.84) can be described in the Preisach operator form as

$$\epsilon = g(0,\sigma) + \int_0^{\infty} g(r, F_r[\sigma])dr \tag{6.92}$$

$$\mathsf{U} = G(0,\sigma) + \int_0^{\infty} G(r, F_r[\sigma])dr. \tag{6.93}$$

Let us assume that $g(r,0) = 0$ for all r and that the partial derivative in the second argument $\partial_\xi g(x,\xi)$ is non-negative, measurable and sufficiently regular. Brokate-Sprekels [1996b] have shown that

$$\dot{q} = \left|\frac{d}{dt}\int_0^{\infty} rg(r, F_r[\sigma])dr\right|, \tag{6.94}$$

which motivates us to define the **dissipation operator**

$$H_d[\sigma] = \int_0^{\infty} r\; g(r, F_r[\sigma])dr, \tag{6.95}$$

where H_d is a Preisach operator. By integrating (6.94), we obtain the total dissipation of energy as

$$d(t) = Var_{[0,t]}H_d[\sigma]. \tag{6.96}$$

Thus, we see that the total dissipation of energy can be expressed in terms of the variation of a Preisach operator in a way similar to the accumulated damage. This result is very important in view of the discussion in the introduction regarding hysteresis loops.

6.5 Hysteresis in the wave equation

Equation (6.7) can be written in the form of a wave equation with hysteresis as

$$\frac{\partial^2 u}{\partial t^2} = \frac{\partial}{\partial\, x}\left(H\left[\frac{\partial\, u}{\partial\, x}\right]\right), \tag{6.97}$$

or, equivalently, in the form

$$\frac{\partial}{\partial t}\left(H^{-1}\left[\frac{\partial z}{\partial t}\right]\right) = \frac{\partial^2 u}{\partial z}, \tag{6.98}$$

for a variable z representing the time integral of the stress,

$$\begin{cases} z(x,t) = \int_0^t \sigma(x,\tau)d\,\tau + z^0(x), \\ z^0(x) \quad = -\int_x^1 \frac{\partial}{\partial t}u(\xi,0)d\,\xi\,. \end{cases} \tag{6.99}$$

For a solution of (6.95) with appropriate boundary conditions, see Brokate [Th 7.2, 1994].

6.6 Problems.

Problem 6.1. (i) Show that the play operator is Lipschitz continuous on $W^{11}(0,T) = H^1(0,T)$.

(ii) Show that the hysteresis operator is Lipschitz continuous on $W^{11}(0,T) = H^1(0,T)$.

Problem 6.2. (i) What do you understand by the **height function**?

(ii) State Masing's law and the Ramberg-Osgood equation of cyclic plasticity equation.

Problem 6.3. Show that the composition of the two Prandtl operators is also a Prandtl operator. What can you say about the inverse of a Prandtl operator?

Problem 6.4. Write down Maxwell's equations with hysteresis and discuss their solutions.

Problem 6.5. Discuss the accumulated damage in the Safing sensor's process.

Problem 6.6. Discuss the solutions of the Helmholtz equation with hysteresis.

Problem 6.7. Discuss a weak solution of the wave equation with hysteresis. Under what conditions, the solution is unique?

Problem 6.8. Solve the Poisson equation with hysteresis by the boundary element method.

Problem 6.9. Explain the concept of the shape memory effect.

Chapter 7

Appendix

7.1 Introduction to mathematical models and simulation

A_1 Introduction.

In every problem related to any aspect of technological and industrial development, three main steps are involved:

1. Mathematical modelling of the physical phenomenon under consideration.

2. Discretization of the model into a discrete algebraic form amenable to numerical solutions. In other words, writing the model in an appropriate from which can be written in a language acceptable to computers.

3. Evaluation of the programme by the computer and visualization of the result on the monitor and subsequent realization and interpretation.

According to the McGraw-Hill Encyclopedia of Science and Technology, a **model** is an entity to represent some other entity for some well defined purpose. A system is a collection of entities that interact. It is a proxy for a particular situation.

Examples of models.

1. A picture or drawing, such as a map used to record geological data or a solid model to design a machine component.

2. An idea (mental model), such as the internalized model of a person's relationships with the environment used to guide behaviour.

3. A description (linguistic model) of the protocol for a biological experiment or the transcript of a medical operation to guide and improve procedures.

4. A physical object like an electronic circuit, airfoil used in the wind tunnel testing of a new aircraft design.

5. A system of equations and logical expressions (mathematical model or computer simulation) like the mass and energy balance equations that predict the end-product of a chemical reaction, or a computer program that simulates the flight of a space vehicle.

Models are built and developed to help hypothesize, define, explore, understand, simulate, predict, design and communicate some aspect of the original entity for which the model is a substitute. Models provide principal support for the study of every scientific field and new areas of technology and industrial development also use models extensively. Study of a system, physical situation and phenomenon through its models is more economical. Changes in the structure of a model are easier to implement, and changes in the behaviour are easier to isolate, understand and communicate to others. A model can be used to achieve a thorough and deep understanding when direct experimentation with an actual system is very dangerous, disruptive or demanding. A model can be used to study those properties of a system that have not been observed or built or cannot be built or observed with the present technologies. The discrete algebraic forms are often called the **simulation model** which, when expressed as a sequence of computer instructions, provides the computer simulation program. The evaluation of the physical phenomenon with the help of a program and computer is known as **computer experiments**. Vast literature exists on the three steps mentioned above which are known as the mathematical modelling, algorithms and programming, and computer simulation, respectively see, for example, Hockney and Eastwood [1981], Gaylord and Wellin [1994], Wolfram [1990], Crandall [1991], Bahder [1995], Ross [1995], Vvendensky[1994], Eriksson [1995] and Heath [1997]. Software like **MATLAB**, **Mathematica**, **Femlab** are now-a-days quite popular for scientific computations.

A_2 Mathematical models.

The description of a physical situation (phenomenon) can be termed as a model. Modelling is an attempt to describe a phenomenon in a logical way. In this process, we identify generalizable principles and processes and frame a mathematical structure (equations – algebraic, difference, differential, partial differential, stochastic differential, integro-differential, matrix, operator; inequalities – variational inequality, boundary and initial value problems, iterative formulae, etc.) which is used to investigate and simulate new possibilities leading to the understanding of the intricacies of nature and solutions of problems in different disciplines of the emerging technologies and related industrial development. Models may also consist of stories, fables and analogies or computer programmes in which they also permit the simulation of dynamic behaviour. As simulation games, they comprise a set of rules guiding behaviour, gaming simulations for tackling new and unusual situations. The examples of models may range from cars, rail engines, bridges, aircraft, robots, epidemics, electronic equipment, economic equilibrium, pharmaceutical compounds, chemical reactors, economic growth, traffic control, paper, glass and steel industries, social process, urban development, population explosion, wars, and environmental pollution to global climate change. Models can be very useful for scientific understanding of phenomenon, systematic development of technologies, management and planning. It may be understood clearly that a good model is the best possible representation of a given physical situation or phenomenon; however its solution may reproduce a limited set of behaviour of the original due to certain unavoidable factors. A phenomenon may have different models and objectives according to the purpose of the modelling. Classical science emphasizes observation and description while the model developers stress the understanding of the processes and their changing behaviour. The validity of a model and its logical admissibility (well-posedness) must be carefully examined. The validity of a model can be categorized into four categories: behavioral validity, structural validity, empirical validity and application validity. While modelling a phenomenon of a specific technology or industry, a fairly good understanding of that process and physical situation must be acquired. The objectives of the model must be specified in the first place as the mathematical structures and variables involved will depend upon the objectives. The determination of physical laws and identification of parameters may not be an easy task; therefore, multi-disciplinary discussion becomes essential. The existing information may be very useful in developing a model. The conservation laws of mass, energy, motion, electricity are quite useful for modelling technological and industrial problems. Analytical solutions of models in emerging areas of technologies are rarely possible. Therefore, the model must be discretized for numerical solution which may lead to error. The minimization of error, that is, controlling the process of discretization such that the error is minimum, is a vital question in the modelling processes. Model-building in areas of biomedical sciences, biotechnology, chemical and pharmaceutical industries is a fairly difficult task and the solution of such models may require the knowledge of new mathematical techniques to be invented in future. A material such as silicon, germinimum, gallium arsenide whose electrical conductivity increases with temperature and lies between metals like silver, copper and insulators like glass and fused quartz is known as **semiconductor**. The phenomenon is due to the thermal generation of equal numbers of negative charge carriers, known as electrons and positive charge carrier called **holes**. Semiconductors of different conductivity like n-type, p-type can be brought together to form a variety of junctions which are the basis of semiconductor devices used as electronic components. The term semiconductor is often used for the devices themselves. Advances in the electronic and computer industry are closely related to the semiconductor technology. A systematic study of mathematical modelling of semiconductor devices has been started in the late seventies keeping in view the rapidly growing demand of this field (see, for example, Markowich [1986], Bank et al. [1990] and Bhattacharya [1994]).

Mathematical modelling is essential for appropriate analysis and simulation. A model is called **behaviourally valid** if it produces the same behaviour as the original physical phenomenon under similar conditions. If the numerical solution of an original situation coincides with the numerical solution of the corresponding model, then it is called **empirically valid.** For application validity, we must check that the model delivers correct results in a specific situation set by the model builder (developer). For a comprehensive study of mathematical models, we refer to Taylor [1986], Kapur [1987] and Bellomo and Preziosi [1995].

Algorithm, programming and software. A prescribed set of well defined rules or instructions for the solution of a problem in a finite number of steps is called an **algorithm**, e.g., the rules for finding minima of a real-valued function. Some important algorithms for 'optimization problems' have been discussed in Chapter 2 while some other important algorithms like the fast Fourier transform algorithm, fast wavelet algorithm, the Chaos game algorithm are presented in Chapter 5 and Appendix B, respectively. Expressing an algorithm in a formal notation is an important part of a programme. Let $P = \{0, 1, 2, 3 \cdots\}$ and $f : P^n \longrightarrow P$, n a natural number. f is called **effectively computable** if there is an effective procedure, that is, an algorithm that correctly calculates f. An effective algorithm is one that meets the following specifications:

(i) The finite numbers of instructions be such that there must not be any ambiguity concerning the order in which the instructions are to be carried out.

(ii) For n-tuple $x \in P^n$ after a finite number of steps, the calculation must terminate and output $f(x)$ and n-tuple not in P^n must not output a value.

The study of whether effective algorithms exist to compute a particular quantity forms the basis of the theory of algorithms. Except for the simplest of algorithms, it is difficult to prove whether a particular algorithm is correct or false. In practice, we should be satisfied if the algorithm is valid, that is, it performs the calculation (computation) assigned to it. It involves testing the routine against a variety of instances of the problem and ensuring that it performs satisfactorily for these test cases. If the test set is chosen sufficiently well, there can then be confidence in the algorithm. The study of the performance characteristics of a given algorithm is known as the **algorithm analysis**. Algorithms can be analyzed in terms of their complexity and efficiency. The efficiency can be characterized by its order. If two algorithms for the same problem are of the same order, then they are approximately equally efficient in terms of computation. Algorithm efficiency is a measure of the average execution time necessary for an algorithm to complete work on a set of data. A language or notation used to express clearly an algorithm is known as the **algorithmic language**. It is usually part of the programming language.

A_3 Program.

A program (computer program) is a set of statements that can be submitted as a unit to some computer system and used to direct the behaviour of the system. A procedural program gives a precise definition of the procedure to be followed by the computer system in order to obtain the required results. By contrast, a non-procedural program specifies constraints that must be satisfied by the results that are produced but does not specify the procedure by which these results should be obtained. Such a procedure must be determined by the computer system itself. A programming language is a notation for the precise description of computer programmes or algorithms. Programming languages are artificial languages in which the syntax and semantics are strictly defined. They serve their purpose but do not have the freedom of expression that is characteristic of a natural language. Some important programming languages are Fortran (latest Fortran-90), C, C+, C++ Turbo (Turbo basic, Turbo C, Turbo Pascal and Turbo Prolog) Algol, Pascal, Pascal$^+$, Basic and Wordstar. For details, see Press et al [1990], Cargill [1992] and Dowd [1993].

A_4 Software.

This term stands for those components of a computer system that are intangible rather than physical. It is generally used to refer to the programmes executed by a computer system as distinct from the physical hardware of that computer system, and to encompass both symbolic and executable forms for such programmes. Software library, the same as program library, is a collection of programmes and packages that are made available for common use within similar situations. A typical software library may contain compilers, utility programmes, packages for mathematical operations, etc. Before citing some specific sources of good mathematical software, we give below the desirable characteristics that such software should possess.

- **Reliability:** always works correctly for easy problems.
- **Robustness:** usually works for hard problems, but fails gracefully and informatively when it does fail.
- **Accuracy:** produces results as accurate as warranted by the problem and input data, preferably with an estimate of the accuracy achieved.
- **Maintainability:** is easy to understand and modify.
- **Portability:** adapts with little or no change to new computing environments.
- **Usability:** has convenient and well-documented user interface.
- **Applicability:** solves a broad range of problems.

These properties often conflict, and it is rare for a software to satisfy all of them. Nevertheless, this gives mathematical software users some idea what qualities to look for and developers some worthy goals to strive for. Some general-purpose mathematical software are listed here. The software listed are written in Fortran unless otherwise stated. For more information about available mathematical software, we refer to **URL http://gams.nist.gov** on the Internet's World-Wide Web.

- **IMSL** (International Mathematical and Statistical Libraries): A commercial product of Visual Numerics Inc., Houston, Texas. A comprehensive library of mathematical software; the full library is available in Fortran, and a subset is available in C. See —bf URL http://www.vni.com.

- **NAG** (Numerical Algorithms Group): A commercial product of NAG Inc. Downers Grove, Illinois. A comprehensive library of mathematical software; the full library is available in Fortran, and a subset is available in C. See —bf URL http://www.nag.com.

- **netlib:** A collection of free software from diverse sources available over the Internet. See URL **http://www.netlib.org**, or send email containing the request "send index" to **netlibornl.gov**, or **ftp** to one of several websites, such as **netlib.bell-labs.com** or **netlib2.cs.utk.edu.**

- **NR** (Numerical recipes): A collection of software accompanying the book **numerical Recipés,** by Press et al. [1992], available in C and Fortran editions.

Scientific computing environments. The software libraries mentioned above contain subroutines that are meant to be called by user-written programmes, usually in a conventional programming language such as Fortran or C. An increasingly popular alternative for scientific computing is interactive environment that provides powerful, conveniently accessible, built-in mathematical capabilities, often combined with sophisticated graphics and a very high-level programming language designed for rapid prototyping of new algorithms. One of the most popular computing environments is MATLAB. Another family of interactive computing environments is based primarily on symbolic computation rather than numeric, often called **computer algebra**. These packages, which include Axiom, Derive, Macsyma, Maple, Mathematica and Reduce, provide many of the same mathematical and graphical capabilities, in addition to providing symbolic differentiation, integration, equation solving, polynomial manipulation, etc. We present here some details of Mathematica and MATLAB; for others, see Heath [1997] and references therein.

A_5 Mathematica in science and technology.

The main purpose of this section is to draw the attention of mathematicians to a very powerful programming language named 'Mathematica' invented by Wolfram Research, Inc. with its formal announcement on 23rd June, 1988. Until very recently, computer languages Fortran, C, C^{++} were mainly used by scientists and engineers for their professional work. These languages are best suited for numerical calculations and development of large software programmes and do not directly support mathematical calculations as carried out by scientists and engineers. Mathematica handles both types of problems. All important functions of mathematical physics and matrix functions are already built into Mathematica. Its tremendous graphics capability for visualizing results of calculations has made it very popular. Some very significant calculations in different fields, especially solid state physics, genetics and population biology, which could not be carried out by the classical languages were made possible through this language. In "Mathematica for the Sciences", programmes in the mathematical language carry the main message, with words, diagrams and formula providing support. A traditional scientific formula might give the solution to a particular kind of equation, but cannot describe the process by which such a solution is found. In a Mathematica formulation, however, one can describe the complete computational process that is required. To reproduce computations, we are only required to execute the Mathematica programmes on a computer directly to the problems under consideration. The writing of a program in Mathematica requires a small fraction of the time required to write the corresponding program in other classical scientific languages like Fortran and C. Mathematica has symbolic computation capabilities that are lacking in Fortran and C. However, Mathematica is more time-consuming than these programmes in certain situations and this shortcoming can be overcome with a bridge known as Math-Link. The Math-Link system of communication is a general scheme where two running programmes can exchange data. Math-Link can be used to call Mathematica functions from inside a Fortran or C program. Since most of Mathematica's functions are capable of taking complex arguments, it may be convenient to use Mathematica as a function library by a Fortran or C program. Math-Link can also be used to call a Fortran subroutine or C function can be installed into Mathematica and made to behave as if it were a built-in Mathematica function. There are at least two situations where communication between Mathematica, Fortran and C will be appropriate: (i) when a complex program is written in a scientific language like Fortran or C and it works efficiently and is of high quality, and (ii) when Mathematica takes more time. The Mathematica system of communication enables us to exchange data. Computations can be carried out in Fortran or C and shipped back to Mathematica via Math-Link. For a solution of the Maxwell equation, Schröding equation, *KdV* equation *L-C* tank, heat and wave equations, we refer to Bahder [1995] and Crandall [1991].

A_6 MATLAB

MATLAB is a proprietary commercial product of The MathWorks, Inc. (see **URL http://www.mathwork.com)**. **MATLAB**, which stands for Matrix Laboratory, is an interactive system that integrates extensive mathematical capabilities, especially in linear algebra, with powerful scientific visualization, a high-level programming language, and a variety of optional "toolboxes" that provide specialized capabilities in particular applications, such as signal processing, image processing, control, system identification, optimization, and statistics. There is also a **MATLAB** interface for the **NAG** mathematical software library. **MATLAB** is available for a wide variety of personal computers, workstations, and supercomputers, and comes in both professional and inexpensive student editions. If **MATLAB** is not available on your computer system, there are similar, though less powerful, packages that are available by **ftp**, including **octave (ftp.che.utexas.edu/pub/octave)** and **RLaB (evans.ee.adfa.oz.au/pub/RLab)**. Other similar commercial products include **GAUSS,HiQ,IDL,Mathcad**, and **PV-WAVE**. Users program will be much shorter. The environment often has built-in functions for many of the problems we may encounter, which greatly simplifies the interface with such routines because much of the necessary information (array sizes, etc.) is passed implicitly by the environment. An additional bonus is built-in graphics, which avoids having to do this separately in a post-processing phase. Even if one intends to use a standard language such as C or Fortran in the long run, one may still find it beneficial to learn a package such as **MATLAB** for its usefulness as a rapid prototyping environment in which new algorithms can be tried out quickly and later recorded in a standard language, if necessary, for greater efficiency or compatibility. For acquiring a good knowledge of **MATLAB**, we refer to Acton [1996], Biran and Breiner [1995], Nakamura [1996], Part-Enander[1996], Redfen and Cambell [1996].

A_7 Simulation.

The general meaning of simulation is imitation of some existing system or systems to be manufactured or created. It may be applied in areas like communication network designing, designing of sophisticated equipment, exploration of traffic pattern and prediction of weather forecasting. It has major applications in the fast-growing computer industry. The simulation can be classified into two categories: (i) discrete simulation, and (ii) continuous simulation. For a discrete-event simulation, one considers all significant changes to the state of the system as distinct events that occur at specific points in time; the simulation then achieves the desired behaviour by modelling a sequence of such events treating each event individually. In the continuous case, changes occur gradually over a period of time and the progress of changes is tracked. Simulation is nothing but examination of problem without experimentation. The simulation of a device is usually taken to be the process of representing the characteristics of the device of examining the operation of analogous systems or processes. Therefore, by providing a description of the device and its operating conditions, a simulation can predict how the actual device would behave. The discrete algebraic equations describe the simulation model which, when expressed as a sequence of computer instructions, provides the computer simulation program. Computer simulation may be regarded as the theoretical exercise of numerically solving a boundary value problem. Simulation can be used to predict accurately the working of complex devices in order to have many variations to be evaluated before expensive technology is employed in manufacturing or constructing the optimal choice. It can also be used to obtain information not available at hand, for example, the study of galaxies and submicron electronic devices. Steps involved in simulation (computer simulation) are:

1. Mathematical modelling of the physical situation or phenomenon.

2. Discretization of the model.

3. Computer programming.

4. Implementation of the program on an appropriate computer (PC's/Work-station/Laptop etc).

7.2 Fractal image compression

Fractal image compression. Image compression techniques based on fractals have been developed in the last few years and promise better compression performance. These techniques are being developed on the recognition that fractals can describe natural scenes better than shapes of traditional geometry. There are three main fractal image compression techniques: (i) fractal image compression based on the iterated function systems (IFS), (ii) The segment based coding, and (iii) The yardstick coding. Here, we confine ourselves to the first technique where images are compressed into compact IFS codes at the encoding stage, and fractal images are generated to approximate the original image at the decoding stage. For this technique, we refer to Barnsley [1988] and [1996], Barnsley and Hurd [1993], Fischer [1994] and Lu [1997]. This topic is also discussed in Chapter 5.

The word fractal was coined by Benôt Mandelbrot from the Latin word fractus, meaning broken, for describing objects that were too irregular to fit into the traditional geometrical setting. There are several definitions of fractals but we shall treat fractals as fixed points of mappings on the metric space of all compact subsets of a complete metric space into itself. Cantor set, Sierpinski gasket, Sierpinski carpet, Von Koch curves are examples of fractals. An image can be approximated by these objects and some numbers/quantity, characteristic of these objects will be transmitted or stored and the original image can be retrieved from these characteristic numbers/quantity. For example, the fern image can be represented by 24 integers which can be represented by 192 bits (if each integer is represented by 8 bits). On the other hand, if we store fern pixels, we need 307200 bits (supposing that the image size is 480 lines by 640 pixels). Therefore, we achieved compression ratio of 1600 to 1 by the IFS technique of fractal image compression. Here, we briefly introduce the fractal objects, IFS Theorem, Collage Theorem, and methods for finding IFS for an image and image for an IFS. Commercial software to implement the process are available.

Cantor set: The Cantor set C is a subset of the metric space $X = [0,1]$, which is obtained by successive deletions of middle third open subintervals, as follows:
$I_0 = [0,1]$,
$I_1 = [0, \frac{1}{3}] \cup [\frac{2}{3}, \frac{3}{3}]$
$I_2 = [0, \frac{1}{9}] \cup [\frac{2}{9}, \frac{3}{9}] \cup [\frac{6}{9}, \frac{7}{9}] \cup [\frac{8}{9}, \frac{9}{9}]$
$I_3 = [0, \frac{1}{27}] \cup [\frac{2}{27}, \frac{3}{27}] \cup [\frac{6}{27}, \frac{7}{27}] \cup [\frac{8}{27}, \frac{9}{27}] \cup [\frac{18}{27}, \frac{19}{27}] \cup [\frac{20}{27}, \frac{21}{27}] \cup [\frac{24}{27}, \frac{25}{27}] \cup [\frac{26}{27}, \frac{27}{27}]$
$I_4 = I_3$ minus the middle open third of each interval in I_3
...
...
...
$I_N = I_{N-1}$ minus the middle open third of each interval in I_{N-1}.
The cantor set C is defined as
$C = \bigcap_{n=0}^{\infty} I_n$, where
$I_0 \supseteq I_1 \supseteq I_2 \supseteq I_3 \supseteq I_{N-1} \supseteq I_N \supseteq I_{N+1}$.

$C \neq \emptyset$ as $0 \in C$. C is a perfect set.

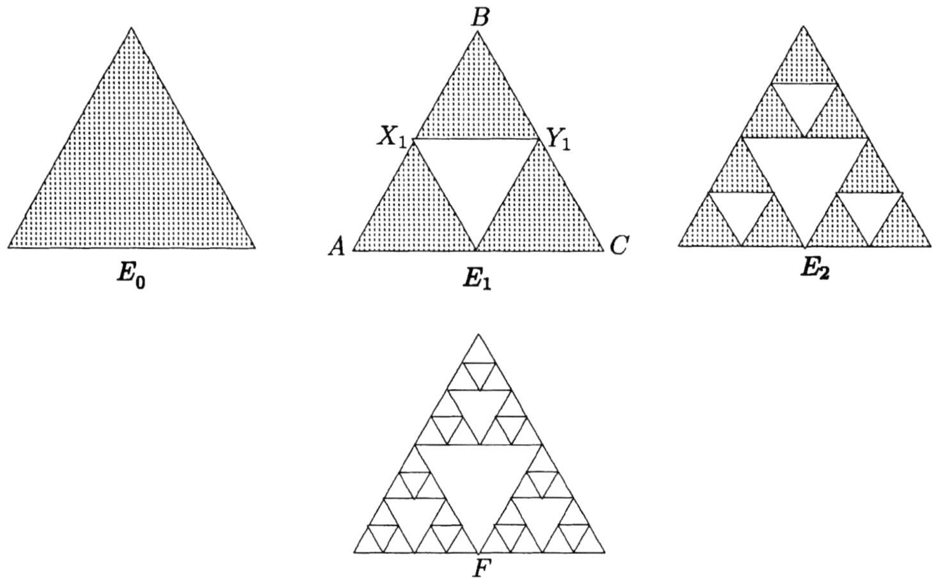

Figure B1. Sierpinski gasket.

In Figure B1, X_1, Y_1, Z_1 are middle points of the side AB, BC and AC, respectively. Remove the triangle with vertices X_1, Y_1 and Z_1.

X_1'', X_1''', Y_1' are middle points of the triangle AX_1Z_1 and so on. Remove the central triangles of three left-over triangles of E_1.

Continue this process. Whatever is left is called the Sierpinski gasket or Sierpinski triangle. For an interesting account of the Sierpinski gasket, we refer to Stewart [95]. It may be observed that there is a close connection between Chaos and fractals.

Iterated function system.

Theorem B1 (Banach contraction fixed point theorem). *Let X be a complete metric space and let $T : X \longrightarrow X$ be a contraction mapping, that is, $d(T(x), T(y)) \leq \alpha d(x, y), 0 \leq \alpha < 1$. Then T has a unique fixed point, that is, $\exists$ unique $u \in X$ such that $Tu = u$. Furthermore, $T^{0n}(y) \longrightarrow u$ as $n \longrightarrow \infty$ where $T^{0n}(\cdot)$ is defined as follows:*

$$T^{00}(x) = x$$
$$T^{01}(x) = T(x), T^{02}(x) = T(T(x)) = T(T^{01}(x))$$
$$T^{03}(x) = T(T^{02}(x) \cdots , T^{0n}(x) = T(T^{0^{(n-1)}}(x)\ .$$

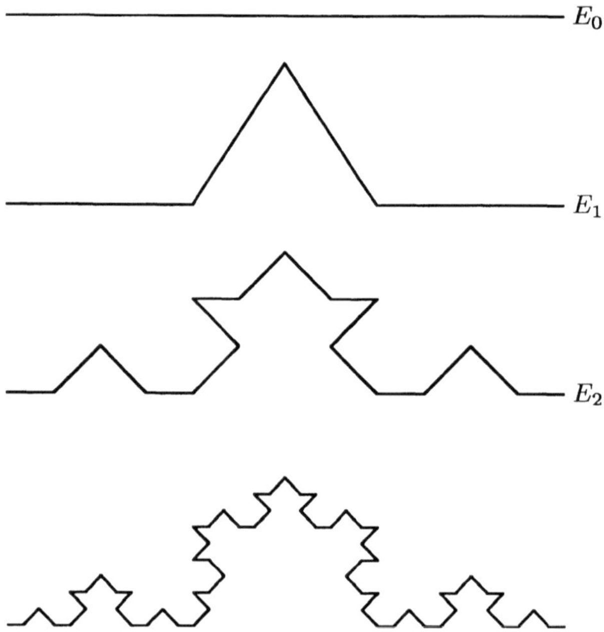

Figure B2. Von koch curve.

This theorem is a key to the fractal image compression.

Definition B1. *Let (X, d) be a complete metric space. X together with a finite set of contraction mappings $W_n, n = 1, 2 \cdots, N$ with contractivity factors $s_n, n = 1, 2, \cdots, N$ is called an "iterated function system" abbreviated as IFS, where*

$$s = \max_n s_n.$$

An IFS will be denoted by

$$\{X, W_n, n = 1, 2, \cdots N, s\}, s = \max_n s_n.$$

Example B1. $\{R, W(x) = 0, W(x) = \frac{2}{3}x + \frac{1}{3}\}$ is an IFS. Let (X, d) be a complete metric space and $H(X) = \{K \subseteq X/K\text{compact}\}$. For $B \in H(X)$ and $x \in X, d(x, B) = \min\{d(x, y)/y \in B\}$,

$$\begin{aligned} d(A, B) &= \max\{d(x, B)/x \in A\} \\ d(A, B) &\neq d(B, A) \\ h(A, B) &= \max\{d(A, B), d(B, A)\}, \end{aligned}$$

where $h(.,.)$ is metric on $H(X)$, $(H(X), h)$ is a complete metric space, and $h(.,.)$ is called the **Hausdorff metric**.

Let $\{R^2, W_n, n = 1, 2, 3 \cdots N,\ s\}$ be an IFS and

$$W_n \begin{bmatrix} x \\ y \end{bmatrix} = \begin{bmatrix} a_n & b_n \\ c_n & d_n \end{bmatrix} \begin{bmatrix} x \\ y \end{bmatrix} + \begin{bmatrix} e_n \\ f_n \end{bmatrix} = A_n + t_n$$

IFS $\{R^2, W_1, W_2, W_3 \cdots W_N\}$ can be expressed in the table as follows:

Table B1.

W	a	b	c	d	e	f	b
1	a_1	b_1	c_1	d_1	e_1	f_1	b_1
2	a_2	b_2	c_2	d_2	e_2	f_2	b_2
3	a_3	b_3	c_3	d_3	e_3	f_3	b_3
4	a_4	b_4	c_4	d_4	e_4	f_4	b_4
.		,	,	,	,	,	,
:	,	,	,	,	,	,	,
.	,		,	,	,		
N	a_N	b_N	c_N	d_N	e_N	f_N	b_N

where

$$p_i \simeq |a_i d_i - b_i c_i| / \sum_{i=1}^{N} |a_i d_i - b_i c_i| .$$

This table is known as the IFS code table. It can be seen that for IFS of fern leaf

$$\begin{aligned} N = 4,\ & a_1 = 0, \quad a_2 = .2, \quad a_3 = .85,\ a_4 = .85, \\ & b_1 = 0, \quad b_2 = .26,\ b_3 = .04,\ b_4 = .04, \\ & c_1 = 0, \quad c_2 = .23,\ c_3 = .04,\ c_4 = .04, \\ & d_1 = .16,\ d_2 = .22,\ c_3 = .85,\ c_4 = .85, \\ & e_1 = 0, \quad e_2 = 0, \quad e_3 = 0, \quad e_4 = 0, \\ & f_1 = 0, \quad f_2 = .2, \quad f_2 = .2, \quad f_3 = .2, \quad f_4 = .2 \end{aligned}$$

Theorem B2 (The Iterated Function Theorem–IFS). *Let $\{X, W_n,\ n = 1, 2, \cdots, N\}$ be an iterated function system with contractivity factor s. Then the transformation defined by*

$$W(B) = W_1(B) \cup W_2(B) \cup \cdots \cup W_N(B) \ \ \textit{for all} \ \ B \in H(X)$$

is a contraction mapping on the complete metric space $(H(X), h(.))$ with contractivity factor s; that is,

$$h(W(B), W(C)) \leq \ sh(B, C) \ \forall\ B, C \in H(X).$$

W has a unique fixed point, $A \in H(X)$ satisfying

$$A = W(A) = \bigcup_{n=1}^{N} W_n(A),$$

and is given by

$$A = \lim_{n \leftarrow \infty} W^{0n}(B) \quad \textit{for any} \quad B \in H(X).$$

Definition B2. *The fixed point $A \in H(X)$ described in the IFS theorem is called the attractor of the IFS. The attractor is also called* **deterministic fractal or fractal**.

Let $W : R^2 \longrightarrow R^2$ be a contraction map, then A and $W(A)$ are shown in Figure B3.

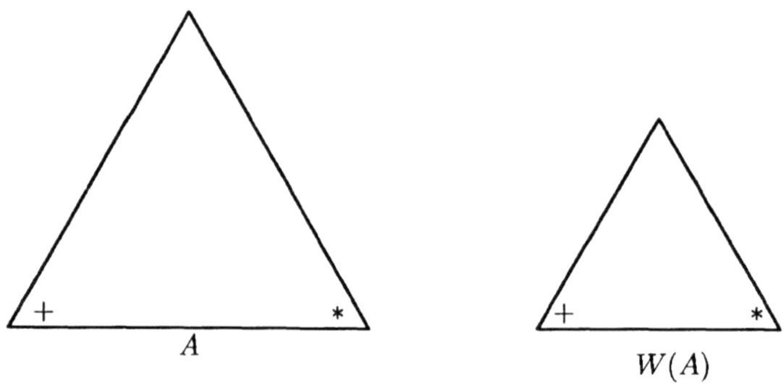

Figure B3. Contraction map.

Relation between mappings on X and $H(X)$.

Lemma B1. *Let W_1 be a continuous mapping on the metric space (X, d) into itself. Then W_1 maps $H(K)$ into itself.*

Lemma B2. *Let $V : X \longrightarrow X$ be a contraction mapping on the metric space (X, d) with contractivity factors s. Then $W : H(X) \longrightarrow H(X)$ defined by $W(B) = \{V(x)/x \in B, \forall B \in H(X)\}$ is a contraction mapping on $(H(X), h(.))$ with the same contractivity factor.*

Lemma B3. *Let (X, d) be a metric space and $\{W_n, n = 1, 2, 3, s, \cdots, N\}$ be an IFS. Let the contractivity factor for W_n be s_n. Define $W : H(X) \longrightarrow H(X)$ by*

$$W(B) = W_1(B) \cup W_2(B) \cup \cdots \cup W_N(B) = \bigcup_{n=1}^{N} W_n(B)$$

for each $B \in H(X)$. *Then* W *is a contraction mapping with contractivity factor* $s = \max\{s_n : n = 1, 2, \cdots, N\}$.

Example B3. (a) $X = [0,1], W_1(x) = \frac{1}{3}, W_2(x) = \frac{1}{3}x + \frac{2}{3}$. W_1 and W_2 are contraction maps on X into itself. The deterministic fractal or fractal of the IFS system $\{X, W_n, n = 1, 2\}$ is the cantor set.

(b) $X = [0,1] \times [0,1]$

$$W_i(x) = T_i \begin{pmatrix} x \\ y \end{pmatrix} + b_i, i = 1, 2, 3,$$

$$W_1 = \begin{pmatrix} 1/2 & 0 \\ 0 & 1/2 \end{pmatrix}, b_1 = \begin{pmatrix} 0 \\ 0 \end{pmatrix}$$

$$W_2 = \begin{pmatrix} 1/2 & 0 \\ 0 & 1/2 \end{pmatrix}, b_2 = \begin{pmatrix} 1/4 \\ 1/4 \end{pmatrix}$$

$$W_3 = \begin{pmatrix} 1/2 & 0 \\ 0 & 1/2 \end{pmatrix}, b_3 = \begin{pmatrix} 1/2 \\ 0 \end{pmatrix}.$$

The fixed point A of the map $W : H(X) \longrightarrow H(X)$ defined as $W(B) = \bigcup_{n=1}^{N} W_n(B)$ is the Sierpinski gasket.

(c) Sierpinski gasket and Von Koch curve are examples of fractal (deterministic fractal).

Theorem B3 (The Collage Theorem). *Let* B *be an arbitrary target image (an element of* $H(X), X = R^2$*) and* $\{X, W_n = 1, 2 \cdots N\}$ *be an iterated function system with contractivity factor* $s, 0 < s < 1$. *Further, let* $W(B)$ *be as in Theorem B2 such that the Hausdorff distance between* B *and* $W(B)$ *be less than* ξ. *Then the Hausdorff distance between* B *and the attractor* A *of the given IFS system is less than* $(1-s)^{-1}\xi$.

Remark B1.

(i) Theorem B2 (Iterated Function Theorem) tells us that each IFS defines a unique fractal image (attractor). This also says that the attractor of the IFS is exactly the same as the union of the transformations of the attractor.

(ii) Small (affine) deformed copies of the target are arranged so that they cover up the target as exactly as possible. This Collage of deformed copies determines an IFS (Theorem B2). Theorem B3 tells us that the better the Collage, as measured by the Hausdorff distance, the closer will be the attractor of the IFS to the target.

(iii) An interesting consequence of the Collage Theorem is that if the matrix entries in two codes are close, thus attractors of the codes are also close. In other words, small errors in codes lead to small errors in images.

Methods for finding IFS for an image. Let S be an image and we want to find an IFS for S. We can split the whole image into non-overlapping segments whose union covers the whole image. Each segment is similar to the whole image and can be obtained by a transformation of the whole image. If we can find transformations for each segment, the combination of these transformations is the IFS encoding of the original image. In other words, to encode the image into IFS is to find a set of contractive affine transformations $W_1, W_2 \cdots, W_N$, so that the original image S is the union of N subimages:

$$W_1(S) \cup W_2(S) \cup \cdots \cup W_N(S) .$$

The IFS for the Sierpinski triangle can be found as follows: The Sierpinski

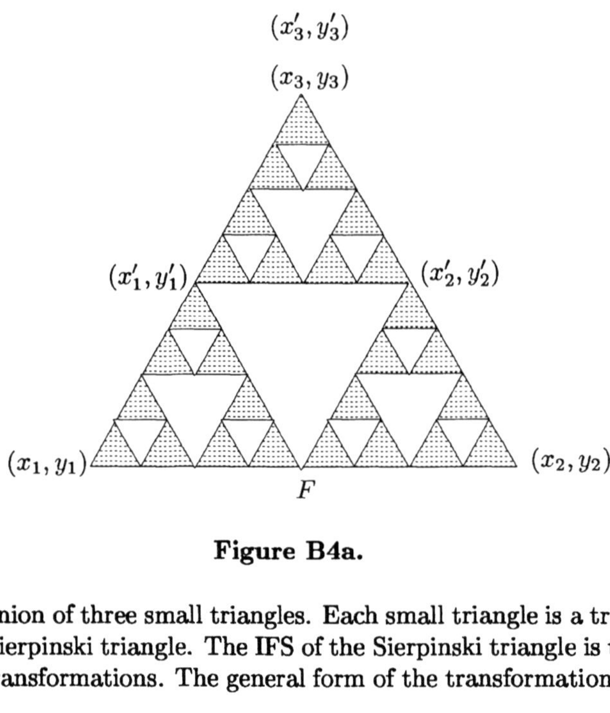

Figure B4a.

triangle is the union of three small triangles. Each small triangle is a transformation of the original Sierpinski triangle. The IFS of the Sierpinski triangle is the collection of these three transformations. The general form of the transformation representing the triangles is:

$$W(x) = W \begin{bmatrix} x \\ y \end{bmatrix} = \begin{bmatrix} a & b \\ c & d \end{bmatrix} \begin{bmatrix} x \\ y \end{bmatrix} + \begin{bmatrix} e \\ f \end{bmatrix} .$$

To determine the transformation, we are required to find values of a, b, c, d, e, f. The top triangle is obtained by the original triangle through transformation of the points (x_1, y_1) to (x'_1, y'_1), (x_2, y_2) to (x'_2, y'_2) and (x_3, y_3) to (x'_3, y'_3). We have the following six linear equations:

$$ax_1 + by_1 + e = x'_1$$

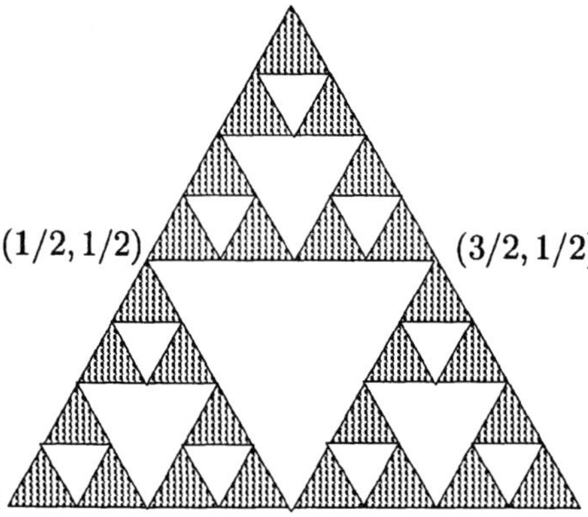

Figure B4b.

$$ax_2 + by_2 + e = x_2'$$
$$ax_3 + by_3 + e = x_3'$$
$$cx_1 + dy_1 + f = y_1'$$
$$cx_2 + dy_2 + e = y_2'$$
$$cx_3 + dy_3 + e = y_3' .$$

Putting the values of the coordinates of Figure B4b and solving this system of equation, we get $a = 0.5, b = 0, c = 0, d = 0.5, e = 0.5, e = 0.5$ and $f = 0.5$. In a similar manner, we can find values of the six variables for the bottom right triangles.

Thus, the IFS of the Sierpinski gasket of Figure B4b is $\{R^2, W_n, n = 1, 2, 3.\}$ where

$$W_1(x)\begin{bmatrix} x \\ y \end{bmatrix} = \begin{bmatrix} 0.5 & 0 \\ 0 & 0.5 \end{bmatrix}\begin{bmatrix} x \\ y \end{bmatrix} + \begin{bmatrix} 0.5 \\ 0.5 \end{bmatrix}$$
$$W_2(x)\begin{bmatrix} x \\ y \end{bmatrix} = \begin{bmatrix} 0.5 & 0 \\ 0 & 0.5 \end{bmatrix}\begin{bmatrix} x \\ y \end{bmatrix} + \begin{bmatrix} 0 \\ 0 \end{bmatrix}.$$

It may be observed that the IFS of an image is not unique. There may be many IFSs for an image. However, an IFS has a unique attractor (fractal)

$$W_3(x)\begin{bmatrix} x \\ y \end{bmatrix} = \begin{bmatrix} 0.5 & 0 \\ 0 & 0.5 \end{bmatrix}\begin{bmatrix} x \\ y \end{bmatrix} + \begin{bmatrix} 1.0 \\ 0 \end{bmatrix}.$$

The IFS codes of Figure B4b are given in Table B2 below.

Table B2.

W	a	b	c	d	e	f	$p=b$
W_1	0.5	0	0	0.5	0.5	0.5	p_1
W_2	0.5	0	0	0.5	0	0	p_2
W_3	0.5	0	0	0.5	1	0	p_3

where

$$p_i \simeq |a_i d_i - b_i c_i| / \sum_{i=1}^{3} |a_i d_i - b_i c_i| \, .$$

Let a target image S lie in the rectangle

$$A = \{(x,y) \in R^2 / a \leq x \leq b, c \leq y \leq d\} \, .$$

Let S be digitized or a polygonalized approximation. S is rendered on a computer graphic monitor. An affine transformation

$$W_1 \begin{bmatrix} x \\ y \end{bmatrix} = \begin{bmatrix} a_1 & b_1 \\ c_1 & d_1 \end{bmatrix} \begin{bmatrix} x \\ y \end{bmatrix} + \begin{bmatrix} e_1 \\ f_1 \end{bmatrix} = A_1(x) + t_1$$

is introduced, with coefficients initialized at $a_1 = d_1 = 0.25, b_1 = c_1 = 0.25$.

The image $W_1(S)$ is displayed on the monitor in a different colour from S. The image $W_1(S)$ is a quarter-sized copy of S, centered closer to the point $(0,0)$. The user now interactively adjusts the coefficients by specifying changes with a mouse or some other interaction technique so that the image $W_1(S)$ is variously translated, rotated, and sheared on the screen. The goal of the user is to transform $W_1(S)$ so that it lies over a part of S. It is important that the dimensions of $W_1(S)$ are smaller than those of S to ensure that W_1 is a contraction. Once $W_1(S)$ is suitably positioned, it is fixed, its coefficients are recorded, and a new affine transformation W_2 and a new supcopy $W_2(S)$ are introduced. The image W_2 is adjusted interactively until $W_2(S)$ covers a subset of those pixels in S that are not in $W_1(S)$. Overlapping is permissible but it should be kept to a minimum.

In this manner, the user determines a set of contractive affine transformations $\{W_1, W_2, W_3, \cdots, W_N\}$ with this property: The original target S, and the set

$$S^{\sim} = \bigcup_{n=1}^{N} W_n(S),$$

are virtually close, while N is as small as possible. The mathematical indicator of the closeness of S and $S^{\sim}$ is the Hausdorff distance between them, $h(S, S^{\sim})$. By 'virtually close', we mean that "$h(S, S^{\sim})$" is small. The coefficients of the transformations $\{W_1, W_2, W_3, \cdots, W_N\}$ thus determined are stored. The Collage Theorem assures us that the attractor A of this IFS will also be virtually close to S. Moreover, if

$$S = S^{\sim},$$

then $A = S$. There are several interesting algorithms for computing the attractor of IFS like:

(i) Photocopy machine Algorithm, (See [Barnsley and Hurd 1993, pp. 98-99]).

(ii) Chaotic game algorithm (See Lu [1993]).

(iii) For more details, we also refer to Fisher [1994].

7.3 Some basic results

C_1 Linear Algebra.

A vector space (linear space) X over a field of real numbers R is a set X together with mappings $(x, y) \longrightarrow x + y$ of $X \times X$ into X (+, operation of addition) and $(\alpha, x) \longrightarrow \alpha x$ of $R \times X$ into X (., operation of scalar multiplication) having the following properties:

(i) $x + y = y + x$,

(ii) $x + (y + z) = (x + y) + z$,

(iii) there exists $0 \in X$ with $x + 0 = x$ for all $x \in X$,

(iv) for each $x \in X$ there exists $x' \in X$ with $x + x' = 0$,

(v) $(\alpha + \beta)x = \alpha x + \beta x$ for all $\alpha, \beta \in R$ and $x \in X$,

(vi) $\alpha(x + y) = \alpha x + \alpha y$ for all $\alpha \in R$ and $x, y \in X$,

(vii) $(\alpha\beta)$ x $= \beta(\alpha x)$ for all $\alpha, \beta \in R$ and $x \in X$,

(viii) $1 \cdot x = x$.

A subset Y of X which is also a vector space with respect to the same operation of addition and scalar multiplication is called a subspace.

Examples of a vector space.

(i) The set of real numbers R is a vector space.

(ii) The set R^n, Euclidean space of n dimension, with elements of the type $x = (x_1, x_2, \cdots x_n)$, $x_i \in R$ is a vector space over R.

(iii) The set of complex numbers C is a vector space over R.

(iv) The set $C[a, b]$ of all continuous functions on $[a, b]$ into R is a vector space over R.

(v) The set of all bounded real sequences is a vector space.

(vi) The set $l_p = \left\{ \{x_n\} / \sum_{n=1}^{\infty}, |x_n|^p < \infty, \quad 1 \le p < \infty \right\}$ is a vector space.

(vii) The set L_p of all measurable functions such that $|f|^p$, $1 \le p < \infty$, is Lebesgue integrable is a vector space.

A set of n linearly independent vectors, $\{e_i\}_{i=1}^n$ is called the basis of a vector space X if every element $x \in X$ can be written as $x = \sum_{i=1}^{n} \alpha_i e_i$. The number of vectors in the basis of X is called the dimension of X. X is called a finite-dimensional vector space of dimension n if the number of elements in the basis is finite, say n. The vector

$$x = \sum_{i=1}^{n} \alpha_i e_i \in X$$

will always be represented by

$$x = \begin{pmatrix} \alpha_1 \\ \alpha_2 \\ \vdots \\ \alpha_n \end{pmatrix},$$

in the matrix notation. It is called the column vector. $x^T = (\alpha_1, \alpha_2 \cdots, \alpha_n)$ is known as the **row vector**. The row vector x^T is the transpose of the column vector. Let X be a vector space of finite dimension over R. The function $\langle \cdot, \cdot \rangle : X \times X \longrightarrow R$, $\langle x, y \rangle = xy^T = x^T y = \sum_{i=1}^{n} \alpha_i \beta_i$ is called the **Euclidean scalar product**. Two vectors x and y of X are orthogonal denoted by $x \perp y$ if $< x, y >= 0$. A vector x of X is called orthogonal to a subset Y of X denoted by $x \perp Y$ if $x \perp y$ for all $y \in Y$. A set $\{x_1, x_2, \cdots x_n\}$ of vectors in X is called orthogonal if $\langle x_i, x_j \rangle = \delta_{ij}$, $1 \le i, j \le k$, where δ_{ij} is the Kronecker delta: $\delta_{ij} = 1$ if $i = j, \delta_{ij} = 0$ if $i \ne j$.

Let X and Y be two vector spaces equipped with the bases $(e_j)_{j=1}^n$ and $(f_i)_{i=1}^m$, respectively. Relative to these bases, a linear operator $\mathcal{A} : X \to Y$ is represented by the matrix having m rows and n columns:

$$A = \begin{pmatrix} a_{11} & a_{12} & \cdots & a_{1n} \\ a_{21} & a_{22} & \cdots & a_{2n} \\ \vdots & & & \\ \vdots & & & \\ a_{m1} & a_{m2} & \cdots & a_{mn} \end{pmatrix},$$

the elements a_{ij} of the matrix A being defined uniquely by the relations

$$\mathcal{A} e_j = \sum_{i=1}^{m} a_{ij} \, fi \quad 1 \le j \le n.$$

Equivalently, the j-th column vector

$$\begin{pmatrix} a_{1j} \\ a_{2j} \\ \vdots \\ a_{mj} \end{pmatrix},$$

of the matrix A represents the vector Ae_j relative to the basis $(f_i)_{i=1}^m$. We call

$$(a_{i1} \;\; a_{i2} \cdots a_{in}),$$

the i-th row vector of the matrix A.

A matrix with m rows and n columns is called a matrix of type (m, n). A column vector is a matrix of type $(m, 1)$ and a row vector a matrix of type $(1, n)$. A matrix A with element a_{ij} is written as

$$A = (a_{ij}) \;\; i = 1, \cdots m, j = 1, 2 \cdots n;$$

where the first subscript i always denotes the row and the second subscript j denotes the column. $A_{mn}(R)$ denotes the space of all $m \times n$ matrices $a_{ij} \in R$. The transpose of the matrix $A = (a_{ij})$, $i \cdots, 1 \cdots m, j = 1, 2 \cdots n$ is the matrix $A^T = (a_{ji})$; equivalently $\langle Ax, y\rangle_m = \langle x, A^T y\rangle$. If $A = (a_{ij})$ *and* $B = (b_{kj})$ are matrices of (m, l) and (l, n)-type, respectively, then their product AB is the matrix of type (m, n) defined by

$$AB = \sum_{k=1}^{l} a_{ik} \, b_{kj} \, .$$

It can be seen that $(AB)^T = B^T A^T$. $A = a_{ij}$ $i = 1, 2 \cdots m, j = 1, 2 \cdots n$ is called a square matrix of order n if $m = n$. $A_{nn}(R)$ is a ring which is called the ring of square matrices of order n. Elements a_{ii} of a square matrix $A = (a_{ij}), i = 1, 2, \cdots n, j = 1, 2, \cdots n$ are called diagonal elements and the elements $a_{ij}, i \neq j$, are called the off-diagonal elements. $I = (\delta_{ij}), i, j = 1, 2 \cdots n$ is called the identity matrix. A square matrix $A = (a_{ij})$ is called **invertible matrix** and its inverse is denoted by A^{-1} if $AA^{-1} = A^{-1}A = I$, otherwise A is called **singular**. If A and B are invertible matrices, then

$$(AB)^{-1} = B^{-1}A^{-1}, \quad (A^T)^{-1} = (A^{-1})^T.$$

A is called

(i) **symmetric** if $a_{ij} = a_{ji}$, that is, $A = A^T$,

(ii) **orthogonal** if $AA^T = A^T A = I$,

(iii) **normal** if $AA^T = A^T A$, and

(iv) **diagonal** if $a_{ij} = 0$ for $i \neq j$.

The trace of A denoted by $\mathrm{tr}(A)$ is defined by $\mathrm{tr}(A) = \sum_{i=1}^{n} a_{ii}$, $\mathrm{tr}(AB) = \mathrm{tr}(BA), \mathrm{tr}(A+B) = \mathrm{tr}(A) + \mathrm{tr}(B)$. The concept of eigenvalues, eigenvector and spectrum for matrices can be derived from discussion of these concepts for linear operators mentioned in the next section. The matrix equation $Ax = b$; that is

$$\begin{pmatrix} a_{11} & a_{12} & \cdots & a_{1n} \\ a_{21} & a_{22} & \cdots & a_{2n} \\ \vdots & & & \\ \vdots & & & \\ a_{n1} & a_{n2} & \cdots & a_{nn} \end{pmatrix} \begin{pmatrix} x_1 \\ x_2 \\ \vdots \\ x_n \end{pmatrix} = \begin{pmatrix} b_1 \\ b_2 \\ \vdots \\ b_n \end{pmatrix} \tag{C.1.1}$$

has a unique solution if

$$\sum_{j=1}^{n} | - a_{ij} + \delta_{ij} | \leq k < 1.$$

Iterative methods for the solution of Ax = b. Given an invertible matrix A and a vector b, one would like to find the solution u of the linear system

$$Ax = b.$$

Assume that a matrix B and a vector c can be found such that $(I - B)$ is invertible and the unique solution of the linear system

$$u = Bu + c$$

is also the solution of $Ax = b$. The form of the system $u = Bu + c$ invokes the definition of an iterative method for $Ax = b$. Choosing an arbitrary initial vector u_0, the sequence of vectors $\{u_k\}$ is defined by

$$u_{k+1} = Bu_k + c, k \geq 0.$$

We say that the iterative method is convergent if the sequence u_k is convergent to u for every initial vector u_0; that is,

$$\lim_{u \to \infty} u_k = u,$$

for very choice of u_0.

The iteration method is convergent if and only if $||B|| < 1$. The matrix B is constructed from the matrix A. There are many methods for constructing B from A, for example, Jacobi, Gauss-Seidel and relaxation methods. For details of these methods, we refer to Ciarlet [1989], Datta [1995], Goulub and van Loan [1996] and Press Teukolsky et al. [1992].

C_2 Banach and Hilbert spaces.

Let X be a non-empty set, then a mapping $d : X \times X \longrightarrow R_+$ defined on $X \times X$ into the set of non-negative real numbers R_+ is called a metric on X if it satisfies the following conditions:

(i) $d(x, y) \geq 0$, d(x,y) $= 0$ if and only $x = y$,

(ii) $d(x, y) = d(y, x)$ for all $x, y \in X$, and

(iii) $d(x, y) \leq d(x, z) + d(z, y)$ for all $x, y, z \in X$.

The metric is a generalization of the notion of distance on the real line. A non-empty set of X equipped with the metric $d(\cdot, \cdot)$ is called a metric space. It is clear that $(R, d(\cdot, \cdot))$ is a metric space with $d(x, y) = |x - y|$ for $x, y \in R$. R^n, $n \geq 1$ is a metric space with respect to

$$d(x, y) = \left(\sum_{i=1}^{n} |x_i - y_i|^p \right)^{\frac{1}{p}}, \quad 1 \leq p < \infty,$$

for every

$$x = (x_1, x_2 \cdots x_n) \text{ and } y = (y_1, y_2 \cdots y_n) \in R^n$$

and

$$d(x, y) = \max_{1 \leq i \leq n} (|x_i|, |y_i|)$$

where

$$x = (x_1, x_2 \cdots x_n) \text{ and } y = (y_1, y_2, \cdots y_n) \in R^n .$$

ay* $C[a, b]$ is a metric space with respect to

$$d(f, g) = \max_{a \leq x \leq b} |f(x) - g(x)|, f, g \in C[a, b],$$

and

$$d(f, g) = \left(\int_a^b |f(x) - g(x)|^2 dx \right)^{\frac{1}{2}}, f, g \in C[a, b].$$

The set of all bounded real sequences, denoted by m, is a metric space with respect to

$$d(x, y) = \sup_i |x_i - y_i|, x = \{x_i\} \text{ and } y = \{y_i\},$$

belonging to m.

$(0, 1]$ is a metric space with respect to $d(x, y) = |x - y|$. Every metric space is a Hausdorff topological space.

Let X be a vector space over a field of real numbers R. A function defined on X into R, denoted by $\|\cdot\|$ is called a norm if it satisfies the following conditions:

(i) $\|x\| \geq 0, \|x\| = 0$ if $x = 0$,

(ii) $\|\alpha x\| = |\alpha|, \|x\|, \alpha \in R, x \in X$, and

(iii) $\|x + y\| \leq \|x\| + \|y\|, x, y \in X$.

$(X, \|\cdot\|)$ is called a real normed linear space or simply normed space. One can define normed linear space over a field of complex numbers and most of the results mentioned here may be true for such spaces. However, here we confine ourselves to the real case. Every normed space is a metric space but the converse may not be true. R, $R^n, C[a,b], l_p, 1 \leq p < \infty$, $m, L_p, 1 \leq p < \infty$ are normed spaces. A normed linear space, in which every Cauchy sequence is convergent; that is

$$\|x_n - x_m\| \to 0 \text{ as } m, n \to \infty \Rightarrow \exists x \in X \text{ such that } \|x_n - x\| \to 0 \text{ as } n \to \infty$$

is called a Banach space. A normed linear space is called finite-dimensional if the underlying vector space is finite-dimensional.

Every finite-dimensional normed space is a Banach space. R^n is Banach space with respect to the norms

$$\begin{aligned} \|x\| &= \left(\sum_{i=1}^{n} |x_i|^2\right)^{\frac{1}{2}} \\ \|x\| &= \left(\sum_{i=1}^{n} |x_i|^p\right)^{\frac{1}{p}}, \quad 1 \leq p < \infty \\ \|x\| &= \max_{1 \leq i \leq n} |x_i|. \end{aligned}$$

$C[a, b]$ is a Banach space with respect to the norm

$$\|f\| = \sup_{a \leq x \leq b} |f(x) - g(x)|.$$

$C[a, b]$ is a normed space with respect to the norm $\|f\| = \left(\int_a^b |f(x)|^2 dx\right)^{\frac{1}{2}}$, but it is not a Banach space.

$l_p, 1 \leq p < \infty$ is a Banach space with respect to the norm

$$x = (x_1, x_2 \cdots x_n \cdots) \in l_p, \|x\| = \left(\sum_{i=1}^{\infty} |x_i|^p\right)^{\frac{1}{p}}, \quad 1 \leq p < \infty,$$

$L_p, 1 \leq p < \infty$ is a Banach space with respect to the norm

$$||f|| = \left(\int_a^b |f(x)|^p dx \right)^{\frac{1}{p}},$$

where $L_p = \{f : [a,b] \to R$ measurable functions$/|f|^p$ is Lebesgue integrable$\}$. We say that two elements f and g of L_p are equal if they are equal almost everywhere.

A mapping T on a normed linear space X into another normed space Y over fields R is called a linear operator if

(i) $T(x+y) = Tx + Ty$ for every $x, y \in X$, and

(ii) $T(\alpha x) = \alpha T(x)$, for every $\alpha \in R$ and $x \in X$,

If $Y = R$, then an operator is called **functional**, that is, functional is a mapping on a normed space into R. A linear operator is bounded if $\exists k > 0$ such that $||Tx|| \leq k||x||$ for all $x \in X$. The concepts of boundedness, continuity and continuity at '0' are equivalent. The set of all bounded linear operators on a normed space X into normed linear space Y denoted by $BL[X,Y]$ is a normed linear space with respect to the following norm:

$$||T|| = \inf \{k/||Tx|| \leq k||x||\}.$$

Operations of addition and scalar multiplications are defined as

$$(T+S)(x) = Tx + Sx$$
$$(\alpha T)(x) = \alpha T(x).$$

$BL[X,Y]$ is a Banach space if Y is a Banach space. Thus the set of all bounded linear functionals is a Banach space. It is denoted by $X^\star$ or X' or X^c and is known as the dual or adjoint or conjugate space of X. We say that X and Y are algebraic and topologically equivalent denoted by $X = Y$ or $X \simeq Y$ if there exists a mapping $T : X \longrightarrow Y$ such that

(i) $T(x+y) = Tx + Ty$,

(ii) $T(\alpha x) = \alpha Tx$,

(iii) $||Tx|| = ||x||$,

(iv) T is $1-1$, and

(v) T is onto.

It can be proved that

$$(R^n)^\star = R^n, (l_p)^\star = l_q, \frac{1}{p} + \frac{1}{q} = 1$$
$$(L_p)^\star = L_q, \frac{1}{p} + \frac{1}{q} = 1.$$

A list of dual spaces of a fairly large number of normed spaces can be found in Siddiqi [1986, Table 1.1, 34–35]. An operator mapping, a continuous function on $[a,b]$ to its integral on $[a,b]$, is linear and bounded functional, that is, $T : C[a,b] \longrightarrow R$ defined by $Tf = \int_a^b f(t)dt$ is linear and bounded.

Differential equations, integral equations and systems of algebraic equations can be written in the form of operator equation $Tx = y$ where T is a linear operator on a normed linear space into itself or into another normed linear space. For example, a system of n linear equations:

$$\begin{aligned} a_{11}x_1 + a_{12}x_2 + \cdots\cdots + a_{1n}x_n &= b_1 \\ a_{21}x_1 + a_{22}x_2 + \cdots\cdots + a_{2n}x_n &= b_2 \\ &\vdots \\ &\vdots \\ a_{n1}x_1 + a_{n2}x_2 + \cdots\cdots + a_{nn}x_n &= b_n, \end{aligned}$$

can be written in the form

$$Tx = y, \text{ where } x = (x_1, x_2 \cdots x_n) \in R^n \text{ and } y = (y_1, y_2, \cdots y_n),$$

and

$$y_i = \sum_{j=1}^{n} (-a_{ij} + \delta_{ij})x_j + b_i,$$

where $i = 1, 2 \cdots n$, $(a_{ij}), i, j = 1, 2 \cdots n$ is an $n \times n$ matrix. The differential equation $\frac{dy}{dx} = f(x,y)$ can be written in the form $Ty = f$, where $y \in C^1[a,b]$ = space of all functions on $[a,b]$ whose first derivative exists and is continuous on $[a,b]$.

Let

$$h(x) = g(x) + \mu \int_a^b K(x,y)f(y)dy,$$

be an integral equation where $K(x,y)$ is a measurable function in the square $A = \{(x,y) | a \le x \le b, a \le y \le b\}$ such that

$$\int_a^b \int_a^b |K(x,y)|^2 dx\, dy < \infty,$$

and $g(x) \in L_2(a,b)$. Then the integral equation can be written in the form $Tf = h$ where $T : L_2(a,b) \longrightarrow L_2(a,b)$ is a linear operator.

The set of $n \times n$ matrices, n finite, is a Banach space; in fact, a Banach algebra with respect to the following norms. For a matrix $A = (a_{ij}), i, j = 1, 2 \ldots, n$

$$\|A\| = \sup_{i,j} |a_{ij}|, \quad \|A\| = \sum_{i=1}^{n} \sum_{j=1}^{n} |a_{ij}|$$

$$\|A\| = \left(\sum_{i=1}^{n} \sum_{j=1}^{n} |a_{ij}|^2 \right)^{\frac{1}{2}} .$$

A subset M of a vector space X is called convex if $\alpha x + \beta y, \in, M$ for α, $\beta \in R, \alpha + \beta = 1, \alpha, \beta \geq 0$ and $x, y \in M$. A functional F defined on a convex subset K of a vector space X is called a convex functional if $F(\alpha x + \beta y) \leq \alpha F(x) + \beta F(y), \alpha, \beta \geq 0, \alpha + \beta = 1$. Every subspace is a convex set but the converse may not be true. Every linear functional is a convex functional but the converse need not be true. An operator T on a normed linear space X into another normed space Y of type $Tx = Sx + b$, for all $x \in X$ and $b \in Y$ (for b fixed) is called an **affine operator** where S is a linear operator on X into Y. If $Y = R$, T is called an affine functional. For details, see Siddiqi [1986]. A vector space equipped with the function $\langle \cdot, \cdot \rangle : X \times X \longrightarrow R$, called inner product, having the following properties:

(i) $\langle x + x', y \rangle = \langle x, y \rangle + \langle x', y \rangle$,

(ii) $\langle \alpha x, y \rangle = \alpha \langle x, y \rangle \alpha \in R, x, y \in X$,

(iii) $\langle x, y \rangle = \langle y, x \rangle$,

(iv) $\langle x, x \rangle \geq 0 \;\; \forall x \in X, \langle x, x \rangle = 0$ if and only if $x = 0$,

is called an **inner product space** on a **pre-Hilbert space**. Every inner product space is a normed linear space with the norm $\|x\|^2 = \langle x, x \rangle$, but the converse may not be true. A pre-Hilbert space which is complete as a normed space, that is, its every Cauchy sequence is convergent, is called a Hilbert space. The Hilbert space has an additional structure like the concepts of an angle and orthogonality between two elements. The following theorems play a very important role in different disciplines like optimization, variational inequalities, optimal control, mathematical economics.

Theorem C1. *Let K be a non-empty, convex, closed subset of a Hilbert space H. Given any element $w \in H$, there exists a unique element Pw such that $Pw \in K$ and $\|w - Pw\| \leq \|w - v\|$ for all $v \in K$; equivalently*

$$\|w - Pw\| = \inf_{v \in K} \|w - v\|. \tag{C2.1}$$

The element $Pw \in K$ satisfies

$$\langle Pw - w, v - Pw \rangle \geq 0 \;\; \textit{for every } v \in K, \tag{C2.2}$$

and, conversely, if an element u satisfies

$$u \in K \text{ and } < u - w, v - u > \geq 0 \text{ for every } v \in K, \tag{C2.3}$$

then $u = Pw$.

The function $P : H \longrightarrow K$ *defined in this manner satisfies the relation*

$$||Pw_1 - Pw_2|| \leq ||w_1 - w_2|| \text{ for every } w_1, w_2 \in H. \tag{C2.4}$$

The function P is linear if and only if the subset K is a subspace of H, in which case (C2.2) *takes the form*

$$\langle Pw - w, v \rangle = 0 \text{ for every } v \in K. \tag{C2.5}$$

Theorem C2. *Let M be a closed subspace of a Hilbert space H, then*

$$H = M \oplus M^{\perp}, \tag{C2.6}$$

where $M^{\perp} = \{y \in H \langle x, y \rangle = 0 \text{ for all } x \in M\}$, *that is, H is the direct sum of M and* $M^{\perp}$.

Theorem C3. *For each* $F \in H^{\star}$ *(dual space of H), there exists a unique element* $y \in H$ *such that*

$$F(x) = < x, y > \text{ for every } x \in H .$$

Definition C1. *The function P defined on Theorem* (C2.1) *on H into K is called the projection operator, and the element Pw is called the* **projection** *onto the set K of the element w.*

Remark C1.

(i) Relation of (C2.1) tells us that Pw is the nearest element of K to w.

(ii) Inequality (C2.2) expresses the necessity for the angle formed by the vectors $Pw - w$ and $v - Pw$ to be less than or equal to π for all elements $v \in K$.

(iii) $w - Pw = 0$ if and only if $w \in K$.

(iv) $H = R^n$ and $K = R^n_+ = \{x = (x_1, x_2 \cdots x_n) | x_i \geq 0 \text{ for all } i\}$, then $(Pw)_i = max\{w_i, 0\}$, $1 \leq i \leq n$, satisfies (C2.2).

Definition C2. (a) *A mapping* $a(\cdot, \cdot) : H \times H \longrightarrow R$, *where H is a Hilbert space, is called a bilinear form if it is linear in both the variables, that is,*

(i) $a(x_1 + x_2, y) = a(x_1, y) + a(x_2, y)$

(ii) $a(\alpha x, y) = \alpha a(x, y)$

(iii) $a(x, y_1 + y_2) = a(x, y_1) + a(x, y_2)$

(iv) $a(x, \beta y) = \beta a(x, y)$.

The bilinear form $a(\cdot,\cdot)$ is called bounded if there exists $k > 0$ such that

$$|a(x, y)| \leq k||x||||y||.$$

(b) $a(\cdot,\cdot)$ *is called*

(i) **symmetric** *if* $a(x, y) = a(y, x)$

(ii) **positive** $a(x, x) \geq 0$ *for all* $x \in H$

(iii) **positive definite** *if* $a(x, x) \geq 0$ *for all and* $a(x, x) = 0 \Rightarrow x = 0$.

(c) $a(\cdot,\cdot)$ *is called* **coercive** *if there exists $\alpha > 0$ such that $a(x, x) \geq \alpha||x||^2$ for all $x \in H$.*

(d) *The norm of a bounded bilinear form $a(\cdot,\cdot)$ is defined by*

$$\begin{cases} ||a|| = \sup\limits_{x\neq 0, y\neq 0} \dfrac{|a(x, y)|}{||x|| \quad ||y||}; \\ \quad = \sup\limits_{x\neq 0, y\neq 0} |a(\dfrac{x}{||x||}, \dfrac{y}{||y||})|; \\ \quad = \sup\limits_{||x||=||y||=1} |a(x, y)|. \end{cases} \tag{C2.7}$$

Theorem C4. *Let T be a bounded linear operator on a Hilbert space H into itself. Then a bounded bilinear form $a(\cdot,\cdot)$ can be defined by the relation*

$$a(x, y) = \langle x, Ty\rangle, \tag{C2.8}$$

and $||a|| = ||T||$. Conversely let $a(.,.)$ be a bounded bilinear form on $H \times H$ into R, then a bounded linear operator T on H into itself can be defined by the relation

$$\langle x, Ty\rangle = a(x, y)$$

such that $||a|| = ||T||$.

Theorem C5 (Lax-Milgram Lemma). *If $a(\cdot,\cdot)$ is a bounded bilinear and coercive form on a* **Hilbert space** *H, then the functional/operator equation*

$$\begin{cases} a(x, y) = F(y) \text{ for all } y \in H; \\ \Leftrightarrow Tx = y, T : H \longrightarrow H, \end{cases} \tag{C2.9}$$

has a unique solution.

Theorem C6 (Fundamental Theorem of Calculus of Variation). *If $a(\cdot,\cdot)$ is a bounded, bilinear, symmetric and coercive form on a Hilbert space H, then finding the solution of* (C2.9) *is equivalent to finding the solution of optimization problem for $J(v) = \frac{1}{2}a(v,v) - F(v)$ where $F \in H^\star$.*

Examples of bilinear form:

(i)

$$H = R^n, a(x,y) = \sum_{i=1}^{n} x_i y_i,$$

where

$$x = (x_1, x_2 \cdots x_n), y = (y_1, y_2 \cdots y_n) \in R^n.$$

(ii) Let

$$H = H^1(0,1) = \left\{ f \in L | \frac{df}{dx} \in L \right\},$$

then

$$a(f,g) = \int_0^1 f(x)g(x)dx + \int_0^1 \frac{df}{dx}\frac{dg}{dx}dx$$

is a bilinear, symmetric and coercive form.

(iii) Let

$$A = (a_{ij}) \quad i,j = 1,2,\cdots,n \text{ be } n \times n,$$

be a symmetric matrix and $H = R^n$, then $a(u,v) = \frac{1}{2}\langle Au, v\rangle$ is a bilinear symmetric bilinear form.

7.4 Results from distribution theory and Sobolev spaces

The distribution theory was developed by the French mathematician Laurent Schwartz around 1950 to resolve the discrepancy created by the Dirac delta function whose value is zero at all points except one while its Lebesgue integral is 1. This function was introduced by the famous physicist P.M. Dirac around 1930 which contradicted the celebrated Lebesgue theory of integral. A class of Lebesgue integrable functions was introduced by the Russian physicist S.L.Sobolev around 1936 which has been found very useful in many areas of current interest, and is now known as the Sobolev space. The results related to this space have provided a sound foundation for modern theory of ordinary and partial differential equations – analytical as well as numerical methods. Some important results of these two topics are mentioned here along with important references. For appreciating and understanding the results of different chapters, especially Chapter 3, these results have been very useful.

Let n be a positive integer. A vector or n-tuple $\alpha = (\alpha_1, \alpha_2 \cdots, \alpha_n)$, where $\alpha_i, i = 1, 2 \cdots, n$, are non-negative integers, is called multi-index of dimension n. The number $|\alpha| = \sum_{i=1}^{n} \alpha_i$, is called the magnitude or length of the multi-index.

For given α, β, $\alpha + \beta$ means $(\alpha_1 + \beta_1, \alpha_2 + \beta_2, \cdots, \alpha_n + \beta_n)$

$$\begin{aligned} \alpha\,! &= \alpha_1\,! \cdots \alpha_n\,! \\ C^\alpha_\beta &= \binom{\alpha}{\beta} = \frac{\alpha!}{\beta\,!(\alpha-\beta)!} \\ x^\alpha &= x_1^{\alpha_1} x_2^{\alpha_2} \cdots x_n^{\alpha_n} \quad \text{where } x = (x_1, x_2, \cdots, x_n) \in R^n. \end{aligned}$$

We say that multi-indices α, β are related by $\alpha \le \beta$ if $\alpha_i \le \beta_i$ for all $i = 1, 2, 3 \cdots, n$.

We are familiar with the concept of the classical partial derivative for a function of n variables $f = f(x), x = (x_1, x_2, \cdots, x_n)$. Laurent Schwartz introduced the concept of multi-index and a new notation of derivatives given below. In the new terminology, D^α will denote the expression $\frac{\partial^{|\alpha|}}{\partial x_1^{\alpha_1} \cdots \partial\, x_n^{\alpha_n}}$. $D^{(\alpha)} f$ will be called a derivative of f of order $|\alpha|$. For $n = 1, \alpha_1 = 1, D^\alpha f = \frac{\partial^1 f}{\partial x_1^1}$ which is the classical derivative of a function of single variable denoted by $\frac{df}{dx}(x_1 = x)$. For $n = 2, \alpha_1 = 1, \alpha_2 = 1, \frac{\partial^2 f}{\partial x_1^1 \partial x_2^1} = D^\alpha f$. This is nothing but the partial derivative of the function of two variables $f(x_1, x_2)$ which is denoted by $\frac{\partial^2 f}{\partial x_1 \partial x_2}$. We also have $D^{(1,1)} f = \frac{\partial^2 f}{\partial x_2 \partial x_1} = \frac{\partial}{\partial x_2}\left(\frac{\partial f}{\partial x_1}\right)$ which is equal to $\frac{\partial^2 f}{\partial x_1 \partial x_2} = \frac{\partial}{\partial x_1}\left(\frac{\partial f}{\partial x_2}\right)$. In distributional derivative to be defined later, we shall not distinguish between $\frac{\partial^2 f}{\partial x_1 \partial x_2}$ and $\frac{\partial^2 f}{\partial x_2 \partial x_1}$.

For $\alpha_1 = 1, \alpha_2 = 2$

$$
\begin{aligned}
\frac{\partial^3 f}{\partial x_1^1\, \partial x_2^2} &= D^{(1,2)}, \alpha_1 = 1, \alpha_2 = 2, \alpha_3 = 1 \\
D^{(1,2,1)} f &= \frac{\partial^4 f}{\partial x_1^1\, \partial x_2^2 \partial x_3}, \quad \alpha = (0,0) \quad D^{(0,0)} f = \frac{\partial^0 f}{\partial x_1^0\, \partial x_2^0} = f, \\
D^{(1,0)} f &= \frac{\partial^1 f}{\partial x_1^1\, \partial x_2^0} = \frac{\partial f}{\partial x_1} \\
D^{(0,1)} f &= \frac{\partial^1 f}{\partial x_1^0\, \partial x_2^1} = \frac{\partial f}{\partial x_2} \\
D^{(0,1,1)} f &= \frac{\partial^2 f}{\partial x_1^0\, \partial x_2^1\, \partial x_3^{'}} = \frac{\partial^2 f}{\partial x_3^2} \\
D^{(0,0,2)} &= \frac{\partial^2 f}{\partial x_1^0\, \partial x_2^0\, \partial x_3^2} = \frac{\partial^2 f}{\partial x_3^2}.
\end{aligned}
$$

The relationship between the multi-index terminology, Laplace operator, and harmonic operator is discussed in Chapter 3 of Rektorys [1980].

All functions are defined on a bounded subset Ω of R^n into R. The boundary of Ω is denoted by Γ or $\partial\, \Omega$ and $\bar{\Omega} = \Omega + \Gamma$. We say that $f \in L_2$ if $|f|^2$ is Lebesgue integrable on Ω. $L_2(\Omega)$ is a Hilbert space with respect to $||f|| = \left(\int_\Omega |f|^2 dx\right)^{\frac{1}{2}}$ or $\langle f, g\rangle = \int_\Omega fg dx$ (Here $\int_\Omega f dx$ stands for $\int \cdots \int_\Omega f(x_1, x_2 \cdots, x_n) dx_1, \cdots dx_n$).

Throughout our discussion, Ω is a bounded subset of R^n with the Lipschitz boundary Γ. A technical definition is slightly complicated; however broadly it means that the boundary will not contain cuspidal points and edges. Examples in two- and three-dimension domains having the boundary Lipschitz condition or Lipschitzian boundary are circles, squares, triangles, spheres, cubes, annuli, etc. In one dimension, $\Omega = (a, b)$ broadly speaking, the function representing Lipschitzian boundary will be smooth or piecewise smooth and will have no singularity. A function f defined on Ω into R is said to satisfy the Hölder condition with exponent $\lambda, 0 < \lambda \leq 1$, if there exists a constant $M > 0$ such that

$$|f(x) - f(y)| \leq M||x - y||^\lambda \;\forall x, y \in \Omega,$$

where $||\cdot||$ is the Euclidean norm on R^n.

Let Ω be a bounded domain and let there exist positive constants $\alpha > 0, \beta > 0$, a finite number of Cartesian coordinate systems $x_1^r \quad x_2^r \cdots x_n^r (r = 1, 2, \cdots\ m)$ and m functions $a_r\left(x_1^r, x_2^r \cdots x_{n-1}^r\right)$ which are continuous in $(n-1)$-dimensional closed cubes E_r

$$|x_i^r| \leq \alpha, i = 1, 2 \cdots n - 1,$$

such that

(i) every point x of Γ can be expressed at least in one of m considered coordinate systems in the form

$$x = \left(x_r^{'}, a_r(x_r^{'})\right); x_r^{'} = \left(x_{n-1}^r \in E_r\right).$$

(ii) points $(x_r^{'}, x_n^r)$ satisfying

$$a_r(x_r^{'}) < x_n^r < a_r(x_r)^{'} + \beta, \quad x_r^{'} \in E_r,$$

lie in Ω ,

(iii) points $(x_r^{'}, x_n^r)$ satisfying

$$a_r(x_r^{'}) - \beta < x_N^r < a_r(x_x^{'}), x_r^{'} \in E_r,$$

lie outside $\bar{\Omega}$. Γ satisfying these conditions is called continuous. It is called Lipschitz continuous if there exists a constant $M > 0$ such that

$$|a_r(y_r^{'}) - a_r(x_r^{'})| \le M \left\{ \sum_{i=1}^{n-1} (y_i^r - x_i^r)^2 \right\}^{\frac{1}{2}} \quad \forall x_r^{'}, y_r^{'} \in E_r, r = 1, 2, \cdots m.$$

$C^0(\Omega)$ = Space of continuous functions on Ω into R,

$C^k(\Omega)$ = Space of all functions defined on Ω into R, whose derivatives upto k-th order exist and are continuous.

$C^\infty(\Omega) = \cap_{k=0}^{\infty} C^k(\Omega)$ = Space of all functions defined on Ω into R having derivatives of all order.

supp f = closure $\{x \in \Omega / f(x) \neq 0\}$ is called the **support** of f. If K is compact, f is said to have compact support.

It can be easily seen that $C_0^\infty(\Omega)$ is a vector space with respect to usual operations.

A sequence $\{\phi_n\}$ in C_0^∞ is said to converge to an element ϕ in $C_0^\infty(\Omega)$; namely, $\phi_n \to \phi$, if (i) there is a fixed compact set $K \subset \Omega$ such that supp $\phi_n \subset K$ for all n, (ii) ϕ_n and all its derivatives converge uniformly to $\phi(x)$ and its derivatives; that is, $D^\alpha \phi_n \to D^\alpha \phi$ for all α uniformly.

$C_0^\infty(\Omega)$ equipped with the topology induced through the convergence is called the space of **test functions** and is often denoted by $D(\Omega)$. A continuous linear functional defined on $D(\Omega)$ is called a **distribution** or **Schwartz distribution** (A functional F on $D(\Omega)$ is continuous if $\phi_n \to \phi \Rightarrow F\phi_n \to F\phi$).

The space of distributions is nothing but the dual space of $D(\Omega)$ and is often denoted by $D^{'}(\Omega)$. If $\Omega = R^n$, we write simply $D^{'}$.

The distributional derivative or the derivative of a distribution is a continuous linear functional defined as:

$$\text{For } F \in D^{'}(\Omega), \langle D^\alpha F, \phi \rangle = (-1)^{|\alpha|} \langle F, D^\alpha \phi \rangle \forall \phi \in D(\Omega).$$

A function $f : \Omega \to R$ is called locally integrable if for every compact set $K \subset \Omega$, $\int_K |f| < \infty$, that is, f is Lebesgue integrable over compact set K.

Every continuous functional is locally integrable. Every Lebesgue integrable function is locally integrable over $[a, b]$.

If $\Omega = S_r$ = ball of radius $r > 0$ and center $(0,0)$ in R_2, then $f(r) = \frac{1}{r}, r = |x|$ is locally integrable on S_r. Locally integrable functions can be identified with distributions. We mention some results for $n = 1$ and $n = 2$. Case $n = 1$ (Distribution theory on real line).

$$\phi(x) = \begin{cases} \exp\left[\dfrac{1}{(x-1)(x-3)}\right], & 1 < x < 3 \\ 0, & \text{outside the open interval } (1,3) \end{cases}$$

$$\text{supp } \phi = [1,3], \phi(x) \text{is a test function}$$

$$\phi(x) = \exp(-x^{-2}), x > 0$$

$$= 0\,,\ x \leq 0\,,$$

$\phi(x)$ is a test function.

(i) $F(\phi) = \phi(0)$, F is linear and continuous on $D(\Omega), \Omega = (a, b)$ and so F is a distribution.

(ii) Let $F(\phi) = \displaystyle\int_a^b f(x)\phi(x)dx$, f is a locally integrable function. F is linear and continuous on $D(\Omega)$ and therefore it is a distribution.

(iii) Let $F(\phi) = \displaystyle\int_a^b |\phi(x)|^2 dx$, F is continuous on $D(\Omega)$ but not linear and, therefore, F is not a distribution.

(iv) The Dirac delta distribution is defined as

$$\langle D_8, \phi \rangle = \phi(0) \quad \forall\, \phi \text{ in } D(\Omega)\,.$$

D_8 is linear and continuous on $D(\Omega)$ and hence a distribution.

A distribution generated by locally integrable functions as in example (ii) is called a regular distribution; otherwise it is called a singular distribution.

The Heaviside function H is defined by

$$H(x) = \begin{cases} 0, & x < 0 \\ \dfrac{1}{2}, & x = 0 \\ 1, & x > 0\,. \end{cases}$$

Let

$$H_1(x) = \begin{cases} 0, x \leq 0 \\ 1, x > 0\,. \end{cases}$$

H and H_1 generate the same distribution and such functions are identified identically in distribution theory.

Two distributions F and G are equal if $\langle F, \phi \rangle = \langle G, \phi \rangle$ for all $\phi \in D(\Omega)$ such that $\text{supp}(\phi) \subset (a, b)$.

For a comprehensive account of one-dimensional theory, see Griffel [1987]. The distributional derivative of $|x|$ is $\text{sgn}(x)$ defined as

$$\begin{aligned} \text{sgn}(x) &= -1, x < 0 \\ &= 1, x > 0 \\ &= 0, x = 0 . \end{aligned}$$

The distributional derivative of H is the Dirac delta distribution.

Meaning of $\int_\Gamma f(x) d\Gamma$. Let Ω be a domain with the Lipschitzian boundary Γ.

Every point of the boundary Γ can be expressed in at least one of the considered coordinate systems in the form

$$x = \left[(x_1^{(r)}, x_2^{(r)} \cdots, x_{N-1}^{(r)}), a_r(x_1^{(r)}, x_2^{(r)}, \cdots x_{N-1}^{(r)}) \right] .$$

The surface element above the element $dx_1^{(r)}\ dx_2^{(r)} \cdots dx_{n-1}^{(r)}$ is

$$dS = \left[1 + \left(\frac{\partial a_r}{\partial x_1^{(r)}} \right)^2 + \cdots\cdots + \left(\frac{\partial a_r}{\partial x_{N-1}^{(r)}} \right)^2 \right]^{\frac{1}{2}} dx_1^{(r)} \cdots dx_{N-1}^{(r)} .$$

If on $\Gamma^{(r)}$, a function $f(S) = f(x_1^{(r)}, x_2^{(r)}, \ldots x_{n-1}^{r}, a_r(x_1^{(r)}, x_2^{(r)}, \ldots x_{n-1}^{r}))$ is given measurable on E_r, we define

$$\int_{\Gamma^{(r)}} f(S) dS = \int_{E_r} f(x_1^{(r)}, x_2^{(r)} \cdots x_{n-1}^{(r)}, a_r(x_1^{(r)}, x_2^{(r)}, \cdots x_{n-1}^{(r)})) dS.$$

It may be observed that values of the function $a_r(x_1^{(r)}, x_2^{(r)}, \cdots x_{n-1}^{(r)})$ characterize the part $\Gamma^{(r)}$ of the boundary above the cube E_r, where the integral on the right is taken in the Lebesgue sense. If $\int_{\Gamma^{(r)}} f^2(S) dS$ exists, we say that $f(S) \in L_2(\Gamma^{(r)})$. If $f^2(S)$ is integrable over $\Gamma^{(r)}$ for every $r = 1, 2 \cdots, m$, we say that $f(S) \in L_2(\Gamma)$. The integral $\int_\Gamma f(S) dS$ can be defined with the aid of a partition of unity as

$$f_r(S) = f(S) \phi_r(S),$$

where the functions $\phi_r(x)$ exist with compact supports in the system of neighbourhoods defined earlier such that

$$\sum_{r=1}^{m} \phi_r(x) = 1 \text{ for each } x \in \Gamma.$$

From here, we get

$$\sum_{r=1}^{m} f_r(S) = f(S) \text{ on } \Gamma.$$

$\int_\Gamma f(S)dS$ is defined by the equation

$$\int_\Gamma f(S)dS = \sum_{r=1}^{m} \int_{\Gamma^{(r)}} f_r(S)dS \ .$$

With this definition of the choice of the m coordinate systems and the choice of functions $\phi_r(r)$, we proceed further.

Let $L_2(\Gamma) = \left\{ f(S) \mid \int_\Gamma |f(S)|^2 \, dS \text{ exists} \right\}$ be a Hilbert space with respect to $\langle f, g \rangle = \int_\Gamma f(S)g(S)dS$, $S \in \Gamma$.

The induced norm and metric are given as

$$\begin{aligned} \|f\|_\Gamma &= \langle f, f \rangle \Gamma \\ d(f,g) &= \|f - g\|_\Gamma \,. \end{aligned}$$

$H^m(\Omega) = \{f \in L_2(\Omega) | D^\alpha \in L_2(\Omega), |\alpha| \leq m\}$, m is any positive integer, is called the **Sobolev space of order m.**

$H^1(\Omega)$ denotes the Sobolev space of order 1 and $H^2(\Omega)$ is the Sobolev space of order m.

$H^m(\Omega)$ is a Hilbert space with respect to the inner product

$$\langle f, g \rangle_{H^m(\Omega)} = \sum_{|\alpha| \leq m} \langle D^\alpha f, D^\alpha g \rangle_{L_2(\Omega)}.$$

For $m = 1, \Omega = (a, b)$,

$$\begin{aligned} D^\alpha f &= \frac{df}{dx} \\ \langle f, g \rangle_{H^1(a,b)} &= \langle f, g \rangle_{L_2(a,b)} + \langle \frac{df}{dx}, \frac{dg}{dx} \rangle_{L_2(a,b)} \\ &= \int_a^b f(x)g(x)dx + \int_a^b \frac{df}{dx}\frac{dg}{dx}dx. \end{aligned}$$

It can be easily verified that $H^1(a, b)$ is a Hilbert space.

For $m = 2, \Omega = S_r = \{(x, y)/x^2 + y^2 \leq r\}$ = circle with origin as the centre; or $\Omega = \{(x, y)/a < x < b, c < y < d\}$ = a rectangle with sides of length $b - a$ and $d - c$.

$$\langle f, g \rangle_{H^2(\Omega)} = |\alpha| \sum_{|\alpha| \leq 2} \langle D^\alpha f, \ D^\alpha g \rangle_{L_2(\Omega)},$$

where

$$\alpha = (0,1),(1,0),(1,1),(0,2),(2,0)$$

$$|\alpha| = \alpha_1+\alpha_2 \le 2,\ \langle D^\alpha f, D^\alpha g\rangle = \langle D^{(0,0)}f, D^{(0,0)}g\rangle_{L_2(\Omega)}$$

$$+ \langle D^{(0,1)}f, D^{(0,1)}g\rangle_{L_2(\Omega)} + \langle D^{(1,0)}f, D^{(1,0)}g\rangle_{L_2(\Omega)}$$

$$\langle D^{(1,1)}f, D^{(1,1)}g\rangle + \langle D^{(2,0)}f, D^{(2,0)}g\rangle_{L_2(\Omega)} + \langle D^{(0,2)}f, D^{(0,1)}g\rangle_{L_2(\Omega)}$$

$$= \langle f,g\rangle_{L_2(\Omega)} + \langle \frac{\partial f}{\partial x_1}, \frac{\partial g}{\partial x_1}\rangle_{L_2(\Omega)} + \langle \frac{\partial f}{\partial x_2}, \frac{\partial g}{\partial x_2}\rangle_{L_2(\Omega)}$$

$$+ \langle \frac{\partial^2 f}{\partial x_1 \partial x_2}, \frac{\partial^2 g}{\partial x_1 \partial x_2}\rangle_{L_2(\Omega)}$$

$$+ \langle \frac{\partial^2 f}{\partial x_1^2}, \frac{\partial^2 g}{\partial x_1^2}\rangle_{L_2(\Omega)}$$

$$+ \langle \frac{\partial^2 f}{\partial x_2^2}, \frac{\partial^2 g}{\partial x_2^2}\rangle_{L_2(\Omega)}$$

$$= \int_a^b \int_c^d fg\, dx_1\, dx_2 + \int_a^b \int_c^d \frac{\partial f}{\partial x_1}\frac{\partial g}{\partial x_1} dx_1 dx_2$$

$$+ \int_a^b \int_c^d \frac{\partial f}{\partial x_2}\frac{\partial g}{\partial x_2} dx_1 dx_2 + \int_a^b \int_c^d \frac{\partial^2 f}{\partial x_1^2}\frac{\partial^2 g}{\partial x_1^2} dx_1 dx_2$$

$$+ \int_g^b \int_a^b \frac{\partial^2 f}{\partial x_2^2}\frac{\partial^2 g}{\partial x_2^2} dx + \int_a^b \int_e^d \frac{\partial^2 f}{\partial x_1 \partial x_2}\frac{\partial^2 g}{\partial x_1\, \partial x_2} dx_1 dx_2$$

$$H^m(\Gamma) = \{f \in L_2(\Gamma)/D^\alpha f \in L_2(\Gamma), |\alpha| \le m\},$$

$H^m(\Gamma)$ is Hilbert space with respect to

$$\langle f,g\rangle_{H^m(\Gamma)} = \sum_{|\alpha|\le m} \langle D^\alpha f, D^\alpha g\rangle_{L_2(\Gamma)}$$

$$\langle f,g\rangle_{H^1(\Gamma)} = \int_\Gamma fg dx + \int_\Gamma \frac{df}{dx}\frac{dg}{dx} dx$$

$$\langle f,g\rangle_{H^2(\Gamma)} = \sum_{|\alpha|\le 2} \langle D^\alpha f, D^\alpha g\rangle_{L_2(\Gamma)}$$

$$= \int_\Gamma fg dS + \int_\Gamma \frac{\partial f}{\partial x_1}\frac{\partial g}{\partial x_2} dS$$

$$+ \int_\Gamma \frac{\partial f}{\partial x_2}\frac{\partial g}{\partial x_2} dS + \int_\Gamma \frac{\partial^2 f}{\partial x_1^2}\frac{\partial^2 g}{\partial x_1^2} dS$$

$$+ \int_\Gamma \frac{\partial^2 f}{\partial x_2^2}\frac{\partial^2 g}{\partial x_2^2} dS$$

$$+ \int_\Gamma \frac{\partial^2 f}{\partial x_1 \partial x_2}\frac{\partial^2 g}{\partial x_1\, \partial x_2} dS.$$

The dual space of $H_0^m(\Omega)$, that is, the space of all bounded linear functionals on

$H_0^m(\Omega)$ is denoted by $H^{-m}(\Omega)$.

The restriction of a function $f \in H^m(\Omega)$ to the boundary Γ is called the trace of f and is denoted by Γ_f, that is, $\Gamma_f = f(S) =$ value of f at Γ.

$$\begin{aligned} H_0^m(\Omega) &= \{f \in H^m(\Omega)/\Gamma_f = 0\} = \text{ closure of } D(\Omega) \text{ in } H^m(\Omega) \\ H^m(R^n) &= H_0^m(R^n), m \geq 0 . \end{aligned}$$

$f \in H^{-m}(\Omega)$ if and only if $f = \sum_{|\alpha| \leq m} D^\alpha g$ for some $g \in L_2(\Omega)$ [Kardestuncer J., 1987].

Sobolev space with a real index. $H^s(\Omega), s \in R_+$. For $0 < \sigma < 1$, we define $H^\sigma(\Omega)$ to be the closure of $C^\infty(\Omega)$ with respect to the norm

$$\|u\|_s^2 = \|u\|_{[2]}^2 + \sum_{|\alpha|=[s]} \|D^\alpha u\|_\theta^2,$$

where

$$\|u\|_\theta^2 = \int_\Omega \int_\Omega \frac{|u(t) - u(\tau)|^2}{\|t - \tau\|^{n+2\theta}} dt d\tau,$$

and n is the dimension of the domain Ω.

For $s > 0$ not integer, write $s = k + \sigma$ where $k = [s]$, that is k is integer ($k = [s]$ and $\sigma = s - k$) and define

$$H^s(\Omega) = \{f \in H^\sigma(\Omega)/D^\alpha f \in H^\sigma(\Omega) \text{ for } |\alpha| \leq 2\},$$

where $H^s(\Omega)$ is a Hilbert space with respect to

$$\begin{aligned} \langle f, g \rangle_{H^s(\Omega)} &= \sum_{|\alpha| \leq 2} \langle D^\alpha f, D^\alpha g \rangle_{L_2(\Omega)} \text{ if } s = k \\ &= \sum_{|\alpha| \leq 2} \Bigg\{ \langle D^\alpha f, D^\alpha g \rangle_{L_2(\Omega)} \\ &\quad + \int_\Omega \int_\Omega \frac{(D^\alpha f(x) - D^\alpha f(y))(D^\alpha g(x) - D^\alpha g(y))}{\|x - y\|n + 2\sigma} dx dy \Bigg\} . \end{aligned}$$

Note: $H_p^m(\Omega), H_p^s(\Omega), H_p^{-m}(\Omega), \frac{1}{p} + \frac{1}{q} = 1, 1 \leq p < \infty$ can be defined replacing $L_2(\Omega)$ by $L_p(\Omega)$.

The dual of $H_0^s(\Omega)$ is denoted by $H^{-s}(\Omega)$. $H^s(\Gamma), H^{-s}(\Gamma)$ can also be defined keeping in mind the definition of $H^m(\Gamma)$ and $H^s(\Gamma)$, see Dautry and Lions [1990, Chapter III] for more details about $H^s(\Gamma)$. $H^s(\Gamma)$ is also a Hilbert space for $s \in R$.

It can be observed that every element of $H^1(a, b)$ is absolutely continuous and hence continuous.

Some important inequalities.

Ostrogradski's formula. For a field $p = (p_1, p_2, \cdots, p_n)$,

$$\int_\Omega \text{div p } dx = \int_\Gamma p.ndS, S \in \Gamma\ .$$

Green's formula for integration by parts.

(i) $\int_\Omega v\Delta u dx + \int_\Omega \text{grad } u \text{grad } v dx = \int_\Gamma v\frac{\partial u}{\partial n}dS$

(ii) $\int_\Omega (u\Delta v - v\Delta u)dx = \int_\Gamma \left(u\frac{\partial v}{\partial n} - v\frac{\partial u}{\partial n}\right)d\gamma.$

It is clear that (i) is a generalization of the integration by parts formula stated below (for $n = 1, \Omega = [a, b]$)

$$\int_a^b u''(x)v(x)dx = u'(b)v(b) - u'(a)v(a) - \int_a^b u'(x)v'(x)dx.$$

The Friedrichs inequality. Let Ω be a bounded domain of R^n with a Lipschitz boundary. Then there exists a constant $k_1 > 0$, depending on the given domain, such that for every $f \in H^1(\Omega)$,

$$||f||^2_{H^1(\Omega)} \le k_1 \left\{ \sum_{j=1}^n \int_\Omega \left(\frac{\partial f}{\partial x_j}\right)^2 dx + \int_\Gamma f^2(S)dS \right\}.$$

The Poincaré inequality. Let Ω be a bounded domain of R^n with a Lipschitz boundary. Then there exists a constant, say, k_2 depending on Ω such that for every $f \in H^1(\Omega)$

$$||f||^2_{H^1(\Omega)} \le k_2 \left\{ \int_\Omega \left(\frac{\partial f}{\partial x_j}\right) dx + \left(\int_\Omega f dx\right)^2 \right\}.$$

The above two inequalities hold for elements of $L_2(\Omega)$. For $f \in H^1(\Omega)$ such that $\int_\Omega f(x)dx = 0$

$$||f||^2 L_2(\Omega) \le k_1 \sum_{j=1}^n || \frac{\partial f}{\partial x_j} ||^2_{L_2(\Omega)}, k_1 > 0 \quad \text{constant depending on } \Omega\ .$$

Let

$$\text{div } v = \sum_{i=1}^n \frac{\partial v_i}{\partial x_i} \quad \text{for all } v = (v_1, v_2, \cdots v_n) \in (\mathcal{D}'(\Omega))^n = D'(\Omega)^n).$$

Thus divergence, written in brief div., is an operator on $D'(\Omega)$ into $\mathcal{D}'(\Omega)^n$, usually from application point of view $n = 2$ or 3.

For case $n = 3$ and $v = (v_1, v_2, v_3) \in \mathcal{D}'(\Omega)^3$, we put

$$\text{curl } v = \left(\frac{\partial v_3}{\partial x_2} - \frac{\partial v_2}{\partial x_3}, \frac{\partial v_1}{\partial x_3} - \frac{\partial v_3}{\partial x_1}, \frac{\partial v_2}{\partial x_1} - \frac{\partial v_1}{\partial x_2}\right),$$

which defines the differential operator, denoted by curl in $\mathcal{D}'(\Omega)^3$. Sometimes, it is also denoted by $\nabla\wedge$ or "rot" (rotation).

In case $n = 2$, we put

$$\text{curl } v = \frac{\partial v_2}{\partial x_1} - \frac{\partial v_1}{\partial x_2} \text{ for all } v = (v_1, v_2) \in D(\Omega)^2,$$

the curl operator defined on $D(\Omega)^2$ into $D(\Omega)$. A curl or $\overline{\text{curl}}$ on $D'(\Omega)^2$ into itself can be defined as follows:

$$\overline{\text{curl}}\ v = \text{curl } v \ = \left(\frac{\partial v}{\partial x_1} - \frac{\partial v}{\partial x_2}\right).$$

For $n = 3$, the diagram

$$D'(\Omega) \overset{\text{grad}}{\longrightarrow} D'(\Omega)^3 \overset{\text{curl}}{\longrightarrow} D'(\Omega)^3 \overset{\text{div}}{\longrightarrow} D'(\Omega)$$
$$D(\Omega) \overset{\text{grad}}{\longrightarrow} D(\Omega)^3 \overset{\text{curl}}{\longrightarrow} D(\Omega)^3 \overset{\text{div}}{\longrightarrow} D(\Omega)$$

is such that curl grad $v \ = \ 0$ for all $v \in \mathcal{D}'(\Omega)^3$.

The following relations are satisfied:

(i) Im grad $\subset$ ker curl.

(ii) Im curl $\subset$ ker div.

(iii) $\langle \text{div } v, \phi\rangle = \langle v, -\text{grad } \varphi\rangle$ for all $v \in \mathcal{D}(\Omega)^3, \phi \in \mathcal{D}(\Omega)$, that is, div is the transpose of $-$grad.

(iv) $\langle \text{curl} v, \phi\rangle = \langle v, \text{curl}\phi\rangle$ for all $v \in \mathcal{D}'(\Omega)^3, \phi \in \mathcal{D}(\Omega^3)$.

For $n = 2$, we have the diagrams

$$D'(\Omega) \overset{\text{grade}}{\longrightarrow} D'(\Omega)^2 \overset{\text{curl}}{\longrightarrow} \mathcal{D}'(\Omega)$$
$$D'(\Omega) \overset{\text{curl}}{\longrightarrow} \mathcal{D}'(\Omega)^2 \overset{\text{div}}{\rightarrow} \mathcal{D}'(\Omega),$$

such that

$$\text{curl grad } v = 0 \text{ for } v \in D'(\Omega)$$
$$\text{div curl } v = 0 \text{ for all } v \in \mathcal{D}'(\Omega)^2.$$

Thus

$$\begin{aligned} &\mathrm{Im\,grad} \subset \ker \mathrm{curl} \\ &\mathrm{Im\,curl} \subset \ker \mathrm{div} \\ &\langle \mathrm{curl}\ v, \phi\rangle = \langle v,\ \mathrm{curl}\ \phi\rangle, \text{for all } v \in D'(\Omega)^2, \phi \in \mathcal{D}(\Omega). \end{aligned}$$

Let Ω' be an open set in R^3, with bounded complement $\overline{\Omega}$.

Let us put

$$B'(\Omega') = \left\{ u \in \mathcal{D}'(\Omega) | \frac{\partial u}{\partial x_i} \in L_2(\Omega') \right\}.$$

$W'(\Omega')$ = closure of $\mathcal{D}(\overline{\Omega'})$ under the semi norm, $\|\mathrm{grad}\,\|_{L_2(\Omega')^n}$. $W'(\Omega')$ is called the Beppolevi space.

7.5 Numerical solutions of linear systems

Here, we present iterative methods required to solve a very large and sparse linear system $Ax = b$. These include the following:

1. The Jacobi method.
2. The Gauss-Seidel method.
3. The successive overrelaxation (SOR) method.

Another important iterative method to solve such system, known as the conjugate gradient method, is discussed in Chapter 2.

E1. Fundamental idea.

The basic idea behind an iterative method is first to write the system $Ax = b$ in an equivalent form

$$x = Bx + d, \tag{E1}$$

then, starting with an initial approximation $x^{(1)}$ of the solution vector x, generate a sequence of approximations $\{x^{(k)}\}$ iteratively defined by

$$x^{(k+1)} = Bx^{(k)} + d, \quad k = 1, 2, \ldots \tag{E2}$$

with the hope that under certain mild conditions, the sequence $\{x^{(k)}\}$ converges to the solution as $k \to \infty$.

To solve the linear system $Ax = b$ iteratively using the idea, we therefore need to know

1. How to write the system $Ax = b$ in the form of (E1).
2. How $x^{(1)}$ should be chosen so that iteration (E2) converges to the limit with any arbitrary choice of $x^{(1)}$.

E2. Stopping criteria for iteration (2).

It is natural to wonder when iteration (E2) can be terminated! When convergence occurs, $x^{(k+1)}$ is a better approximation than $x^{(k)}$, so a natural stopping criterion is as follows:

Stopping Criterion 1

Stop iteration (E2) if

$$\frac{\| x^{(k+1)} - x^{(k)} \|}{\| x^{(k)} \|} < \epsilon$$

for a prescribed small positive number ϵ (ϵ should be chosen according to the accuracy desired) or if the number of iterations exceeds a predetermined number.

A stopping criterion in terms of the residual is given in the following:

Stoping Criterion 2

Stop iteration (E2) if

$$\| r^{(k+1)} \| \leq \epsilon(\| A \| \cdot \| x^{(k+1)} \| + \| b \|)$$

where

$$r^{(k+1)} = Ax^{(k+1)} - b$$

or, if the number of iterations exceeds a predetermined number. ϵ is the tolerance to be chosen such that $\mu < \epsilon < 1$, μ is the machine precision.

E3. The Jacobi method.

The system

$$Ax = b$$

or

$$\begin{aligned}
a_{11}x_1 + a_{12}x_2 + \cdots + a_{1n}x_n &= b_1 \\
a_{21}x_1 + a_{22}x_2 + \cdots + a_{2n}x_n &= b_2 \\
&\cdots \\
a_{n1}x_1 + a_{n2}x_2 + \cdots + a_{nn}x_n &= b_n
\end{aligned}$$

can be written (under the assumption that $a_{ii} \neq 0,\ i = 1, \ldots, n$) as

$$\begin{aligned} x_1 &= \frac{1}{a_{11}}(b_1 - a_{12}x_2 - \cdots - a_{1n}x_n) \\ x_2 &= \frac{1}{a_{22}}(b_2 - a_{21}x_1 - \cdots - a_{2n}x_n) \\ &\vdots \\ x_n &= \frac{1}{a_{nn}}(b_n - a_{n1}x_1 - \cdots - a_{n,n-1}x_{n-1}). \end{aligned}$$

In matrix notation,

$$\begin{pmatrix} x_1 \\ x_2 \\ \vdots \\ x_n \end{pmatrix} = \begin{pmatrix} 0 & -\frac{a_{12}}{a_{11}} & \cdots & \cdots & -\frac{a_{1n}}{a_{11}} \\ -\frac{a_{21}}{a_{22}} & 0 & -\frac{a_{23}}{a_{22}} & \cdots & -\frac{a_{2n}}{a_{22}} \\ \vdots & \ddots & \ddots & \ddots & \vdots \\ & & \ddots & \ddots & -\frac{a_{n-1,n}}{a_{n-1,n-1}} \\ -\frac{a_{n1}}{a_{n,n}} & \cdots & \cdots & -\frac{a_{n,n-1}}{a_{n,n}} & 0 \end{pmatrix} \begin{pmatrix} x_1 \\ x_2 \\ \vdots \\ x_n \end{pmatrix} + \begin{pmatrix} \frac{b_1}{a_{11}} \\ \frac{b_2}{a_{22}} \\ \vdots \\ \frac{b_n}{a_{nn}} \end{pmatrix},$$

or

$$x = Bx + d.$$

If we write the matrix A in the form

$$A = L + D + U,$$

where

$$L = \begin{pmatrix} 0 & 0 & 0 & \cdots & 0 \\ a_{21} & 0 & 0 & \cdots & 0 \\ \vdots & \ddots & \ddots & & \vdots \\ \vdots & & \ddots & \ddots & 0 \\ a_{n1} & \cdots & \cdots & a_{nn-1} & 0 \end{pmatrix},$$

$D = \text{diag}(a_{11}, \ldots, a_{nn})$, and

$$U = \begin{pmatrix} 0 & a_{12} & a_{13} & \cdots & a_{1n} \\ 0 & 0 & a_{23} & \cdots & a_{2n} \\ \vdots & & \ddots & \ddots & \vdots \\ \vdots & & & \ddots & a_{n-1n} \\ 0 & \cdots & \cdots & \cdots & 0 \end{pmatrix},$$

then it is easy to see that

$$B = -D^{-1}(L+U) = (I - D^{-1}A)$$
$$d = D^{-1}b.$$

(Note that, because of our assumption that $a_{ii} \neq 0,\ i = 1, \ldots, n,\ D$ is nonsingular.)

We shall call the matrix

$$B = -D^{-1}(L+U) = (I - D^{-1}A)$$

the **Jacobi iteration matrix** and denote it by B_j. Similarly, we shall denote the vector $D^{-1}b$ by b_j, which we call **the Jacobi vector**

The Jacobi Iteration Matrix and the Jacobi Vector

Let $A = L + D + U$. Then
$B_j = -D^{-1}(L+U)$
$b_j = D^{-1}b$

With the Jacobi iteration matrix and the Jacobi vector as defined, iteration (E2) becomes the following.

The Jacobi iteration

$$x_i^{(k+1)} = \frac{1}{a_{ii}}\left(b_i - \sum_{j=1}^{n} a_{ij}x_j^{(k)}\right), \quad j \neq i,\ i = 1, \ldots, n.$$

We thus have the following iterative procedure, called the **Jacobi method.**

ALGORITHM 1. The Jacobi method.

Step 1: Choose $x^1 = \begin{pmatrix} x_1^{(1)} \\ x_2^{(1)} \\ \vdots \\ x_n^{(1)} \end{pmatrix}$.

Step 2: For $k = 1, 2, \ldots,$ do until a stopping criterion is satisfied.

$$x^{(k+1)} = B_j x^{(k)} + b_j \tag{E3}$$

or

$$x_i^{(k+1)} = \frac{1}{a_{ii}}\left(b_i - \sum_{j=1}^{n} a_{ij}x_j^{(k)}\right), \quad j \neq i,\ i = 1, \ldots, n. \tag{E4}$$

Note: In practical implementation, we will use Equation (E4).

E4. The Gauss-Seidel method.

In the Jacobi method, to compute the components of the vector $x^{(k+1)}$, only the components of the vector $x^{(k)}$ are used; however, note that to compute $x_i^{(k+1)}$, we could have used $x_1^{(k+1)}$ through $x_{i-1}^{(k+1)}$, which were already available to us. Thus, a natural modification of the Jacobi method will be to rewrite the Jacobi iteration (E4) in the following form:

The Gauss-Seidel Iteration

$$\left. \begin{array}{l} x_i^{(k+1)} = \frac{1}{a_{ii}} \left(b_i - \sum_{j=1}^{i-1} a_{ij} x_j^{(k+1)} - \sum_{j=i+1}^{n} a_{ij} x_j^{(k)} \right), \\ i = 1, 2, \ldots, n \end{array} \right\} . \tag{E5}$$

The idea is to use each new component, as soon as it is available, in the computation of the next component.

The iteration (E5) is known as the **Gauss-Seidel iteration**, and the iterative method based on this iteration is called the **Gauss-Seidel method.**

In the notation used earlier, the Gauss-Seidel iteration is

$$x^{(k+1)} = -(D+L)^{-1}Ux^{(k)} + (D+L)^{-1}b.$$

(Note that the matrix $D + L$ is a lower triangular matrix with $a_{11}, \ldots, a_{nn}$ on the diagonal; because we have assumed that these entries are nonzero, the matrix $(D+L)$ is nonsingular).

We call the matrix

$$B = -(D+L)^{-1}U$$

the **Gauss-Seidel Matrix** and denote it by B_{GS}. Similarly, the Gauss-Seidel vector $(D+L)^{-1}b$ is denoted by b_{GS}.

The Gauss-Seidel matrix and Gauss-Seidel vector

Let $A = L + D + U$. Then

$$\begin{aligned} B_{GS} &= -(D+L)^{-1}U \\ b_{GS} &= (D+L)^{-1}b. \end{aligned}$$

ALGORITHM 2. The Gauss-Seidel method.

Step 1: Choose an initial approximation $x^{(1)}$.

Step 2: For $k = 1, 2, \ldots,$ do until a stopping criterion is satisfied.

$$x^{(k+1)} = B_{GS}x^{(k)} + b_{GS}, \tag{E6}$$

or

$$x_i^{(k+1)} = \frac{1}{a_{ii}}\left(b_i - \sum_{j=1}^{i-1} a_{ij}x_j^{(k+1)} - \sum_{j=i+1}^{n} a_{ij}x_j^{(k)}\right), \quad i = 1, 2, \ldots, n. \tag{E7}$$

On actual computer implementations, it is certainly economical to use equations (E4) and (E7) rather than (E3) and (E6).

It is often hard to make a good guess of the initial approximation $x^{(1)}$. Thus, it will be nice to have conditions that will guarantee the convergence of iteration (E2) for any arbitrary choice of the initial approximation. We have the following results in this direction.

Theorem E1. (Iteration Convergence Theorem) *The iteration*

$$x^{(k+1)} = Bx^{(k)} + c$$

converges to a limit with an arbitrary choice of the initial approximation $x^{(1)}$ if and only if the matrix $B^{(k)} \to 0$ as $k \to \infty$; that is, B is a convergent matrix.

Conditions for convergence of iteration

A necessary and sufficient condition for the convergence of the iteration (E 2), for any arbitrary choice of $x^{(1)}$, is that $\rho(B) < 1$. A sufficient condition is that $\| B \| < 1$, for some norm. Here $\rho(B)$ denotes the spectral radius of B.

This result can be applied to identify classes of matrices for which the Jacobian and/or Gauss-Seidel methods converge for any choice of initial approximation $x^{(1)}$.

The Jacobi and Gauss-Seidel methods for diagonally dominant matrices.

1. If A is row-diagonally dominant, then the Jacobi method converges for any arbitrary choice of the initial approximation $x^{(1)}$.

2. If A is row-diagonally matrix, then the Gauss-Seidel method converges for any arbitrary choice of $x^{(1)}$.

The Gauss-Seidel method for a symmetric positive definite matrix. The following theorem tells us that the Gauss-Seidel method converges, with an arbitrary choice of $x^{(1)}$, for a symmetric positive definite matrix.

Theorem E2. *Let A be a symmetric positive definite matrix. Then the Gauss-Seidel method converges for any arbitrary choice of the initial approximation of $x^{(1)}$.*

Rates of convergence and comparison between the Gauss-Seidel and the Jacobi methods. We have seen that for row-diagonally dominant matrices, both the Jacobi and the Gauss-Seidel methods converge for an arbitrary $x^{(1)}$. The question naturally arises whether this is true for some other matrices as well. Also, when both methods converge, another question arises: which one converges faster?

If A is a matrix having diagonal entries positive and off-diagonal entries negative, then

1. Either one or both the Jacobi and the Gauss-Seidel methods converge or diverge.

2. When both the methods converge, the Gauss-Seidel method converges faster than the Jacobi method.

E5. The successive overrelaxation (SOR) method.

The Gauss-Seidel method is frustratingly slow when $\rho(B_{\text{GS}})$ is close to unity. However, the rate of convergence of the Gauss-Seidel iteration can, in certain cases, be improved by introducing a parameter ω, known as the relaxation parameter. The following modified Gauss-Seidel iteration is known as the successive overrelaxation iteration or, in short, the SOR iteration, if $\omega > 1$.

The SOR Iteration

$$\left.\begin{array}{l} x_i^{(k+1)} = \dfrac{\omega}{a_{ii}}\left(b_i - \sum_{j=1}^{i-1} a_{ij}x_j^{(k+1)} - \sum_{j=i+1}^{n} a_{ij}x_j^{(k)}\right) \\ \qquad +(1-\omega)x_i^k,\ i = 1, 2, \ldots \end{array}\right\}. \tag{E8}$$

From (E8), we note the following:

1. When $\omega = 1$, the SOR iteration reduces to the Gauss-Seidel iteration.

2. When $\omega > 1$, in computing the $(k+1)$ approximation, more weight is placed on the most current value than when $\omega < 1$, with the hope that the convergence will be faster.

In matrix notation, the SOR iteration is

$$x^{(k+1)} = (D+\omega L)^{-1}[(1-\omega)D - \omega U]x^{(k)} + \omega(D+\omega L)^{-1}b, \quad k = 1, 2, \ldots \tag{E9}$$

(Note that because $a_{ii} \neq 0, i = 1, \ldots, n$, the matrix $(D + \omega L)$ is nonsingular.) The matrix $(D+\omega L)^{-1}[(1-\omega)D-\omega U]$ is called the SOR matrix and is denoted by B_{SOR}. Similarly, the vector $(D+\omega L)^{-1}b$ is denoted by b_{SOR}.

The SOR Matrix and the SOR Vector

$$B_{SOR} = (D+\omega L)^{-1}[(1-\omega)D - \omega U]$$
$$b_{SOR} = \omega(D+\omega L)^{-1}b.$$

In matrix notation, the SOR iteration algorithm is as follows:

ALGORITHM 3. The successive overrelaxation method:
Step 1: Choose $x^{(1)}$.
Step 2: For $k = 1, 2, \ldots$ do until a stopping criterion is satisfied

$$x^{(k+1)} = B_{\text{SOR}} x^k + b_{\text{SOR}}$$

or compute $x^{(k+1)}$ using Equation (E8).

Choice of ω in the convergent SOR iteration. It is natural to wonder about the range of ω for which the SOR iteration converges and is the optimal choice of ω. To this end, we mention here the following important results due to William Kahan and Ostrowski-Reich:

THEOREM: Kahan. The SOR iteration cannot converge for an arbitrary initial approximation if ω lies outside the interval (0,2).

THEOREM: Ostrowski-Reich. Let A be a symmetric positive definite matrix and let $0 < \omega < 2$. Then the SOR method will converge for any arbitrary choice of $x^{(1)}$.

For further details of these methods and proofs of theorems, we refer to Datta [1995]. MATLAB can be applied to implement these algorithms. The MATLAB toolkit MATCOM may enable us to study different algorithms for the same problem, comparing efficiency, stability, and accuracy.

7.6 Black-Scholes world of option pricing

F1. Introduction.

This section is based on Wilmott et al., [1995], Wilmot [1998], Siddiqi, Manchanda, Kocvara [1999], Sircar and Papanicolaou [1998]; see the references in Scholes [1997] and Merton [1997] for more details. Option pricing theory is a relatively new discipline within the economics and banking industry which encompasses mathematical finance theory and finance practice. This special field of finance is the study of allocation and deployment of economic resources, both spatially and across time in an uncertain environment. Sophisticated mathematical models and numerical methods described in Chapters 3 and 4 have been included to capture the influence and interaction of time and uncertainty. One of the fastest algorithms for a nonlinear problem in this area will be presented in this appendix along with some basic results.

In October 1997, Myron Scholes and Robert Merton were awarded the Nobel Prize for Economics on their work, which can be expressed in the form of a linear parabolic partial differential equation (diffusion equation) that enables investors to price accurtely their bets on the future. More precisely, it allows investors to calculate the fair price of a derivative security whose value depends on the value of another security, known as the underlying, based on a small set of assumptions on the price behaviour of that underlying. Fisher Black had died in 1995. Black-Scholes pricing is considered as one of the modern financial theory's biggest successes in terms of both approach and applicability. Before publication of their work in 1973, pricing of a derivative was rather a mysterious task owing to their often complex dependencies on the underlying, and they were traded mainly over the counter rather than in large markets, generally with high transaction costs. Opening of the Chicago Board of Trade nearly coincided with the publication of the Black-Scholes model and since then trading in derivatives has become a regular feature. The Chicago Board Options Exchange (CBOE) first created standardized, listed options. Initially, there were just options call on 16 stocks. Put options were introduced in 1977. In the US options are traded on CBOE options, the American Stock Exchange, the Pacific Stock Exchange and the Philadelphia Stock Exchange. Worldwide, there are more than 50 exchanges on which options are traded. Extensive use of the Black-Scholes model has significantly influenced the market. This feedback effect has been incorporated in the model by Sircar and Papanicolaou [1998] where we get a nonlinear parabolic partial differential equation. Some generalized forms of the Black-Scholes model are discussed by Wilmott [1998]. Basic theory of derivatives comprising production and markets, derivatives, the random behaviour of assets, elementary stochastic calculus, the Black-Scholes equation and its generalizations and their sensitivity analysis, American options, and multi-asset options is known as the *Black-Scholes World of Option Pricing*. Options are certain kinds of contracts, many of which have been named as European, American, Asian and Russian but they have nothing to do with the context of the origin; rather they refer to a technicality in the option contract. For general financial news, one may visit www.bloomberg.com, www.cnnfn.com, ww.wsj.com, www.ft.com, www.fow.com, www.cbot.com. For information about training programms, one may visit www.wilmott.com.

F2. Basic concepts related to European and American options.

Before introducing European and American options, we briefly mention the commonly used terms like Asset or Underlying Asset, Equity, Derivative, Equity Derivative, Expiry, Option Pricing, Call Option, Put Option, Strike Price (Exercise Price), Risk Management, Volatility.

By **underlying asset**, often called only **underlying** or **asset**, we mean commodity, exchange, shares, stocks and bonds, etc. **Equity** is a share in the ownership of a company which usually guarantees the right to vote at meetings and a share in the dividends (payment to shareholders as return for investment in the corporation). **Derivative** refers to either a contract or a security whose pay-off or final value is dependent on one or more features of the underlying equity. In many cases it is the price of the underlying equity which determines to a large extent the value of the equity derivative or derivative based on equity, although other factors like interest rates, time to maturity and strike price can also play a significant role. The termination time of a derivatives contract, usually when the final pay-off value is calculated and paid, is called **expiry**. **Option pricing** or **options** are some kind of contracts: cts: The right to the holder (owner) and an obligation to the seller (writer) of a contract either to buy or to sell an underlying asset at a fixed price for a premium. In **call options** the holder has the right but not the obligation to buy the underlying asset at the strike price. Options in which the right to sell for the holder and the obligation to buy for the writer at a strike price E for the payment of a premium is guaranteed, are called **put options**. **Strike price** or Exercise price is the price at which the underlying asset is bought in options. **Risk management** is the process of establishing the type and magnitude of risk in a business enterprise and using derivatives to control and shape that risk to maximize the business objective. **Volatility** is a measure of the standard deviations of returns. In practice it is understood as the average daily range of the last few weeks or average absolute value of the daily net change of the last few weeks. Dividends are lump sum payments disbursed every quarter or every six months to the holder of the stock. **Commodities** are usually raw products such as precious metals, oil, food products, etc. Prices of these products may depend on seasonal fluctuations, scarcity of the product and are difficult to predict. Most trading is done on the futures market, making deal to buy or sell the commodity sometime in the future. The **exchange rate** is the rate at which one currency can be exchanged for another. This is the world of foreign exchange (FX). Although fluctuations in exchange rates are unpredictable, there is a link between exchange rates and the interest rate of two countries.

F3. Modelling of European options.

A **European call option** is a contract with the following conditions: At a prescribed time in the future, known as the expiry date, the owner of the option may purchase a prescribed asset, called underlying asset, for a prescribed amount (strike price or exercise price). Similarly, a **European put option** is a contract in which at a prescribed time in the future the owner (holder) of the option may sell an asset for a prescribed amount.

Let $V(S,t)$ denote the value of an option which is the function of the underlying asset S ant time t. Black and Scholes [1973] proved that V is a solution of the parabolic partial differential equation:

$$\frac{\partial V}{\partial t} + \frac{1}{2}\sigma^2 S^2 \frac{\partial^2 V}{\partial S^2} + rS\frac{\partial V}{\partial S} - rV = 0, \tag{F1}$$

where σ and r are volatility and interest rate, respectively.

Let $C(S,t)$ and $P(S,t)$ denote the value of $V(S,t)$, respectively, when it is a call option and put option. It has been shown (see, for instance, Wilmott, et al. [1993] that a European call option $C(S,t)$ is a solution of the following boundary value problem:

$$\frac{\partial C}{\partial t} + \frac{1}{2}\sigma^2 S^2 \frac{\partial^2 C}{\partial S^2} + rS\frac{\partial C}{\partial S} - rC = 0 \tag{F2}$$

$$C(S,T) = \max(S-E,0) \tag{F3}$$

$$C(0,t) = 0 \tag{F4}$$

$$C(S,t) \to S \text{ as } S \to \infty \tag{F5}$$

where S, σ, r are as above, and E and T are the exercise price and expiry time, respectively.

On the other hand, a European put option $P(S,t)$ is a solution of the following boundary value problem:

$$\frac{\partial P}{\partial t} + \frac{1}{2}\sigma^2 S^2 \frac{\partial^2 P}{\partial S^2} + rS\frac{\partial P}{\partial S} - rP = 0 \tag{F6}$$

$$P(S,T) = \max(E - S, 0) \tag{F7}$$

$$P(0,t) = Ee^{-r(T-t)} \tag{F8}$$

if r is independent of time

$$P(0,t) = Ee^{-\int_t^T r(\tau)d\tau}$$

if r is time dependent.

As $S \to \infty$, the option is unlikely to be exercised and so

$$P(S,t) \to 0 \text{ as } S \to \infty. \tag{F9}$$

Equations (F2)–(F5) and (F6)–(F9) are known as **Black-Scholes model** for call and put options, respectively.

The Black-Scholes call option model can be transformed into the diffusion equation:

$$\frac{\partial u}{\partial \tau} = \frac{\partial^2 u}{\partial x^2} \quad \text{for } -\infty < x < \infty,\ \tau > 0 \tag{F10}$$

with

$$u(x,0) = \max(e^{\frac{1}{2}(k+1)x} - e^{\frac{1}{2}(k-1)x}, 0), \tag{F11}$$

by putting $S = Ee^x, t = T - \tau/\frac{1}{2}\sigma^2$ and

$$C(S,t) = Ee^{-\frac{1}{2}(k-1)x - \frac{1}{4}(k+1)^2\tau}u(x,\tau),$$

where $k = \frac{r}{\frac{1}{2}\sigma^2}$.

The Black-Scholes put option model can analogously be written in the form of the diffusion equation.

If it is assumed that the asseet receives a continuous and constant dividend D, then the Black-Scholes equation takes the form

$$\frac{\partial V}{\partial t} + \frac{1}{2}\sigma^2 S^2 \frac{\partial^2 V}{\partial S^2} + (r - D)S\frac{\partial V}{\partial S} - rV = 0. \tag{F12}$$

Currency options are modelled by the following modified form of the **Black-Scholes equation**

$$\frac{\partial V}{\partial t} + \frac{1}{2}\sigma^2 S^2 \frac{\partial^2 V}{\partial S^2} + (r - rf)S\frac{\partial V}{\partial S} - rV = 0, \tag{F13}$$

where rf denotes interest at the foreign rate of interest. If we want to model commodity options, then the relevant feature of commodities, requiring that they have a cost of carry, must be taken into account in the Black-Sholes equation. This is just like having a negative dividend and so we get

$$\frac{\partial V}{\partial t} + \frac{1}{2}\sigma^2 S^2 \frac{\partial^2 V}{\partial S^2} + (r+q)S\frac{\partial V}{\partial S} - rV = 0, \tag{F14}$$

where $q = -D$ denotes the fraction of the value of a commodity that goes towards paying the cost of carry. For details about the models mentioned above see Wilmott [1998]. The following feedback pricing model for European call option has been studied by Sircar and Papanicolaou [1998]:

$$\frac{\partial C}{\partial t} + \frac{1}{2}\left[\frac{\left(1-\rho\frac{\partial C}{\partial S}\right)}{1-\rho\frac{\partial C}{\partial S} - \rho S\frac{\partial^2 C}{\partial S^2}}\right]^2 \sigma^2 S^2 \frac{\partial^2 C}{\partial S^2} + r\left(S\frac{\partial C}{\partial S} - C\right) = 0,$$

$$t < T - \varepsilon \tag{F15}$$

$$C(S, T-\varepsilon) = C_{BS}(S, T-\varepsilon) \tag{F16}$$

$$C(0,t) = 0 \tag{F17}$$

$$\lim_{S\to\infty} \left|C(S,t) - (x - Ke^{-r(T-t)})\right| = 0 \tag{F18}$$

with

$$C(S,t) = C_{BS}(S,t) \tag{F19}$$

for $T - \varepsilon \leq t \leq T$, where C_{BS} denotes the Black-Scholes price and the last condition means that C is equal to C_{BS} in some small interval of time.

The generalized Black-Scholes model incorporating feedback is the following nonlinear parabolic equation

$$\frac{\partial C}{\partial t} + \frac{1}{2}\left[\frac{\left(1-\rho\frac{\partial C}{\partial S}\right)}{1-\rho\frac{\partial C}{\partial S} - \rho S\frac{\partial^2 C}{\partial S^2}}\right]^2 \sigma^2 S^2 \frac{\partial^2 C}{\partial S^2} + r\left(S\frac{\partial C}{\partial S} - C\right) = 0. \tag{F20}$$

F4. Modelling of American options.

American options are those options which can be exercised by any time prior to expiry time. American call and put options are related to buying and selling, respectively. The valuation of American options leads to a free boundary problems. Typically, at each time t, there is a valuation of S which marks the boundary between two regions; namely, to one side one should hold the option and to the other side one should exercise it. Let us denote this boundary by $S_f(t)$ (generally, this critical asset value varies with time). Since we do not know $S_f(t)$ apriori, we are lacking one piece of information compared with the corresponding European option problem. Thus with American options we do not know apriori where to apply boundary conditions. This situation resembles the obstacle problem and can be effectively tackled by methods of variational inequalities (see Chapters 2 and 3). Wilmott, Dewynne and Howison [1993] have shown that American call option and American put option can be formulated as the following boundary value problems and equivalent variational inequalities.

American call option is modelled by the following boundary value problem

$$\frac{\partial u}{\partial \tau} - \frac{\partial^2 u}{\partial x^2} \geq 0 \qquad \text{(F21)}$$

$$u(x,\tau) - g(x,\tau) \geq 0 \qquad \text{(F22)}$$

$$\left(\frac{\partial u}{\partial \tau} - \frac{\partial^2 u}{\partial x^2}\right) \cdot (u(x,\tau) - g(x,\tau)) = 0 \qquad \text{(F23)}$$

$$u(x,0) = g(x,0) \qquad \text{(F24)}$$

$$u(a,\tau) = g(a,\tau) \;\; = \;\; 0 \qquad \text{(F25)}$$

$$u(b,\tau) = g(b,\tau), \qquad \text{(F26)}$$

where

$$g(x,\tau) = e^{\frac{1}{4}(k+1)^2\tau} \max(e^{\frac{1}{2}(k+1)x} - e^{\frac{1}{2}(k-1)x}, 0). \qquad \text{(F27)}$$

The financial variables S, t and the option value C are again computed by putting $S = Ee^x$, $t = T - \tau/\frac{1}{2}\sigma^2$ and

$$C(S,t) = Ee^{-\frac{1}{2}(k-1)x - \frac{1}{4}(k+1)^2\tau} u(x,\tau).$$

In order to avoid technical complications, the problem is restricted to a finite interval (a,b) with $-a$ and b large enough. In financial terms, we assume that we can replace the exact boundary conditions by the approximations that for small values of S, $P = E - S$, while for large values, $P = 0$.

Let us denote by u_τ the function $x \mapsto u(x,\tau)$. The equivalent parabolic variational inequality is as follows:

Find $u = u_\tau \in K_\tau$ (τ runs over $[0, \frac{1}{2}\sigma^2 T]$) such that

$$\begin{aligned} &\left(\frac{\partial u}{\partial \tau}, \varphi - u\right)_2 + a(u, \varphi - u) \geq 0 \\ &\text{for all } \varphi \in K_\tau, \text{ a.e. } \tau \in (0, \frac{1}{2}\sigma^2 T), \; u(x,0) = g(x,0), \end{aligned} \qquad \text{(F28)}$$

where

$$K_\tau := \{v \in H^1(a,b) \mid v(a) = g(a,\tau), v(b) = g(b,\tau), v(x) \geq g(x,\tau)\}$$

and $(\cdot,\cdot)_2$ denotes the inner product on $L^2(a,b)$. With

$$W(0, \frac{1}{2}\sigma^2 T) := \left\{v \mid v \in L^2(0, \frac{1}{2}\sigma^2 T; H^1(a,b)), \frac{\partial v}{\partial \tau} \in L^2(0, \frac{1}{2}\sigma^2 T; H^{-1}(a,b))\right\}$$

$$W_0(0, \frac{1}{2}\sigma^2 T) := \left\{v \mid v \in W(0, \frac{1}{2}\sigma^2 T), v(0) = g(\cdot, 0)\right\}$$

and

$$\mathcal{K} := \left\{ v \mid v \in W(0, \frac{1}{2}\sigma^2 T), v_\tau \in K_\tau \text{ for a.e. } \tau \in [0, \frac{1}{2}\sigma^2 T] \right\}$$
$$\mathcal{K}_0 := \left\{ v \mid v \in W_0(0, \frac{1}{2}\sigma^2 T), v_\tau \in K_\tau \text{ for a.e. } \tau \in [0, \frac{1}{2}\sigma^2 T] \right\}$$

we can formulate an equivalent variational inequality:

Find $u \in \mathcal{K}_0$ such that

$$\int_0^{\frac{1}{2}\sigma^2 T} \left(\frac{\partial u}{\partial \tau}, \varphi - u \right)_2 \mathrm{d}\tau + \int_0^{\frac{1}{2}\sigma^2 T} a(u, \varphi - u)\, \mathrm{d}\tau \geq 0 \quad \text{for all } \varphi \in \mathcal{K}. \tag{F29}$$

American put option is modelled by a boundary value problem that only differs in the boundary conditions and the (transformed) pay-off function g:

$$\left(\frac{\partial u}{\partial \tau} - \frac{\partial^2 u}{\partial x^2} \right) \geq 0 \tag{F30}$$
$$(u(x,\tau) - g(x,\tau)) \geq 0 \tag{F31}$$
$$\left(\frac{\partial u}{\partial \tau} - \frac{\partial^2 u}{\partial x^2} \right) \cdot (u(x,\tau) - g(x,\tau)) = 0 \tag{F32}$$
$$u(x,0) = g(x,0) \tag{F33}$$
$$u(a,\tau) = g(a,\tau) \tag{F34}$$
$$u(b,\tau) = g(b,\tau) \;=\; 0, \tag{F35}$$

where

$$g(x,\tau) = e^{\frac{1}{4}(k+1)^2\tau} \max(e^{\frac{1}{2}(k-1)x} - e^{\frac{1}{2}(k+1)x}, 0). \tag{F36}$$

The equivalent variational inequality is formulated analogously to (F29) or (F28), with the only change in the boundary conditions and the function g.

It may be remarked that in American call option, $C(S,t)$ lies above the pay-off $\max(S-E,0)$; in the transformed variables, this condition takes the form $(u(x,\tau) - g(x,\tau)) \geq 0$. The condition $\left(\frac{\partial u}{\partial \tau} - \frac{\partial^2 u}{\partial x^2} \right) \geq 0$ means that the return from the risk-free delta-hedged portfolio is less than the risk-free interest rate r.

The numerical method to solve the complementarity problems modelling American call and put options will be the same. In Wilmott et al. [1995], it is discussed in quite detail how the projected Successive Overrelaxation (SOR) method can be employed for numerical simulation of American options. In the next section we describe a new, more efficient algorithm, its application to American option pricing and present numerical comparison with the projected SOR method.

F5. Numerical simulation of American call option by a two-step algorithm.

We discretize the boundary value problem (F21)–(F26) by the finite difference method and solve it using the Crank-Nicholson scheme; see, e.g., Wilmott et al.

[1993] or Glowinski [1984]. At each time step we need to solve a linear complementarity problem (LCP):

Find $u^{m+1} \in \mathbb{R}^n$ such that

$$\begin{aligned} Cu^{m+1} \geq b^m, \qquad u^{m+1} \geq g^{m+1} \\ (u^{m+1} - g^{m+1})^T (Cu^{m+1} - b^m) = 0. \end{aligned} \tag{F37}$$

Here C is an $n \times n$ real symmetric positive definite matrix given below:

$$C = \begin{pmatrix} 1+\alpha & -\frac{1}{2}\alpha & 0 & \cdots & 0 \\ -\frac{1}{2}\alpha & 1+\alpha & -\frac{1}{2}\alpha & & \vdots \\ 0 & -\frac{1}{2}\alpha & \ddots & \ddots & 0 \\ \vdots & & \ddots & 1+\alpha & -\frac{1}{2}\alpha \\ 0 & \cdots & 0 & -\frac{1}{2}\alpha & 1+\alpha \end{pmatrix}$$

with $\alpha = \frac{\delta\tau}{(\delta x)^2}$, $\delta\tau$ and δx being the time and space discretization parameters, respectively. Vectors u^{m+1} and g^{m+1} are the discrete counterparts of $u(x, \tau)$ and $g(x, \tau)$ from (F21)–(F26) at a time step $(m + 1)\delta\tau$ and b^m is a "right-hand side" vector containing information from the previous time step $m\delta\tau$.

Note that the matrix C is large and sparse and that the problem (F37) has to be solved repeatedly in each time step. Thus we need a fast LCP solver (in this application, literally, time is money). Several algorithms were proposed for LCPs with large sparse symmetric positive definite matrix. These algorithms are either based on methods for solving linear systems, like the successive overrelaxation (SOR) method or the preconditioned conjugate gradient (PCG) method, or on optimization methods for convex quadratic programs, like the gradient projection method or interior-point methods.

Recently, a two-step algorithm has been proposed by Kočvara and Zowe [1994]. This algorithm, based on ideas of multigrid methods, combines the efficiency of the PCG method for solving linear systems with the ability of SOR to smooth down the solution error. The smoothing property of a variant of SOR with a projection on a feasible set, called **SORP**, enables to detect fast the active set

$$I(u) := \{i \mid u_i = g_i\}.$$

In the above definition and in the rest of this section we skip the time step index and write (F37) as

Find $u \in \mathbb{R}^n$ such that

$$\begin{aligned} Cu \geq b, \qquad u \geq g \\ (u - g)^T (Cu - b) = 0. \end{aligned} \tag{F38}$$

Instead, we will use the upper index to denote the successive iterate; i.e., u^k will be the kth iterate of a particular method. By u^* we denote the (unique) solution to

(F38). We further denote by C_{ij} the $(i,j)-$component of the matrix C. Finally, let us define the feasible set of (F38):

$$S := \{v \in \mathbb{R}^n \mid v_i \geq g_i,\ i = 1, 2, \ldots, n\}.$$

The two-step algorithm mentioned above proved to be very efficient for large LCPs. The examples in Kocvara and Zowe [1994] even indicate that the algorithm, based on PCG, is asymptotically as fast as PCG for linear systems. It is our strong belief that the algorithm fits naturally to our LCP and significantly improves the efficiency of the overall time stepping procedure. In the following text we describe the algorithm in detail.

We first recall the definition of SORP:

SORP

Choose $x^0 \in \mathbb{R}^n$ and put for $k = 0, 1, 2, \ldots$

$$x_i^{k+1} = \max\left\{x_i^k - \omega\frac{1}{C_{ii}}\left(\sum_{j<i} C_{ij}x_j^{k+1} + \sum_{j\geq i} C_{ij}x_j^k - b_i\right), c_i\right\}, \quad i = 1, 2, \ldots, n, \tag{F39}$$

where $\omega \in (0, 2)$ is a relaxation parameter.

We will work with a symmetric version of SORP, called SSORP; one SSORP step consists of one forward and one backward SORP step:

$$\begin{aligned} \tilde{u}_i^{k+1} &= \max\left\{u_i^k - \omega\frac{1}{C_{ii}}\left(\sum_{j<i} C_{ij}\tilde{u}_j^{k+1} + \sum_{j\geq i} C_{ij}u_j^k - b_i\right), g_i\right\}, \quad i = 1, 2, \ldots, n \\ u_i^{k+1} &= \max\left\{\tilde{u}_i^{k+1} - \omega\frac{1}{C_{ii}}\left(\sum_{j\leq i} C_{ij}\tilde{u}_j^{k+1} + \sum_{j>i} C_{ij}u_j^{k+1} - b_i\right), g_i\right\}, \quad i = n, n-1, \ldots, 1. \end{aligned} \tag{F40}$$

We shall denote by $SSORP^m(x; C, b, g)$ the value which we obtain in m steps (F40). If $m = 1$ we skip the superscript.

The new algorithm can be viewed as a variant of the active set strategy. That means, at each iteration step one has to solve a linear system with a matrix of similar structure as that of C. This system, however, does not have to be solved exactly, particularly when the actual active set $I(u^k)$ is far away from $I(u^*)$. The idea is to perform just a few steps of a preconditioned conjugate gradient method. For completeness we give below the definition of the PCG algorithm for the solution of system

$$Au = b$$

with a symmetric and positive definite matrix A.

PCG

Let M be a symmetric positive definite matrix (the preconditioner).
Choose $x^0 \in \mathbb{R}^n$ and $\varepsilon > 0$.
Set $r^0 = b - Ax^0, p^0 = z^0 = M^{-1}r^0$ and do for $k = 0, 1, 2, \ldots$:

$$\begin{aligned}
&\alpha_k = \langle r^k, z^k\rangle / \langle p^k, Ap^k\rangle \\
&x^{k+1} = x^k + \alpha_k p^k \\
&r^{k+1} = r^k - \alpha_k A p^k \\
&z^{k+1} = M^{-1} r^{k+1} \\
&\text{if } \langle z^{k+1}, r^{k+1}\rangle \geq \varepsilon, \text{ then if } \|r^{k+1}\| \geq \varepsilon, \text{ continue} \\
&\beta_k = \langle r^{k+1}, z^{k+1}\rangle / \langle r^k, z^k\rangle \\
&p^{k+1} = z^{k+1} + \beta_k p^k
\end{aligned}$$

We denote by $PCG^s(u; A, b)$ the point which we reach in s PCG steps starting from u.

We are now going to explain the new approach. Assume that we have an approximation u of the solution u^* to (F38). We again denote by $I(u)$ the active set with respect to the constraint $u \geq g$, i.e.,

$$I(u) := \{i \mid u_i = g_i\}.$$

Let $p(u)$ be the cardinality of $I(u)$ and $P_{I(u)} : \mathbb{R}^n \to \mathbb{R}^{n-p(u)}$ the operator which assigns a vector $v \in \mathbb{R}^n$ the reduced vector in $\mathbb{R}^{n-p(u)}$ obtained by omitting the components of v with indices from $I(u)$. We skip u in this notation if this does not lead to confusion. Further we write I^* for $I(u^*)$.

The basic idea of the new method is the following: We try to identify the active set by a few steps of SSORP; we get an approximation of I^*. With this approximation, we perform several steps of PCG. Then, again some steps of SSORP to improve the approximation of I^*, and so on. Therefore we call the method SSORP-PCG. One iteration of the SSORP-PCG algorithm consists of two steps:

SSORP–PCG

Choose $u^0 \in \mathbb{R}^n, m \in \mathbb{N}, s \in \mathbb{N}$ and do for $k = 0, 1, 2, \ldots$:
Step 1: Perform m SSORP steps (F40) and put

$$u^{k+1/2} := SSORP^m(u^k; C, b, g)$$

Step 2: Determine P_I with $I := I(u^{k+1/2})$ and compute

$$r^k = b - Cu^{k+1/2}.$$

Perform s PCG steps and define with

$$z^k := PCG^s(0, P_I C P_I^T, P_I r^k)$$

the next iterate

$$u^{k+1} := u^{k+1/2} + \gamma P_I^T z^k,$$

where γ is the largest real number such that $\gamma \leq 1$ and $u^{k+1} \in S$.

It was proved in Kocvara and Zowe [1994] that the sequence of iterates $\{u^k\}_{k\in\mathbb{N}}$ produced by SSORP–PCG converges to the solution x^* of LCP (F38).

Just as for multigrid methods, the number of SSORP steps (parameter m in Step 1) can be chosen small; already for $m = 2$ we obtained good results. The number of PCG steps (parameter s in Step 2) is more problem-dependent. Generally speaking, s should grow with the condition number of the matrix C. We recommend to take $s = 5$ for well-conditioned problems and $s = 10$ otherwise.

Concerning the preconditioner, it is well known that efficient preconditioning matrices for elliptic problems are those based on incomplete factorization. We implemented the $MIC(0)^*$ algorithm.

Below we will compare SSORP-PCG with SORP. To guarantee equal conditions, we have chosen the following stopping criteria:

- The SORP method with relaxation parameter $\omega = 1.9$ was stopped when $\|u_{k+1} - u_k\|_2$ became less than 10^{-9} the first time.
- The stopping criterion in SSORP-PCG guarantees an accuracy comparable to the one in the SORP implementation: if we applied SORP after stopping SSORP-PCG, then we typically had $\|u_{k+1} - u_k\|_2 \leq 10^{-9}$.

Numerical Results

In this section we present results of an example computed by the new algorithm and, for comparison, also by the plain SORP method. We would like to emphasize that the data shown below should not be evaluated from the viewpoint of overall efficiency; the example is academic, the choice of time and space discretization parameters, as well as the parameter α, can be far from being optimal. Our goal is to demonstrate the efficiency of the SSORP-PCG algorithm for solving a particular subproblem (LCP) which, no doubt, is large and has to be solved repeatedly many times.

We have solved an example from Wilmott et al. [1995] in order to get comparable results. This is a problem of computing American put option with interest rate $r = 0.10$, volatility $\sigma = 0.4$ and exercise price $E = 10$. The calculation is carried out with $\alpha = 1$ and with the expiry time of three months. The space interval $[a, b]$ is chosen as $[-0.5, 0.5]$. We carried out the computation for three different space discretization steps: $\delta x = 0.01$, $\delta x = 0.001$ and $\delta x = 0.0001$.

Table 1

δx	$\delta\tau$	n	N
0.01	10^{-4}	99	200
0.001	10^{-6}	999	20 000
0.0001	10^{-8}	9999	2 000 000

Table 1 shows the corresponding values of the time step $\delta\tau$, the size n of the $n \times n$ matrix C and the number of time steps N saying how many LCPs we have to solve.

These three problems were solved using the two-step algorithm SSORP-PCG. Table 2 shows the overall CPU time needed to solve the problem, as well as the time needed to solve one LCP. These times are compared with the solution times of the plain SORP method. Table 2 also shows the (average) number of active constraints. Note that for large number of active constraints (compared to the problem size n) the SORP methods becomes very efficient. This is observed on the number of SORP iterations, shown in the last column of Table 2; with increasing number of variables, the number of iteration decreases. This is very untypical behaviour caused by the particular data of the problem.

Table 2

δx	S-P one step CPU	S-P overall CPU	SORP one step CPU	SORP overall CPU	#active constr.	#SORP iter.
0.01	0.0021	0.38	0.0089	1.45	20	75
0.001	0.016	390.4	0.041	783.9	400	32
0.0001	0.44	*	1.05	*	8500	28

The second factor which influences the behaviour of the algorithms is that in both, SORP and SSORP-PCG, the solution from the previous time step was taken as an initial approximation for the current time step. This technique, in fact, favours the SORP algorithm. This is clearly seen from Table 3 which shows the CPU times obtained with the initial approximation for each time step taken as zero vector. In this situation, SSORP-PCG is a clear winner.

Table 3

δx	S-P one step CPU	S-P overall CPU	SORP one step CPU	SORP overall CPU	#SORP iter.
0.01	0.0023	0.43	0.013	2.39	109
0.001	0.016	486.3	0.11	2221.1	87
0.0001	0.44	*	2.89	*	74

Our results show that the new algorithm certainly outperforms the SORP method, even though the problem data favour the latter.

All numerical experiments were carried out on a Sun Ultral m140 computer running the operating system Solaris 2.6. The CPU times are in seconds. Study of American options with feedback is an interesting open problem.

Bibliography

[1] Acton, F.S., *Real Computing Made Real,* Princeton University Press, Princeton, NJ. 1996.

[2] Albrecht, J., Collatz, L. et al. (ed.), *Numerical Treatment of Eigenvalue problems,* vol.5, Birkhäuser, Basel-Boston-Berlin, 1991.

[3] Alonso, A. and Valli, A., *A domain decomposition approach for heterogeneous time-harmonic Maxwell equations*, Comput. Math. Appl. Mech. Engrg., 143(1997), 97–112.

[4] Alonso, A. and Valli, A., *An optimal domain decomposition preconditioner for low-frequency time-harmonic Maxwell equations*, Math. Comp., 68(1999), 607–631.

[5] Alvarez, L., Lions, P.L. and Morel, J.M., *Image Selective Smoothing and edge detection by nonlinear diffusion II*, SIAM J. Numer. Anal., 29(1992), 845–866.

[6] Amaratunga, K. and Williams, R., *Wavelet based Green's function approach to 2D PDEs*, Engineering Computations, 10(1993), 349–367.

[7] Antes, H. and Panagiotopoulos, *P.P., The Boundary Integral Approach to Static and Dynamic Contact problems,* Birkhäuser (ISNM 108, International Series of Numerical Mathematics), 1992.

[8] Arino, M.A. and Vidakovic, B., *On wavelet scalograms and their applications in time series*, Preprint 1999.

[9] D'Attellis, C.E. and Fernándoz-Berdaguer, E.M., (eds.) *Wavelet Theory and Harmonic Analysis in Applied Sciences*, Birkhäuser, Boston, Basel, 1997.

[10] Argyris, J.H., *Energy theorems and structural analysis, Aircraft Engg.*, 26(1954), 347–356, 383–387, 394.

[11] Aubin, J.P., *Approximation of Elliptic boundary problems,* Wiley Inter science, New York, 1972.

[12] Babovsky, H., Gopengiesser, F., Neunzert, H., Struckmeier, J. and Wiesen, B., *Low discrepancy methods for the Boltzmann equation*, 16th Internat. Symppos. on Rarified Gas Dynamics, Pasadena, Ca., July 1988.

[13] Babovsky, H. and Illner, R., *A convergence proof of Nanbu's simulation method for full Boltzmann equation*, SIAM J. Num. Anal., 26(1989), 45–65.

[14] Bahader, T. B., *Mathematica for Scientists and Engineers*, Addison Wesley Publishing Company Inc. Reading, Massachusetts, 1995.

[15] Balder, R. and Zenger, C., *The solution of multidimensional real Helmholtz equations on sparse grids*, SIAM J. Sci. Comput., 17(1996), 631–646.

[16] Banerjee, P.K., *The Boundary element methods in Engineering*, McGraw Hill Book Company, London, 1994.

[17] Bank, R.E., Bulirsch, R., Merten, K., *Mathematical Modelling and Simulation of Electrical Circuits and Semi conductor Devices*, Birkhäuser, Basel, 1990.

[18] Bari, N.K., *Treatise on Trigonometric Series, Vol. I, II*, Pergamon Press, Oxford, 1961.

[19] Barnsley, M.F., *Fractal Everywhere*, Academic Press, San Diago, CA, 1988.

[20] Barnsley, M.F., *The Desktop Fractal Design Handbook*, Academic Press, New York, 1989.

[21] Barnsley, M.F., *Fractal Image Compression*, Notices of the AMS, 43(1996), 657–662.

[22] Barnsley, M.F., Hurd, L.P., *Fractal Image Compression*, AK Peters, Ltd, Boston, 1993.

[23] Beauchamp, K.G., *Transforms for Engineers, A Guide to Signal Processing*, Oxford University Press, 1987.

[24] Bellomo, N. and Preziosi, L., *Modelling Mathematical Methods and Scientific Computation*, CRC Press, London, 1995.

[25] Berger, M.D. etal, *ICAOS'96 12th Internal Conference on Analysis and Optimization systems Images, Wavelates and PDEs*, Paris, June 26-28, 1996, Springer Verlag, Paris, 1996.

[26] Berkner, K., *A Wavelet-based Solution to the Inverse Problem for Fractal Interpolation Functions in Vahel et al (eds.), Fractals in Engineering*, International Conference on Fractals, Springer, Paris, 1997.

[27] Bertoluzza, S., *Some Error Estimates for Wavelet Expansion*, Mathematical Models and Methods in Applied Sciences, 2(1992), 489–506.

[28] Beste, A., Brokate, M., Dressler, K., *Kann man berechnen, wie lange ein Auto hält*, In Bachem, Jünger, Schröder, Mathematik in der Praxis, Springer-Verlag, 1995.

[29] Berz, M., Bishof, C., Corliss, G. and Griewank, A. (eds.) *Computational Differentiation Techniques, Applications and Tools*, SIAM, Philadelphia, 1996.

[30] Beylkin, G., Coifman, R. and Rokhlin, V., *Fast wavelet transforms and numerical algorithms*, Comm. Pure Appl. Math., 44(1991), 141–183.

[31] Bhattacharya, P., *Semiconductor Optoelectronic devices*, Prentice Hall Inc. International edition, 1994.

[32] Binder, K.,*The Monte Carlo Method in condensed matter Physics*, Springer Verlag, New York-Berlin, 1992.

[33] Biran, A. and Breiner, M., *MATLAB for Engineers*, Addison-Weseley, Reading, MA, 1995.

[34] Bird, G.A., *Molecular Gas Dynamics*, Clarendon Press, Oxford, 1976.

[35] Black, F. and Scholes, M., *The pricing of options and corporate liabilities*, J. Pol. Econ., 81(1973), 637–659.

[36] Blanchard, P. and Brüning E., *Variational Methods in Mathematical Physics*, Springer Verlag, Texts and Monographs in Physics, Berlin, Heidelberg, 1992.

[37] Bossavit, A., *Differential forms and the computation of fields and forces in electromagnetism*, Eur. J. Mech, B., 10(1991), 474–488.

[38] Bossavit, A., *Electromagnetism, en vue de la modelisation (Electromagnetism in view of modeling)*, Springer Verlag, Paris, 1993.

[39] Bossavit, A., *On the homogenization of Maxwell Equations*, COMPEL, 14(1995), 23–26.

[40] Boyce, W. E. (ed.) *Case Studies in Mathematical Modelling*, Pitman Advanced Publishing Program, Boston, 1981.

[41] Bramble J.H. and Nitsche, J.A., *Generalised Ritz-Least square method for Dirichlet Problems*, SIAM J. Numer. Anal., 10(1973), 81–93.

[42] Bramble J.H., *Multigrid Methods*, Pitman Research Notes in Mathematics, v. 124, John Wiley and Sons, 1993.

[43] Bratley, P., Fox, B., Bennet L., Schrage, L. E., *A guide to Simulation*, Springer Verlag, New York, Berlin, 1984.

[44] Bratley, P., Fox, B.L., Niederreiter, H., *Algorithm-738-Programs to Generate Niederreiters Law, Discrepancies,* ACM Transactions on Mathematical Software, 20(1994), 494–495.

[45] Brebbia, C.A., *The Boundary Element Methods for Engineers,* Pentech Press, London, 1978.

[46] Brebbia, C.A., *Topics in boundary element research.* Vol. 1, Springer, 1984.

[47] Brebbia, C.A., *Topics in boundary element research,* Vol. 2, Springer, 1985.

[48] Brebbia, C.A. (ed.), *Boundary element X,* Springer Verlag, Berlin, 1988.

[49] Brebbia, C.A. (ed.), *Boundary Element Technology VI* Elsevier (Appl. Science), London, 1991.

[50] Brebbia, C.A. and Gipson, G.S. (eds.), *Boundary Elements XIII,* Springer-Verlag, Berlin, 1991.

[51] Brebbia, C.A., Tanaka, M. and Honna, T. (eds.), *Boundary elements XII,* Springer Verlag, Berlin, 1990.

[52] Brebbia, C.A., Walker, S., *Boundary Element Techniques in Engineering,* Newnes-Butterworths, London, 1980.

[53] Brebbia, C.A., Telles, J.C.F., Wrobel, L.C., *Boundary Element Techniques,* Springer Verlag, Berlin-Heidelberg, 1984 (Theory and Applications in Engineering).

[54] Brenner, S.C. and Scott, L.R., *The Mathematical theory of finite element methods*, Springer Verlag, New York, Berlin, 1994.

[55] Brezis, H., *Operators maxumus monotone at semigroupes dans les espaces de Hilbert*, North Holland, Amsterdam, 1973.

[56] Brezzi, F., Fortin, M., *Mixed and Hybrid Finite Element Methods,* Springer Verlag, New York, 1991.

[57] Brokate, M., *Some BV properties of the Preisach hysteresis operator, Applicable Analysis,* 32(1989), 229–252.

[58] Brokate, M., *Hysteresis operators, in Phase Transitions and Hysteresis in Visintin, A.(ed.), Lecture Notes in Mathematics* No.1584, Springer Verlag Berlin, Heidelberg, 1994.

[59] Brokate, M., Dressler, K. and Krejči, P., *Rain-flow counting and energy dissipation for hysteresis models in elastoplasticity*, European J. Mechanics A/Solids, 15(1996), 705–737.

[60] Brokate, M. and Siddiqi, A.H. (eds.), *Functional Analysis with current Applications to Science, Technology and Industry,* Pitman Research Notes in Mathematics, Longman, U.K., 1997.

[61] Brokate, M. and Sprekels, J., *Hysteresis and Phase Transitions, Springer Verlag,* Berlin-Heidelberg, 1996.

[62] Brokate, M., and Visintin, A., *Properties of the Preisach model for hysteresis,* J. Reine Angev. Math., (1989), 1–40.

[63] Brokate, M., Beste A., *Dressler, K., Kann man berechenen, wie lange ein Auto hält,* In: Mathematik in der Praxis-Fallstudien aus Industrie, Wirtschaft, Naturwissenschaften und Medizin, 3–24 in (eds.) A. Bachem, M. Jünger, R. Schrader, Springer Verlag, Berlin, 1995, 3–24.

[64] Bruns, W., Motoc, I., O'Driscoll, K.F., *Monte Carlo Applications in Polymer Science,* Lecture Notes in Chemistry, Springer Verlag, 1981.

[65] Burnett, D.S., *Finite Element Analysis From concepts to Applications,* Addison-Wesely, Pub.Comp. Reading, Massachusetts, 1987.

[66] Byrd, R.H., Nocedal, J., *A tool for the analysis of quasi Newton methods with application to unconstrained minimization,* SIAM J. Numer. Anal., 25(1989), 727–739.

[67] Caflisch, Russel E., *Monte Carlo and quasi-Monte Carlo methods,* Numerica, (1998, pp. 1–49, Cambridge University Press, 1999.

[68] Canuto, C. and Cravero, I., *A Wavelet-based Adaptive Finite Element Method for Advection Diffusion Equations,* Mathematical Models and Methods in Applied Sciences, 7(1997), 265–289.

[69] Caracciolo, S. and Fabrocini, A.(ed.) *Monte Carlo Methods in Theoretical Physics,* ETS Editrice, Pisa, 1991.

[70] Cargill, T., *C++ Programming Style,* Addison-Wesely, Rouding, New York, 1992.

[71] Ce'*a*, J., *Approximation Variationnelle de problemes aux limits,* Ann. Inst. Fourier (Grenoble), 14(1964), 345–444.

[72] Cercignani, C., *The Boltzmann Equation and Its Applications,* Springer, Berlin, 1989.

[73] Chambolle, A., DeVor, R.A., Lee, N.Y. and Lucier, B., *Nonlinear wavelet image processing: Variational problems, compresssion, and noise removal through wavelet shrinkage, IEEE Trans. Image Processing,* 7(1998), 319–335.

[74] Chari, M.V.K., and Silvester, P.P. (eds.), *Finite Elements in Electrical and Magnetic field problems,* John Wiley and Sons, Chichester, 1980.

[75] Chavent, G. and Jaffré, J., *Mathematical Models and Finite Elements for Reservoir Simulation,* North Holland, Amsterdam, 1986.

[76] Chen, G. and Zhou, J., *Boundary Element Methods*, Academic Press, Cambridge, 1992.

[77] Chipot, M., March, R., and Vitulano, D., *Numerical Analysis of Oscillations in a Nonconvex Problem Related to Image Selective Smoothing*, Preprint 1999, CNR, Rome, Italy.

[78] Chui, C.K., *An Introduction to Wavelets,* Academic Press, San Diego, CA 1992.

[79] Chui, C.K. (ed.), *Wavelets: A Tutorial in Theory and Applications*, Academic Press, SanDiego, CA, 1992.

[80] Churchil, R.V., *Fourier Series and Boundary value Problems*, McGraw Hill, New York, 1963.

[81] Ciarlet, P.G., *The finite element methods for elliptic problems,* North Holland, Amsterdam, 1978.

[82] Ciarlet, P. G., *Introduction to numerical linear algebra and Optimization*, Cambridge University Press, 1989.

[83] Ciarlet, P.G. and Lions, J.L., *Hand book of Numerical Analysis*, Elsevier Science Publishers, North Holland, 1991.

[84] Clough, R.W., *The Finite Element Method in plane stress analysis,* in Proceedings 2nd ASCE Conference on Electronic Computation, Pittsburg, PA.

[85] Coifman, R.R. Meyer, Y, and Wickerhauser, M.V., *Size Properties of Wavelet packets* in Ruskai, Marry Beth (ed.) et al., Wavelets and their Applications, Boston, MA, 1992, 453–470.

[86] Coifman, R.R., and Wickerhauser, M.V., *Entropy based algorithms for best basis selection,* IEEE Trans. Inform. Theory, 38(1992), 713–718.

[87] Colton, D. and Kress, R., *Integral equation methods in scattering theory,* John Wiley & Sons, New York, 1983.

[88] Cooley, J.W., and Tukey, J.W., *An Algorithm for the Machine Calculation of Complex Fourier series*, Mathematical Computations, 19, 1965.

[89] Crandall, R. E., *Mathematica for the Sciences,* Addison-Weseley Publishing Company, Inc., U.S.A., 1991.

[90] Dahmen, W., *Wavelet and Multiscale Methods for Operator Equations*, Acta Numerica, Cambridge, 1998.

[91] Dai, X., Larson, D.R. and Speegle, D.M., *Wavelet sets in R^n II*, 15–30, Contemporary Mathematics, Vol. 216, AMS (Aldroub and Lin eds.), 1997.

[92] Datta, B.N., *Numerical Linear Algebra*, Brooks/cole publishing companny, Bonn, 1995.

[93] Daubechies, I., *Ten lectures on wavelets,* Society for Industrial and Applied Mathematics (SIAM), Philadelphia, PA, 1992.

[94] Daubechies, I., Grossman, A. and Meyer, Y., *nonorthogonal expansions,* J. Math, Phys., 27(1986), 1271–1283.

[95] Daubechies, I., Mallat, S. and Willsky, (ed.), *A special issue on wavelet transforms and multi resolution signal analysis,* IEEE Trans. Inform. Theory, 38(1992), 529–840.

[96] Dautry, R. and Lions, J., *Mathematical Analysis and Numerical Methods for Science and Technology,* Vols. 1,2,3,4,5,6, Springer Verlag, Berlin, Heidelberg, New York, 1990-1995.

[97] Davies, A.J., *The Finite Element Method, A first approach,* Oxford Applied Mathematics and Computing Sciences, Clarendon Press, Oxford, 1980.

[98] Davis, P.J. and Rabinowitz, P., *Methods of Numerical Integration,* Academic Press, New York, 1984.

[99] Dennis, Jr J.E. and Schnabel, Robert R., *Numerical Methods for unconstrained optimization and nonlinear equations*, Prentice-Hall Inc., New Jersey, 1983 (Reproduction of SIAM, Philadelphia, 1996).

[100] DeVore, R.A., *Nonlinear application,* Acta Numerica, 1998, 51–150, Cambridge University Press.

[101] DeVore, R.A., Jawerth, B. and Lucier, B.J., *Image compression through wavelet transform coding,* IEEE, Trans. Inform., 38(1992), 719–746.

[102] Douglas, J. Jr., Dupont, T., *A Finite Element Collocation Method for Quasi linear Parabolic Equations, Math. Comput.*, 27(1973), 17–28.

[103] Dowd, K., *High Performance Computing,* Oreilly and Associates, Sebastopol, CA, 1993.

[104] Dressler, K. and Hack, M., *Fatigue life time estimation based on rain-flow counted data using the local strain approach,* Eur. J. Mech. A/Solids, 15(1996), 955–968.

[105] Dressler, K. and Hack, M., *Fatigue lifetime estimation based on rainflow counted data using the local strain approach*, Eur. J. Mech. A/Solids, 15(1996), 955–968.

[106] Dressler, K., Hack, M. and Krüger, W., *Stochastic Reconstruction of Loading Histories from a Rain-flow Matrix*, ZAMM. Z. angew. Math. Mech., 77(1997), 217–266.

[107] Duvaut, G., Lions, J.L., *Inequalities in Mechanics and Physics,* Springer, Berlin, 1976.

[108] Eirola, T., Sobolev *characterization of solutions of dilation equation,* SIAM J. Math. Anal., 23(1992), 1015–1030.

[109] Eriksson, K., *Introduction to Adaptive Methods for Differential Equations,* Acta Numerica, 1995, 105–158.

[110] Evans, G., *Practical Numerical integration,* John Wiley and Sons, Chichester, New York, 1993.

[111] Falk, R.S., *Error Estimates for the Approximation of Class of Variation Inequalities,* Math. Comp., 28(1974), 963–971.

[112] Fasman, G. D., *Prediction of Protein Structure and the principles of Protein Conformation,* Fasman, G. D. (ed.), Plenum Press, New York and London 1989, 2nd edition, 1990.

[113] Felippa, *Finite element and Finite Difference energy Techniques for the Numerical solution of partial Differential Equations,* Summer Computer Simulation Conference Proc. Montreal, 1973, Vol. 1, pp.1–14.

[114] Finlayson, B.A., *The method of Weighted Residuals and Variational Principles,* Academic Press, New York, 1972.

[115] Fisher,Y., *Fractal image compression,* Springer Verlag, New York, 1995.

[116] Fletcher, R., *An optimal positive definite update for sparse Hessian matrices,* SIAM, J. Optim., 1(1991), 18–21.

[117] Forte, B. and Vrscay, E.R., *Solving the Inverse Problem for Function and Image Approximation Using Iterative Function Systems, Dynamics of Continuous, Discrete and Impulsive Systems*, 1(1995), 177–231.

[118] Fredholm, I., *Sur une classe d'Equations Functionelles,* Acta Mathematica, 27(1903), 365–390.

[119] Friedman, A., *Mathematics in Industrial problems, for IMA volumes in Mathematics and its Applications,* V. 16, Springer Verlag, 1988.

[120] Gaylord R. J., Wellin, P. R., *Computer Simulations with Mathematica Explorations in Complex Physics and Biological Systems*, Springer Verlag, 1994.

[121] Gilbert, J.C. and Nocedal J., *Global convergence properties of conjugate gradient Method for optim.*, SIAM J. Optimaization, 2(1992), 21–42.

[122] Gill, P. E., Murry, W. and Wright, M. H., *Practical Optimization*, Academic Press, London, 1981.

[123] Glowinski, R., Marroco, A., *Analyse numérique du champ magnetique d' un alternateur par ele'ments finis et sur-relaxation ponctuelle nonlinear*, comput. Methods Appl. Mech. Engg., 3(1974), 55–85.

[124] Glowinski, R., *Numerical Methods for nonlinear variational problems,* Springer Verlag, 1984.

[125] Glowinski, R., Lions, J.L. and Tre'moliereres, *Numerical analysis of variational inequalities, Studies in Mathematics and its applications*, Vol.8, North Holland Publishing Comp., Amsterdam, 1981.

[126] Glowinski, R. et al., *Wavelet solutions of linear and nonlinear elliptic, parabolic and hyperbolic problems in one space dimension* in Glowinski et al. (eds.), Computing Methods in Applied Sciences and Engineering, SIAM, Philadelphia, 1990.

[127] Goldfarb, D., *Algorithms for unconstrained optimization, A Review of Recent Developments, Proc. of Symp. in Appl. Math.*, 48, 1994 Amer. Math. Soc., (0160-7634/94).

[128] Goulub, G.H., Van Loan, C.F., *Matrix Computations, 3rd edition,* The Johns Hopkins University Press, Boltimore and London, 1996.

[129] Gonzalez, R. C. and Woods, R.E, *Digital Image Processing,* Addison - Wesley Publishing Comp., Reading, Massachusetts, 1993.

[130] Goodman, J.W., *Introduction to Fourier Optics*, McGraw Hill, New York, 1968.

[131] Griewank, A. and Corliss, G. F., *Automatic differentiation of Algorithms, Theory Implementation and Application*, SIAM, Philadelphia, 1991.

[132] Griffel, D. H., *Applied Functional Analysis,*, Ellis Horwood Ltd., New York, 1981.

[133] Hack, P., *Quality Control of Artificial Fabrics, Report* AGTM Nr. 45, Kaiserslautern, Germany, 1990.

[134] Hack, M., *Life Time Estimation in the Car Industry, Report* AGTM Nr. 105, 1994.

[135] Hackbusch, W.,*Multigrid methods and applications,* Springer, Berlin 1985.

[136] Hackbusch, W. (ed.), *Robust Mutligrid method,* Notes on numerical fluid mechanics, Vol. 23, Vieweg, Braunschweig/Wiesbaden, 1989.

[137] Hackbusch, W., *Elliptic Differential Equations, Theory and Numerical Treatment,* Springer Verlag, 1992.

[138] Hackbusch, W., *Iterative solution of large sparse system of equations,* Springer Verlag, 1994.

[139] Hackbusch, W., *Integral Equations, Theory and Numerical Treatment,* Basel-Boston-Berlin, Birkhäuser, 1995.

[140] Halton,J.H., *A retrospective and prospective survey of the Monte Carlo Method,* SIAM Rev., 12(1970), 1–63.

[141] Hammersley, J.M. and Handscomb, D.C., *Monte Carlo Methods,* reprint 1967, London Methuen and Co. Ltd. (first print 1964).

[142] Hammond, P., *Energy Methods in Electromagnetism,* Clarendon Press, Oxford, 1981.

[143] Hammond, P., *Electromagnetism for Engineers,* Pergamon Press, Oxford, New York, 3rd edition, 1986.

[144] Hanselman, D. and Littlefield, B., *Mastering MATLAB,* Prentice Hall, Upper Saddle River, NJ, 1996.

[145] Heath, M.T., *Scientific Computing; An Introduction Survey,* The McGraw Hill Complies, Inc., New York, 1997.

[146] Hockney, R. W. and Eastwood James W., *Computer Simulation using particles,* McGraw Hill Inc., U.S.A. 1981.

[147] Hoppe, R. H.W., and Wohlmuth, B., *A Penalty Method for the Approximation Solution of Stationary Maxwell Equations,* Numer. Math., 36(1981), 389–403.

[148] Hoppe, R. H.W., and Wohlmuth, B., *Multilevel iterative solution and adaptive mesh refinement for mixed finite element discretization,* Applied Numerical Mathematics, 23(1997), 97–117.

[149] Hua, L.K. and Wang, Y., *Applications of Number Theory to Numerical Analysis,* Springer Verlag, Berlin-Heidelberg, 1981.

[150] Huebner, K.H., *The Finite Element Method for Engineers,* John Wiley and Sons, New York, London, 1975.

[151] Itzykson, C. Z., Zuber, J.B., *Quantum Field Theory,* McGraw Hill, New York, 1980.

[152] Jackson, J.D., *Electrodynamics of Continuous Media,* Pergamon, Oxford, 1960.

[153] Jackson, J.D., *Classical Electrodynamics,* Wiley, New York, 1975.

[154] Jacoboni C. and Lugli, P. *The Mote Carlo Method for Semiconductor Device Simulation,* Springer Verlag, Wien, New York, 1989.

[155] Jacquin, A.E., *Fractal Image Coding,* A Review, Proc. IEEE, 81(1993), 1451–1465.

[156] Jaswon, H.A. and Symm, G.T., *Integral Equation Methods in Potential Theory and Elastostatics,* Academic Press, London, 1977.

[157] Jin, Jianming, *The Finite Element Method in Electromagnetics,* John Wiley and Sons, INC, New York, 1993.

[158] Johnson, G., Ne'de'lec, J.C. *On the coupling of the Boundary integral and Finite Element Methods,* Math. Comput., 35(1980), 1063–1079.

[159] Kaiser, G., *A friendly guide to wavelets,* Birkauser, Boston-Basel-Berlin, 1994.

[160] Kantorovich, L. V., *Mathematics in Economics:Achievements, Difficulties, Perspectives, Nobel Memorial Lecture,* Dec. 11, 1975.

[161] Kapur, J.N., *Mathematical Modelling,* John Wiley and Sons, New York, 1987.

[162] Kardestuncer, H. and Norrie, D. H., *Finite Element Handbook,* McGraw Hill Book Company, 1987.

[163] Kearsley, S. K. and Smith, M. G., *An Alternative Method for the Alignment of Molecular Structures, Maximising Electrostatic and Steric Overlap,* Tetrahedron Computer Methodology, 3(1990), 615–633.

[164] Kellog, O.D., *Foundation of Potential Theory,* Springer Verlag, Berlin 1929 (also in Dover publ., New York, 1953).

[165] Kelly, S., Kon, M. and Raphael, L. *Pointwise Convergence of Wavelet Expansions,* J. Funct. Anal., 126(1994), 102–138.

[166] Kelly, S.E., Kon, M.A. and Raphael, L.A., *Convergence: Fourier series versus wavelet expansion,* Proc. of NATO Advanced Study Institute: Wavelets and their Applications, J. Byrnes ed., 1994.

[167] Kesavan, S., *Topics in Functional Analysis and Applications,* John Wiley and Sons, New York, 1988.

[168] Kikuchi, N., and Oden, J.T., *Contact problems in Elasticity, A Study of Variational Inequalities and Finite Element Methods,* SIAM Publ. Philadelphia, 1988.

[169] Kinderlehrer, D. and Stampacchia, G., *Introduction to Variational Inequalities and their Applications*, Academic Press, New York, 1980.

[170] Kobayashi, M., *Wavelets and their applications, Case studies*, SIAM, Philadelphia, PA, 1998.

[171] Kocvara, M. and Zowe, J., *An iterative two-step algorithm for linear complementary problems*, Numer. Math., 68(1994), 95–106.

[172] Koenderink, J., *The structure of images, Biological Cybernectics*, Springer Verlag, New York, 1985.

[173] Koonin, S.E. and Mevedith, D. C., *Computational Physics, Fortran version*, Addison Wesely Pub. Comp., U.S.A., 1990.

[174] Koopmans, T. C., *Concepts of optimality and their uses, Nobel Memorial Lecture*, Dec. 11, 1975.

[175] Koorwinder, T.H. ed., Wavelets: *An Elementary Treatment of Theory and Applications*, World Scientific, Singapore, 1993.

[176] Kovačevič, J. and Daubechies, I.(ed.), *Special issue on Wavelets*, Proc. IEEE, 84(1996), 507–614.

[177] Krasnosel'skii, M.A., Pokrovskii A.V., *Systems with hysteresis*, Springer Verlag, Heidelberg 1989.

[178] Kraus, J. D., *Electromagnetics*, McGraw Hill, International, 1992.

[179] Krejčí, P., *On Maxwell equations with the Preisach hysteresis operator: The one dimensional time periodic case*, Appl. Math., 34(1989), 364–374.

[180] Krejčí, P., *Vector hysteresis models*, Europ.J. Appl. Math., 2(1991)(a), 281–292.

[181] Krejčí, P., *Hyseresis memory preserving operators*, Applications of Math., 36(1991), 305–326.

[182] Krejčí, P., *Hysteresis, convexity and dissipation in hyperbolic equations*, Gakkotosho, Tokyo, 1996.

[183] Krizek, M. and Neittaanmäki, P., *Mathematical and Numerical Modeling in Electrical Engineering Theory and Applications*, Kluwer Academic Publisher, Dordrecht/Boston, 1996.

[184] Kröner, D., *Numerical schemes for conservative laws*, Wiley Teubner, Chichester, 1997.

[185] Krüger, W. Scheutzow, M., Beste, A., Petersen, J., *Markov und Rainflow Rekonstruktion stochastischer Beanspruchungszeitfunktionen.* VDI-Report, Serie 18, Nr. 22, 1985.

[186] Kuipers, L. and Niederreiter, *Uniform Distribution of Sequences,* John Wiley, New York, 1974.

[187] Kupradze, V.D., *Potential Methods in the Theory of Elasticity,* Israel Scientific publ. Jerusalem, 1968.

[188] Landau, L., Lipschitz, E., *Classical Theory of Fields,* Pergamon, Oxford, 1959.

[189] Landau, L., *Electrodynamics of Continuous Media,* Pergamon, Oxford, 1960.

[190] Lĕcot, C., *A Quasi-Monte Carlo Method for the Boltzmann equation,* Math. of Comp., 56(1991), 621–644.

[191] Lĕcot, C., *Error Bounds for Quasi-Monte Carlo integration with nets,* 65(1996), 179–187.

[192] Lemaitre, J. and Chaboche, J.L., *Mechanics of solid materials,* Cambridge university Press, Cambridge, 1990.

[193] Lemarié, P.G., Les ondelettes en 1989, *Lecture Notes in Mathematics,* No. 1438, Springer Verlag, Berlin, Heidelberg, 1989.

[194] Lim, J.S. and Oppenheim, A.V., *Advanced Topics in Signal Processing,* Prentice Hall, New Jersey, 1987.

[195] Liu, C.S., and Chan, A.K., *Wavelet Toolware,* Academic Press, London, 1998.

[196] Locker, J., Prenter, P.M., *Optimal L_2 and L_∞ Error estimates for continuous and Discrete Least Squares Methods for Boundary value problems,* SIAM J. Numer. Anal., 15(1978), 1151–1160.

[197] Louis, A.K., Maaβ, P., Rieder, A., *Wavelets Theory and Applications,* Chichester, John Wiley and Sons, 1997.

[198] Lu, G., *Fractal Image Compression, Signal Processing:* Image Communication, 5(1993), 327–343.

[199] Lu, Ning, *Fracal Imaging,* Academic Press, San Diego, London, 1997.

[200] Maaβ P., Stark, H.G., *Wavelets and digital image processing, Surveys on Mathematics for Industry* Springer-Verlag, Wien-New York, 4(1994), 195–235.

[201] Madelung, E., *Über Magnetisierung durch schnellverlaufende Ströme und die Wirkungsweise des Rutherford-Marconischen Magnetdetektors, Ann. Phys.,* 17(1905), 861–890.

[202] Mallat, S., *A theory for multiresolution signal decomposition the Wavelet representation*, IEEE. Pattern Anal.Mech.Intell., 11(1989), 674–693.

[203] Mallat, S., *Multi resolution and wavelet orthonormal bases of* $L_2(R)$, *Trans.* Amer. Math. Soc., 315(1989), 68–87.

[204] Mallat, S., *Multi frequency Channel decompositions of images and wavelet models,* IEEE Trans. Acoust. Speech. Signal Process, 37(1989), 2091–2110.

[205] Mallat, S., *Wavelets for a vision*, Proc. IEEE, 89(1996), 604–614.

[206] Mallat, S. and Huang, W.L., *Singularity detection and processing with wavelets,* IEEE Trans. Inform. Theory, 38(1992), 617–643.

[207] Mallat, S. and Zhong, S., *Characterization of signals from Multiscale edges,* IEEE Trans. Patt. Anal. Mach. Intell., 14(1992), 710–732.

[208] Manchanda, P., Mukheimer, A.A.S. and Siddiqi, A.H., *Certain results concrning the iterated function system,* Preprint 1998 to appear in Numer. Functional Analysis and Optimization, FAAC, Kuait University Issue.

[209] Manchanda, P., Mukheimer, A.A.S. and Siddiqi, A.H., *Pointwise convergence of two-dimensional wavelet expansions incorporating rotation*, Preprint 1999.

[210] Mangin, G.A., *The Thermomechanics of plasticity and fracture,* Cambridge University Press, Cambridge, 1992.

[211] March, R. and Dozio, M., *A variational method for the recovery of smooth boundaries,* Image and Vision Computing, 15(1997), 705–712.

[212] Markowich, P. A., *The Stationary Semiconductor device equations,* Springer Verlag, Wien, New York. 1986.

[213] Mayergoyz, I.D., *Mathematical Models of hysteresis,* Springer Verlag, 1991.

[214] Mendivil, F. Vrscay, E.R., *Correspondence Between Fractal-Wavelet Transforms and Iterated Functions System with Grey Level Maps, in Vehel et al (eds.),* Fractal in Engineering, Springer Verlag, INRIA, June 1997.

[215] Merton, R.C., *Applications of option-pricing theory: Twentyfive years later,* The American Economic Review, June 1998, pp. 50–76 (Nobel Prize Memorial Lecture delivered in Stockholm, December 9, 1997).

[216] Metropolis, N. and Ulam, S.M., *The Monte Carlo Method,* J. Amer. Statist. Assoc., 44(1949), 35–341.

[217] Meyer, Y., *Wavelets and Operators,* Cambridge studies in advanced mathematics, Cambridge University Press, 1992.

[218] Meyer, Y., (Translated and revised by Robert D. Ryan), *Wavelets - Algorithms and Applications,* Society for Industrial and Applied Mathematics, Philadelphia, 1993.

[219] Mikhailov, G.A. *Minimization of computational costs of Non-analogue Monte Carlo Methods,* World Scientific, Singapore, 1991.

[220] Mikhlin, S.G., *Integral Equations,* Pergamon Press, Oxford, 1957.

[221] Mikhlin, S.G., *Approximate Solutions of Differential and Integral Equations,* Pergamon Press, Oxford, 1965.

[222] Morel, J.M., Solimini S., *Variational Methods in Image Segmentation with seven image processing experiments,* Birkhäuser, Boston, 1995.

[223] Morokoff, W.J. and Caflisch, R.E., *A quasi-Monte Carlo approach to particle simulation of the heat equation,* SIAM J. Numer. Anal., 30(1993), 1558–1573.

[224] Morokoff, W.J. and Caflisch, Russel, E., *Quasi-Monte Carlo Integration,* Journal of Computational Physics, 122(1995), 218–230.

[225] Mumford, D. and Shah, J. (i) *Boundary detection by minimizing functionals,* IEEE Conference on Computer Vision and Pattern Recognition, San Francisco 1985. (ii) *Boundary detection by minimizing functionals, Image understanding,* ed. S. Ullman and W.Richards 1988. (iii) *Optimal Approximations by Piecewise Smooth Functions and Associated Variational Problems,* Communications on Pure and Applied Math., XLII (1989).

[226] Murakami, Y. (ed.), *The rainflow Method in fatigue,* Butterworth and Heinemann, Oxford, 1992.

[227] Muskhelishvili, N.I., *Singular Integral Equations,* Noordhoff Publishing Co., Groningen, 1953.

[228] Nakamura, S., *Numerical Analysis and Graphic Visualization with MATLAB,* Printice Hall, Upper Saddle River, NJ, 1996.

[229] Nashed, M.Z., *Differentiation and related properties of non- linear operators, Non-linear Functional Analysis and its Applications (ed.)* L.B. Rall, Academic Press, New York, 1971.

[230] Neuber, H., *Theory of Stress Concentration for Shear-strained Prismatical Bodies with Arbitrary Non-linear Stress-strain law,* Trans. ASME, J. Appl. Mech., 28(1961), 544–550.

[231] Neunzert, H., *Lectures on Mathematical Modelling, Case Studies from the Laboratory of Technomathematics,* Kaiserslautern university, Germany, 1995.

[232] Neunzert, H., Struckmeier, J., *Particle Methods for the Boltzman Equation*, Acta Numerica, 1995, 417–457.

[233] Neunzert, H., Klar, A., Struckmeier, J., *Particle Methods - Theory and Applications* in Kirchgässner, K., Mahrenholtz D., Mennicken R. (ed.) ICIAM95, Proc. Third International congress on Industrial and Applied Mathematics, Akademie Verlag, Berlin, Vol. 87, 1996.

[234] Neunzert, H. and Struckmeier, J., *Boltzmann Simulation by Particle Methods*, Accademia Nationale dei Lincei, Boltzmann's Legacy 150 years after his birth, Roma, 1997.

[235] Neunzert, H. and Siddiqi, A. H., *Industrial Mathematics - Some ideas*, in Brokate, M. and Siddiqi, A. H.(eds.), Functional Analysis with Current Application to Science, Technology and Industry, Pitman Research Notes in Mathematics, Longman, U.K., 1997.

[236] Niederreiter, H., *Quasi-Monte Carlo Methods and Pseudo random number*, Bull. Amer. Math-Sec., 84(1978), 957–1041.

[237] Niederreiter, H., *Random Number Generators and Quasi Monte Carlo Methods*, SIAM, Philadelphia, Vol. 63, 1992.

[238] Niederreiter, H. in Hlawka, E. and Tichy, R.F. (eds.) *Number Theoretic Analysis*, Lect. Note, 1452, Springer Verlag, 1990.

[239] Nievergelt, Y., *Wavelets Made Easy*, Birkhäuser, Boston, 1999.

[240] Ninomiya, S. and Tezuka, S., *Toward real-time pricing of complex financial derivatives*, Appl. Math. Finance, 3(1996), 11–20.

[241] Nussboumer, H.J., *Fast Fourier Transform and Convolution Algorithms*, New York, Springer Verlag, 1982.

[242] Oden, J. Tinsely, in P.G. Ciarlet and J.L. Lions(eds.), *An Introduction, in Handbook of Numerical Analysis, Vol.II, Finite Element Methods (Part I)* 3–15, Elsevier Science Publishers, North Holland, 1991.

[243] Ogden, R.T., *Essential wavelets for statistical applications and data analysis*, Birkhäuser, Boston, 1997.

[244] Oppenheim, Alan V., *Applications of Digital Signal Processing*, Prentice - Hall, New Jersey, 1978.

[245] Outrata, J., Kočvara, M. and Zowe, J., *Non-smooth approach to optimization problems with equilibrium constraints: Theory, applications and numrical results*, Kluwer Academic Publishers, Boston/Dordrecht/London, 1998.

[246] Papageorgiou, A. and Traub, J.F., *Beating Monte Carlo*, Risk, 9(1996), 63–65.

[247] Part-Enander, E., et al, *MATLAB Handbook*, Addison Wesely, Reading, MA, 1996.

[248] Paskov, S. and Traub, J.F., *Faster valuation of financial derivatives*, J. Portfolio Management, 22(1995), 113–120.

[249] Perona, P. and Malik, J., *Scale space and edge detection using anisotropic diffusion*, IEEE Trans. Patt. Anal. and Mach. Intell., 12(1990), 629–639.

[250] Pfreundt, F. and Hackh, P., *Algorithmen zum Molekülmatching vin Wirkstoffdesign.* Abschlußbericht, Sept. 1993, Fachbereich Mathemak, Universität Kaiserslautern, Germany.

[251] Polak, E., *Optimization Algorithms and Consistent Approximations*, Applied Mathematical Sciences Series, Springer, NY, Berlin, 1997.

[252] Polyak, B. T., *Introduction to Optimization,* Optimization Software Inc., Publications Division, New York, 1987.

[253] Powell, M.J.D., *Convergence properties of Algorithms for non-liner optimization*, SIAM Review, 28(1986), 487–500.

[254] Pozrikidis, C., *Boundary Integral and Singularity Methods for linearized viscous flow,* Cambridge Text in Applied Mathematics, Cambridge University Press, Cambridge, 1992.

[255] Prandtl, L., *Ein Gedankenmodell zur kinetischen Theorie der festen Körpor,* ZAMM, 8(1928), 85-106.

[256] Preisach, F., *Über die magnetische Nachwirkung, Z. Physik,* 94(1935), 277–302.

[257] Prenter, P.M., Russel, R.D., *Orthogonal Collocation for Elliptic Partial Differential Equation,* SIAM J. Numer. Anal., 13(1976), 923–939.

[258] Press, W. H., Teukolsky, S.A., Vetterling, W.T. and Flannery, B. P., *Numerical Recipes in Fortran - The Art of Scientific Computing* Cambridge University Press, 1992.

[259] Rabiner, L.R., Bernard, G., *Theory and Application of Digital Signal Processing,* Prentice - Hall, New Jersey, 1975.

[260] Raviart, P. A., *An Analysis of Particle Methods* in Brezzi, F. (eds.), Numerical Methods in Fluid Dynamics, Lect. Notes in Math. Springer Verlag, Heidelberg, 1983.

[261] Reddy, J.N., *An introduction to the Finite Element Method,* McGraw Hill, International Editions, 1985.

[262] Reddy, J.N., *Applied Functional Analysis and Variational Methods in Engineering,* McGraw Hill, 1985.

[263] Redfen, D. and Cambell, C., *The MATLAB Handbook*, Springer Verlag, New York, 1996.

[264] Reissel, M., *3D Eddy current computation using Krylo subspace Methods,* Preprint, Department of Mathematics, Kaiserslautern University, 1995.

[265] Rektorys, K., *Variational Techniques, D. Reidel Publishing Company,* Dordrecht, Holland, 1980.

[266] Richtmyer, R.D., *On the evaluation of definite, integrals and a Quasi Monte Carlo Method based on properties of algebraic numbers,* Report LA-1342, Los Alamos Sci. Lab. Almos, N.M. 1951.

[267] Rjasanow, S., *Parameterschatzverfahren zur Bestimmung der Bahnpametre liner geradlininger, gleichförnubgen Bewegung,* Fachbereich mathematik, universität Kaiserslautern, Germany.

[268] Ross, C. C., *Differential equations an introduction with Mathematica (R),* Springer Verlag, New York, 1995.

[269] Ruskai, M.B., et al, (eds.), *Wavelets and their Applications* Jones and Bartlet, Boston, 1992.

[270] Sarkar, P.K. and Prasad, M.A., *A comparative study of pseudo and quasi-random sequence for the solutions of integral equations,* Journal of Computational Physics, 68(1987), 66–68.

[271] Saupe, D., et al, *Fractal Image Compression - An Introductory Overview,* Fractal Models for Image Synthesis, Compression and Analysis in Saupe, D. and Hart, J. (eds.) ACMSIG – Graph's 96 Course Notes 27, New Orleans, Louisiana, 1996.

[272] Schaltz, A. H., Thome'e, V., Wendland W. L., *Mathematical Theory of Finite and Boundary Finite Element Methods,* Birkhäuser-DMV Seminar Bond 15, Basel, Boston, Berlin, 1990.

[273] Scholes, M.S., *Derivatives in a dynamical environments,* The American Economic Review, June 1998, 15–35 (Nobel Prize Memorial Lecture delivered in Stockholm, December 9, 1997).

[274] Schulenberger, J.R., *The Debye Potential, A Scalar Factorization for Maxwell's equations,* J. Math. Anal. Appl., 63(1978), 502–520.

[275] Schumaker, L.L. and Webb, G.(eds.), *Recent Advances in wavelet Analysis, Academic Press,* Boston San Diego, New York, Inc., 1994.

[276] Schreiner, W., *Particle Verfahren für Kinetische Schemata zu den Eulargleichungen,* Ph.D. Thesis, University of Kaiserslautern, 1994.

[277] Sendov, B., Dimov, I. (eds.) *Monte Carlo Methods and Parallel Algorithms,* World Scientific, Singapore, 1989.

[278] Shreider, Yu. A., *The Monte Carlo Method,* Pergamon Press, Oxford-London, 1966.

[279] Siddiqi, A.H., *Walsh Function,* AMU Press, Aligarh, 1978.

[280] Siddiqi, A.H., *Functional Analysis with Applications, Tata McGraw Hill,* 1986.

[281] Siddiqi, A.H., *Approximation by Walsh Series, Mathematical Analysis and its Applications,* in Mazhar, S.M., Hamoui, A. and Faur, N.S. (ed.), Pergamon Press, 1987, 43-51.

[282] Siddiqi, A.H.,(ed.), *Recent Developments in Applicable Mathematics,* Mcmillan, India, 1994.

[283] Siddiqi, A.H., *Certain current Developments in Variational Inequalities* in Lau and Tweddle (eds.), Pitman Research Notes in Mathematics Series Nr 316, Longman Scientific and Technical, 1994, 219–238.

[284] Siddiqi, A.H., *Lecture Notes on Variational Inequalities with Applications,* Industrial Mathematics, Department of Mathematics, Kaiserslautern University, 1997.

[285] Siddiqi, A.H., *Fractal-wavelet methods in image processing,* Invited Talk International Conference on Fourier Analysis and Applications, Kuwait University, Kuwait, 1998.

[286] Siddiqi, A.H. and Ahmad, K., *Wavelet Methods in Differential equations,* Preprint AMU-95.

[287] Siddiqi, A.H., Ahmad, M. K., and Mukheimer, A. *Current Developments in Fractal Image Compression,* in Brokate, M. and Siddiqi, A. H.(eds.), Functional Analysis with Current Application to Science, Technology and Industry, Pitman Research Notes in Mathematics, Longman, Vol. 377, U.K., 1997.

[288] Siddiqi, A.H., Manchanda, P. and Kočvara, M., *An iterative two-step algorithm for American option pricing,* Preprint 1998, Erlangen University to be published in IMA Journal of Mathematics Applied to Business and Industry.

[289] Siddiqi, A.H. and Ahmad, M.K., *Distortion measure through Sobolev and total variation norm metrics,* Preprint, 1998.

[290] Siddiqi, A.H. and Ahmad, M.K., *Sharp operator and classification of images,* Preprint 1998.

[291] Sircar, K.R. and Papanicolaou, G., *General Black-Sholes models accounting for increased market volatility from heding strategies*, Applied Mathematical Finance, 5(1998), 45–82.

[292] Silvester, R.P. and Ferrari, R.N., *Finite Elements for Electrical Engineers*, Cambridge University Press, 1983 (edition I), 1990 (edition II).

[293] Sobol, I.M., *The Monte Carlo Method*, Mir Publishers, Moscow, 1975.

[294] Stark, H.G., *Multiscale analysis, wavelets and texture quality*, Reports of AGTM, Kaiserslautern University, Germany, Nr. 41, 1990.

[295] Sokolowski, J. and Zolesio, J. P., *Introduction to shape optimization, shape sensitivity Analysis*, Springer-Verlag, Berlin, 1992.

[296] Stewart, I., *Four encounters with Sierpinski Gasket*, Mathematical Intelligencer, 17(1995), 52–64.

[297] Strang, G., *Wavelets and dilation equations - A brief introduction*, SIAM Review, 31(1989), 614-627.

[298] Strang, G., *Wavelet transforms versus Fourier transforms*, Bull. Amer. Math. Soc., 28(1993), 288-305.

[299] Strömberg, J.O., *A modified Franklin system and higher order spline systems on R^n as unconditional bases for Hardy spaces*, in Conference on Harmonic Analysis in Honour of Antoni Zygmund, etal (ed.) Vol. II, University of Chicago Press, Chicago, 1981, 475–494.

[300] Stroud, A.H., *Numerical Quadrature and Solution of Ordinary Differential Equations*, Springer Verlag, New York-Heidelberg-Berlin, 1974.

[301] Sweldons, W. and Piessens, R., *Quadrature formulae and asymptotic error expansions for wavelet approximations of smooth functions*, SIAM J. Numer. Anal., 31(1994), 1240–1264.

[302] Tapia, R.A., *The differentiation and integration of non-linear operators, Nonlinear Functional Analysis and Applications (edit.)* B. Rall, Academic Press, 1971.

[303] Taylor, A.B., *Mathematical Models in Applied Mechanics*, Clarendom Press, Oxford, 1986.

[304] Teo, P.C. and Heeger, D.J., *Perceptual image distortion*, SPIE, 2179(1994), 127–141,

[305] Tezuka, S., *Financial applications of Monte Carlo and quasi-Monte Carlo methods*, in Hellekalek and Larcher, Lect. Notes in Statistics, Springer-Verlag, 1998, pp. 303–332.

[306] Thomas, E.G. and Meadows, A.J., *Maxwell's Equations and their Applications,* Adam Hilger Ltd., Bristol and Boston, 1985.

[307] Tiwari, S, and Rjasanow, S., *Sobolev norm as a criterion of local thermal equilibrium,* to appear in Eur. J. Mech. B/Fluids.

[308] Tricomi, F., *Integral Equations,* Interscience publ., London, New York, 1957 Reprint Dover Publ. New York, 1985.

[309] Turner, M.J., Clough, R.W., Martin, H.C. and Topp, L.J., *Stiffness and deflection analysis of complex structures,* J. Aero, Sci., 23(1956), 805-823.

[310] van Loan, C. *Computational Frameworks for the Fast Fourier Transforms,* SIAM Philedelphia, 1992.

[311] Visintin, A. (ed.) *Models of hysteresis,* Pitman Research Notes in Math. Series No. 286, Longman Scientific and Technical, 1993.

[312] Visintin, A., *Differential Models of hysteresis,* Springer Verlag, 1994.

[313] Visintin, A. (ed.), *Phase transition and Hysteresis,* Lecture Notes in Mathematics, Vol. 1584, Springer Verlag, Berlin, 1994.

[314] Vvedensky, D., *Partial differential Equations with Mathematica* Addison-Wesley Publishing Company, 1994.

[315] Wahlbin, L.B., *Superconvergence in Galerkin Finite Element Methods Lecture Notes in Mathematics,* Springer Verlag, Vol. 1605, Bonn- Heidelberg, 1995.

[316] Wait, R. and Mitchell, A.R., *Finite Element Analysis and Applications,* John Wiley and Sons, Chichester, New York, 1985.

[317] Walker, J.S., *Fourier analysis and wavelet analysis,* Notices of the Amer. Math. Soc., 44(1997), 658–670.

[318] Wallace, G.K., *Overview of the JPEG (ISO/CCITT) Still image compression standard,* SPIE, 1244(1990), 220–233.

[319] Walter, G.G., *Approximation of the delta function by wavelets,* J. Approx. Theory, 71(1992), 329–343.

[320] Walter, G.G., *Pointwise convergence of wavelet expansions,* J. Approx. Theory, 80(1995), 108-118.

[321] Wang, Xin-Hua, *Finite Element Methods for non-linear optical wave guides,* Advances in non-linear Optics, Vol. 2, Gordon and Breach publishers, Amsterdam, 1995.

[322] Weaver, H. Joseph, *Applications of Discrete and Continuous Fourier Analysis,* John Wiley & Sons, Inc. 1983.

[323] Weaver, H. Joseph, *Theory of Discrete and Continuous Fourier Analysis,* John Wiley and Sons, 1989.

[324] Weickert, J., *A model for the cloudiness of Fabrics, Preprint Industrial Math. Lab.* Kaiserslautern university, Germany, 1995, Proc. 8th EC MI, Conference, Wiley-Teubner 1995.

[325] Wendt, J. F. (ed.), *Computational Fluid Dynamics,* Springer Verlag, Berlin, 1991.

[326] Weyl, H., *Überdie Gleichverteilung von Zahlen modulo* Eins, Math. Ann., 77(1917), 313–352.

[327] Wheeler, M.F., *An elliptic Collocation Finite Element Method with Interior penalties,* SIAM J. Numer. Anal., 15(1978), 152–161.

[328] Whiteman, J. (Ed.), *The mathematics of finite elements and applications, I, II, III,* Proc. Conf. Brunel University, Academic Press, (1973, 1976, 1979).

[329] Wickerhauser, M.V., *Adapted wavelet analysis from theory to software* Wallesley, MA: Peters, 1994.

[330] Wilmott, P., *Derivatives, the theory and practice of financial engineering,* John Wiley and Sons, Chichester, New York, 1998.

[331] Wilmott, P., Howison, S.D. and Dewynne, J.N., *The mathematics of financial derivatives,* Cambridge University Press, 1995.

[332] Wilmott, P., Dewynne, J.N. and Howison, S.D., *Option pricing: Mathematical models and computation,* Oxford Financial Press, 1993.

[333] Wojtaszczyk, P., *A mathematical introduction to wavelets,* London Mathematical Society Student Text 37, Cambridge University Press, 1997.

[334] Wolfram, S., Mathematica^{TM}, *A system for doing Mathematics by computer,* Addison-Wesley Publishing Company, 1990.

[335] Zayed, A.I., *Advances in Shannon's Sampling Theory,* CRC Press, London, 1993.

[336] Zenisek, A., *Nonlinear Elliptic and Evolution Problems and their finite element approximations, computational Mathematics and Applications,* Academic Press, Harcourt Brace Jovanovich Publishers, London, 1990.

[337] Zienkiewicz, O.C. and Cheung, Y.K., *The Finite Element Method in Structural and Continuum Mechanics,* McGraw Hill, New York, 1967.

[338] Zlamal, M., *On the Finite Element Method,* Numer. Math., 12(1968), 394-409.

[339] Zygmund, A., *Treatise on Trigonometric Series,* $Vol.I, II$; Oxford Press, 1959.

Symbols

$\rightarrow$	arrows of mapping
$\Rightarrow$	implication sign
$\forall$	for all
$\Longleftrightarrow$	if and only if
N, Z, R	denote the sets of natural numbers, integers and real numbers
ϕ	the empty set
R^n	Euclidean space of dimension n
$\subset$	subset
$\inf A$	the infinimum of $A \subset R$
$\sup A$	the supremum of $A \subset R$
diam A	the diameter of $A \subset R^n$
$d(A, B)$	the distance between $A, B \subset R^n$
$[a, b]$	$= \{x \in R \mid a \leq x \leq b\}$
(a, b)	$= \{x \in R \mid a < x < b\}$
$(a, b]$	$= \{x \in R \mid a < x \leq b\}$
$[a, b)$	$= \{x \in R \mid a \leq x < b\}$
R_+	$= (0, \infty)$
R^+	$= [0, \infty)$
$\|\cdot\|$	norm

support of f = closure of $\{x \in \Omega \mid f(x) \neq 0\}$

$C_0^\infty(u) = D(\Omega)$ the space of all infinitely differentiable functions with compact support.

$D'(\Omega)$ the space of all Distributions = the space of all bounded linear functionals defined on $D(\Omega)$.

$H^m(\Omega) = \{f \in L_2(\Omega) \mid D^\alpha \in L_2(\Omega), |\alpha| \leq m\}$, m is any positive integer, Sobolev space of order m.

$$\begin{aligned} H_0^m(\Omega) &= \{f \in H^m(\Omega)/\Gamma_f = 0\} = \text{ closure of } D(\Omega) \text{ in } H^m(\Omega) \\ H^m(R^n) &= H_0^m(R^n), m \geq 0\,. \end{aligned}$$

$f \in H^{-m}(\Omega)$ if and only if $f = \sum_{|\alpha| \leq m} D^\alpha g$ for some $g \in L_2(\Omega)$.

$$div v = \sum_{i=1}^{n} \frac{\partial v_i}{\partial x_i} \text{ for all } v = (v_1, v_2, \cdots v_n) \in (\mathcal{D}'(\Omega))^n.$$

$$curl\ v = rot\ v = \nabla \times v = \left(\frac{\partial v_3}{\partial x_2} - \frac{\partial v_2}{\partial x_3}, \frac{\partial v_1}{\partial x_3} - \frac{\partial v_3}{\partial x_1}, \frac{\partial v_2}{\partial x_1} - \frac{\partial v_1}{\partial x_2} \right)$$

for $v = (v_1, v_2, v_3) \in \mathcal{D}'(\Omega)^3$.

$$\begin{aligned} \chi_A(x) &= 1, \quad x \in A \\ &= 0, \quad x \notin A \end{aligned}$$ characteristic function or indicatrix function.

Index

Applied Optimization

1. D.-Z. Du and D.F. Hsu (eds.): *Combinatorial Network Theory.* 1996
ISBN 0-7923-3777-8
2. M.J. Panik: *Linear Programming: Mathematics, Theory and Algorithms.* 1996
ISBN 0-7923-3782-4
3. R.B. Kearfott and V. Kreinovich (eds.): *Applications of Interval Computations.* 1996
ISBN 0-7923-3847-2
4. N. Hritonenko and Y. Yatsenko: *Modeling and Optimimization of the Lifetime of Technology.* 1996
ISBN 0-7923-4014-0
5. T. Terlaky (ed.): *Interior Point Methods of Mathematical Programming.* 1996
ISBN 0-7923-4201-1
6. B. Jansen: *Interior Point Techniques in Optimization.* Complementarity, Sensitivity and Algorithms. 1997
ISBN 0-7923-4430-8
7. A. Migdalas, P.M. Pardalos and S. Storøy (eds.): *Parallel Computing in Optimization.* 1997
ISBN 0-7923-4583-5
8. F.A. Lootsma: *Fuzzy Logic for Planning and Decision Making.* 1997
ISBN 0-7923-4681-5
9. J.A. dos Santos Gromicho: *Quasiconvex Optimization and Location Theory.* 1998
ISBN 0-7923-4694-7
10. V. Kreinovich, A. Lakeyev, J. Rohn and P. Kahl: *Computational Complexity and Feasibility of Data Processing and Interval Computations.* 1998
ISBN 0-7923-4865-6
11. J. Gil-Aluja: *The Interactive Management of Human Resources in Uncertainty.* 1998
ISBN 0-7923-4886-9
12. C. Zopounidis and A.I. Dimitras: *Multicriteria Decision Aid Methods for the Prediction of Business Failure.* 1998
ISBN 0-7923-4900-8
13. F. Giannessi, S. Komlósi and T. Rapcsák (eds.): *New Trends in Mathematical Programming.* Homage to Steven Vajda. 1998
ISBN 0-7923-5036-7
14. Ya-xiang Yuan (ed.): *Advances in Nonlinear Programming.* Proceedings of the '96 International Conference on Nonlinear Programming. 1998
ISBN 0-7923-5053-7
15. W.W. Hager and P.M. Pardalos: *Optimal Control.* Theory, Algorithms, and Applications. 1998
ISBN 0-7923-5067-7
16. Gang Yu (ed.): *Industrial Applications of Combinatorial Optimization.* 1998
ISBN 0-7923-5073-1
17. D. Braha and O. Maimon (eds.): *A Mathematical Theory of Design: Foundations, Algorithms and Applications.* 1998
ISBN 0-7923-5079-0

Applied Optimization

18. O. Maimon, E. Khmelnitsky and K. Kogan: *Optimal Flow Control in Manufacturing.* Production Planning and Scheduling. 1998 ISBN 0-7923-5106-1
19. C. Zopounidis and P.M. Pardalos (eds.): *Managing in Uncertainty: Theory and Practice.* 1998 ISBN 0-7923-5110-X
20. A.S. Belenky: *Operations Research in Transportation Systems:* Ideas and Schemes of Optimization Methods for Strategic Planning and Operations Management. 1998 ISBN 0-7923-5157-6
21. J. Gil-Aluja: *Investment in Uncertainty.* 1999 ISBN 0-7923-5296-3
22. M. Fukushima and L. Qi (eds.): *Reformulation: Nonsmooth, Piecewise Smooth, Semismooth and Smooting Methods.* 1999 ISBN 0-7923-5320-X
23. M. Patriksson: *Nonlinear Programming and Variational Inequality Problems.* A Unified Approach. 1999 ISBN 0-7923-5455-9
24. R. De Leone, A. Murli, P.M. Pardalos and G. Toraldo (eds.): *High Performance Algorithms and Software in Nonlinear Optimization.* 1999 ISBN 0-7923-5483-4
25. A. Schöbel: *Locating Lines and Hyperplanes.* Theory and Algorithms. 1999 ISBN 0-7923-5559-8
26. R.B. Statnikov: *Multicriteria Design.* Optimization and Identification. 1999 ISBN 0-7923-5560-1
27. V. Tsurkov and A. Mironov: *Minimax under Transportation Constrains.* 1999 ISBN 0-7923-5609-8
28. V.I. Ivanov: *Model Development and Optimization.* 1999 ISBN 0-7923-5610-1
29. F.A. Lootsma: *Multi-Criteria Decision Analysis via Ratio and Difference Judgement.* 1999 ISBN 0-7923-5669-1
30. A. Eberhard, R. Hill, D. Ralph and B.M. Glover (eds.): *Progress in Optimization.* Contributions from Australasia. 1999 ISBN 0-7923-5733-7
31. T. Hürlimann: *Mathematical Modeling and Optimization.* An Essay for the Design of Computer-Based Modeling Tools. 1999 ISBN 0-7923-5927-5
32. J. Gil-Aluja: *Elements for a Theory of Decision in Uncertainty.* 1999 ISBN 0-7923-5987-9
33. H. Frenk, K. Roos, T. Terlaky and S. Zhang (eds.): *High Performance Optimization.* 1999 ISBN 0-7923-6013-3
34. N. Hritonenko and Y. Yatsenko: *Mathematical Modeling in Economics, Ecology and the Environment.* 1999 ISBN 0-7923-6015-X
35. J. Virant: *Design Considerations of Time in Fuzzy Systems.* 2000 ISBN 0-7923-6100-8

Applied Optimization

36. G. Di Pillo and F. Giannessi (eds.): *Nonlinear Optimization and Related Topics.* 2000 ISBN 0-7923-6109-1
37. V. Tsurkov: *Hierarchical Optimization and Mathematical Physics.* 2000 ISBN 0-7923-6175-X
38. C. Zopounidis and M. Doumpos: *Intelligent Decision Aiding Systems Based on Multiple Criteria for Financial Engineering.* 2000 ISBN 0-7923-6273-X
39. X. Yang, A.I. Mees, M. Fisher and L.Jennings (eds.): *Progress in Optimization.* Contributions from Australasia. 2000 ISBN 0-7923-6175-X
40. D. Butnariu and A.N. Iusem: *Totally Convex Functions for Fixed Points Computation and Infinite Dimensional Optimization.* 2000 ISBN 0-7923-6287-X
41. J. Mockus: *A Set of Examples of Global and Discrete Optimization.* Applications of Bayesian Heuristic Approach. 2000 ISBN 0-7923-6359-0
42. H. Neunzert and A.H. Siddiqi: *Topics in Industrial Mathematics.* Case Studies and Related Mathematical Methods. 2000 ISBN 0-7923-6417-1

KLUWER ACADEMIC PUBLISHERS – DORDRECHT / BOSTON / LONDON

9 780792 364177